PRENTICE-HALL SERIES IN SOLID STATE PHYSICAL ELECTRONICS

NICK HOLONYAK, JR., editor

PRENTICE-HALL INTERNATIONAL, INC., London
PRENTICE-HALL OF AUSTRALIA, PTY. LTD., Sydney
PRENTICE-HALL OF CANADA, LTD., Toronto
PRENTICE-HALL OF INDIA PRIVATE LIMITED, New Delhi
PRENTICE-HALL OF JAPAN, INC., Tokyo

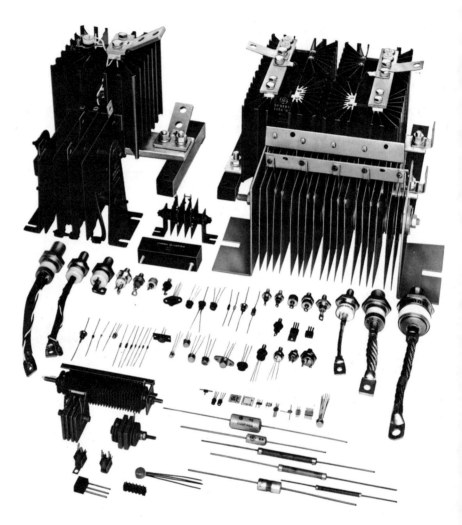

Solid State Devices range in size from tiny multiterminal integrated circuits to large diode rectifiers, as this figure indicates. Some of the devices illustrated here are mounted in chambers with cooling fins, suggesting application at high power levels; others are obviously intended for handling small signals. It is clear that designers and users of modern electronic circuits deal with a wide variety of solid state devices with differing capabilities and limitations. The purpose of this book is to lay the foundation for understanding these devices—their design, fabrication, and application to the requirements of present-day electronics. Photograph courtesy of General Electric Company.

SOLID STATE

ELECTRONIC DEVICES

Ben G. Streetman

*Department of Electrical Engineering
and Coordinated Science Laboratory*

University of Illinois at Urbana-Champaign

PRENTICE-HALL, INC., *Englewood Cliffs, N.J.*

ISBN: 0-13-822023-9

Library of Congress Catalog Card Number 79-173926

10 9 8 7

Drawings by GEORGE E. MORRIS

Printed in the United States of America

CONTENTS

5 p-n Junctions 138

Appendices   **438**

PREFACE

This book is an introduction to solid state electronic devices for undergraduate electrical engineers, other interested engineering and science students, and practicing engineers and scientists whose understanding of modern electronics needs updating. The book is organized to bring students with a background in sophomore physics to a level of understanding which will allow them to read much of the current literature on new devices and applications.

Goals. In my opinion, an undergraduate course in electronic devices has two basic purposes: (1) to provide students with a sound understanding of existing devices, so that their studies of electronic circuits and systems will be meaningful; and (2) to develop the basic tools with which they can later learn about newly developed devices and applications. Perhaps the second of these objectives is the more important in the long run; it is clear that engineers and scientists who deal with electronics will continually be called upon to learn about new devices and processes in the future. For this reason, I have tried to incorporate into the book the basic principles of conduction processes in solids which arise repeatedly in the literature when new devices are explained. Some of these concepts (particularly carrier statistics and energy bands) are often omitted in introductory courses, on the basis that they are unnecessary for understanding the principles of junctions and transistors. I believe this view neglects the important goal of equipping students for the task of understanding a new device by reading the current literature. Therefore, in this text most of the commonly used semiconductor terms and concepts are introduced and related to a broad range of devices.

Reading Lists. As a further aid in developing techniques for independent study, I have omitted the usual bibliography at the end of each chapter and substituted reading lists. These lists contain articles which students can read comfortably as they study this book. These articles have been selected from periodicals such as *Scientific American, Electronics,* and *IEEE Spectrum,* which specialize in introductory presentations. Other articles chosen from books and the professional literature provide a more quantitative treatment of the material. I do not expect that students will read all articles recommended in the reading lists; nevertheless, some exposure to periodicals is useful in laying the foundation for a career of constant updating and self-education.

Units. Consistent with the goals described above, examples and problems are stated in terms of units commonly used in the semiconductor literature. The units are a hybrid of several standard systems, resulting from the fact that semiconductor technology has grown out of the work of engineers, physicists, chemists, and metallurgists. The basic system of units is rationalized MKS, although cm is often used as a convenient unit of length. Similarly, electron volts (eV) are often used rather than joules (J) to measure the energy of electrons. Other variations have grown out of practical considerations; for example, it is not uncommon to find sample dimensions expressed in mils (thousandths of an inch), simply because micrometer gauges and other instruments are often calibrated in this unit. As a result of this rather rich mixture, students new to the field must be careful that the units used in problems are self-consistent. This usually calls for balancing units in formulas by using appropriate conversion factors (Appendix II).

Of course, these complications could be avoided by using MKS throughout; but the result would be extremely unrealistic relative to the semiconductor literature. I believe students can use the commonly accepted system of units with only a little extra work, and the benefits are worth the effort.

Presentation. In presenting this material at the undergraduate level, one must anticipate a few instances which call for a phrase such as "It can be shown . . ." This type of phrase is always disappointing; on the other hand, the alternative is to delay study of solid state devices until the graduate level, where statistical mechanics, quantum theory, and other advanced background can be freely invoked. Such a delay would result in a more elegant treatment of certain subjects, but it would prevent undergraduate students from enjoying the study of some very exciting devices.

The first four chapters of the book provide background on the nature of semiconductors and conduction processes in solids. Included is a brief introduction to quantum concepts (Chapter 2) for those students who do not already have this background from other courses. Chapters 5 and 6 describe the p-n junction and some of its applications. One could argue that Chapter 7, a description of lasers, should appear after the transistor discussion. Such rearrangement would not detract from the presentation. I placed the laser chapter after p-n junctions for a very simple reason: students usually find lasers particularly exciting, and this chapter gives some students a welcome lift near the middle of the course. Chapters 8 and 9 deal with the principles of transistor operation, and 10, 11, and 12 apply the theory of junctions and conduction processes to integrated circuits, switching devices, and microwave devices. All of the devices covered are important in today's electronics; furthermore, learning about these devices should be an enjoyable and rewarding experience. I hope this book provides that kind of experience for its readers.

Acknowledgments. This book was developed during a five-year period of teaching junior-level electrical engineers at the University of Illinois. Portions of the material have been used (in varying degrees of completion) by hundreds of students and more than a dozen teachers. Criticism and suggestions from many students and staff are gratefully acknowledged. I would particularly like to thank George Anner, Don Holshouser, and Marvin Krasnow for useful comments growing out of their teaching of parts of the material. I am also grateful to Harold Korb of Bell Telephone Laboratories, Carl Meyer of Delco Electronics, and Jon Rossi of Lincoln Laboratory for reading and commenting on portions of the manuscript. I would like to thank my wife, Ann, for careful editing and proofreading of the manuscript, and George Morris for creative treatment of the illustrations. Special thanks go to Nick Holonyak, Jr., who generously offered suggestions throughout the development of the book. I am particularly grateful for his careful reading and correction of the final manuscript. Finally, I would like to thank Bell Labs, Delco Electronics, General Electric, Monsanto, Motorola, NASA/JPL, Texas Instruments, and Zenith for generously providing photographs and illustrations of devices and fabrication processes.

Ben G. Streetman

CRYSTAL PROPERTIES AND
GROWTH OF SEMICONDUCTORS 1

In studying solid state electronic devices we are interested primarily in the electrical behavior of solids. However, we shall see in later chapters that the transport of charge through a metal or a semiconductor depends not only on the properties of the electron but also on the arrangement of atoms in the solid. In the first chapter we shall discuss some of the physical properties of semiconductors compared with other solids, the atomic arrangements of various materials, and some methods of growing semiconductor crystals. Topics such as crystal structure and crystal growth technology are often the subjects of books rather than introductory chapters; thus we shall consider only a few of the more important and fundamental ideas which form the basis for understanding electronic properties of semiconductors and device fabrication.

1.1 Semiconductor Materials

Semiconductors are a group of materials having electrical conductivities intermediate between metals and insulators. It is significant that the conductivity of these materials can be varied over wide ranges by changes in temperature, optical excitation, and impurity content. This variability of electrical properties makes the semiconductor materials natural choices for electronic device investigations.

Semiconductor materials are found in column IV and neighboring columns of the periodic table (Table 1-1). The column IV semiconductors, silicon and germanium, are called *elemental* semiconductors because they

TABLE 1-1.   Common semiconductor materials: (a) the portion of the periodic table where semiconductors occur; (b) elemental and compound semiconductors.

(a)	II	III	IV	V	VI
		B	C		
		Al	Si	P	S
	Zn	Ga	Ge	As	Se
	Cd	In	Sn	Sb	Te

(b)	Elemental	IV compounds	III–V compounds	II–VI compounds
	Si	SiC	AlP	ZnS
	Ge		AlAs	ZnSe
			AlSb	ZnTe
			GaP	CdS
			GaAs	CdSe
			GaSb	CdTe
			InP	
			InAs	
			InSb	

are composed of single species of atoms. In addition to the elemental materials, compounds of column III and column V atoms, as well as certain combinations from II and VI, make up the *intermetallic*, or *compound*, semiconductors.

As Table 1-1 indicates, there are numerous semiconductor materials. Among these, Si is used for the majority of semiconductor devices; rectifiers, transistors, and integrated circuits are usually made of Si. The compounds are used most widely in devices requiring the emission or absorption of light. For example, semiconductor light emitters are commonly made of such compounds as GaAs, GaP, and *mixed compounds* such as GaAsP. Fluorescent materials such as those used in television screens usually are II–VI compound semiconductors such as ZnS. Light detectors are commonly made with InSb, CdSe, or other compounds such as the lead salts PbTe and PbSe; Si and Ge are also widely used as infrared and nuclear radiation detectors. An important microwave device, the Gunn diode, is usually made of GaAs. Thus, the wide range of semiconductor materials offers considerable variety in properties and provides device and circuit engineers with much flexibility in the design of electronic functions.

The electronic and optical properties of semiconductor materials are strongly affected by impurities, which may be added in precisely controlled amounts. Such impurities are used to vary the conductivities of semiconductors over wide ranges and even to alter the nature of the conduction processes from conduction by negative charge carriers to positive charge carriers. For example, an impurity density of one part per million can change a sample of Si from a poor conductor to a good conductor of electric current. This process of controlled addition of impurities, called *doping*, will be discussed in detail in subsequent chapters.

To investigate these useful properties of semiconductors, it is necessary to understand the atomic structure of the materials. Obviously, if slight alterations in purity of the original material can produce such dramatic changes in electrical properties, then the nature and specific arrangement of atoms in each semiconductor must be of critical importance. Therefore, we begin our study of semiconductors with a brief introduction to crystal structure.

1.2 Crystal Lattices

In this section we discuss the arrangements of atoms in various solids. We shall distinguish between single crystals and other forms of materials and then investigate the periodicity of crystal lattices. Certain important crystallographic terms will be defined and illustrated in reference to crystals having a basic cubic structure. These definitions will allow us to refer to certain planes and directions within a lattice of arbitrary structure. Finally, we shall investigate the diamond lattice; this structure, with some variations, is typical of most of the semiconductor materials used in devices.

1.2.1 Periodic Structures.
A crystalline solid is distinguished by the fact that the atoms making up the crystal are arranged in a periodic fashion. That is, there is some basic arrangement of atoms which is repeated throughout the entire solid. Thus the crystal appears exactly the same at one point as it does at a series of other equivalent points, once the basic periodicity is discovered. However, not all solids are crystals (Fig. 1-1); some have no periodic structure at all (*amorphous* solids), and others are composed of many small regions of single-crystal material (*polycrystalline* solids). The semiconductor materials we shall study are single crystals, with very small departures from perfect periodicity due to the addition of doping atoms for controlling electrical properties.

The periodic arrangement of atoms in a crystal is called the *lattice*. Since there are many different ways of placing atoms in a volume, the distances and orientation between atoms can take many forms. However, in every case the lattice contains a volume, called a *unit cell*, which is represen-

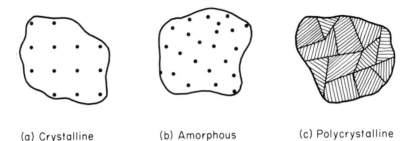

(a) Crystalline   (b) Amorphous   (c) Polycrystalline

FIGURE 1-1.   Three types of solids, classified according to atomic arrangement: (a) crystalline and (b) amorphous materials are illustrated by microscopic views of the atoms, whereas (c) polycrystalline structure is illustrated by a more macroscopic view of adjacent single-crystalline regions, such as (a).

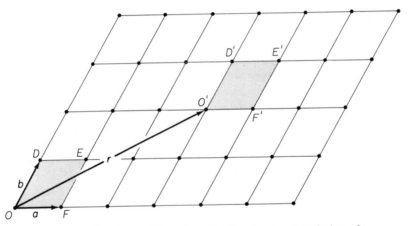

FIGURE 1-2.   A two-dimensional lattice showing translation of a unit cell by $\mathbf{r} = 3\mathbf{a} + 2\mathbf{b}$.

tative of the entire lattice and is regularly repeated throughout the crystal. As an example of such a lattice, Fig. 1-2 shows a two-dimensional arrangement of atoms with a unit cell ODEF. This cell has an atom at each corner shared with adjacent cells. Notice that we can define vectors **a** and **b** such that if the unit cell is translated by integral multiples of these vectors, a new unit cell identical to the original is found (e.g., O'D'E'F'). These vectors **a** and **b** (and **c** if the lattice is three dimensional) are called the *basis vectors* for the lattice. Points within the lattice are indistinguishable if the vector between the points is

(1-1) $$\mathbf{r} = p\mathbf{a} + q\mathbf{b} + s\mathbf{c}$$

where p, q, and s are integers. A unit cell which contains lattice points only at the corners is called a *primitive cell*. In Fig. 1-2 the shaded area ODEF is a primitive cell. In many lattices, however, the primitive cell is not the most convenient to work with. Therefore, we shall deal primarily with unit cells which contain atoms in addition to those at the corners.

The importance of the unit cell lies in the fact that we can analyze the crystal as a whole by investigating a representative volume. For example, from the unit cell we can find the distances between nearest atoms and next nearest atoms for calculation of the forces holding the lattice together; we can look at the fraction of the unit cell volume filled by atoms and relate the density of the solid to the atomic arrangement. But even more important for our interest in electronic devices, the properties of the periodic crystal lattice determine the allowed energies of electrons which participate in the conduction process. Thus the lattice determines not only the mechanical properties of the crystal but also its electrical properties.

1.2.2 Cubic Lattices. The simplest three-dimensional lattice is one in which the unit cell is a cubic volume, such as the three cells shown in Fig. 1-3. The *simple cubic* structure (abbreviated *sc*) has an atom located at each

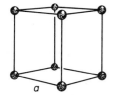 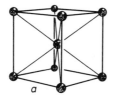 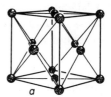

Simple   Cubic   Body-centered   Cubic   Face-centered   Cubic

FIGURE 1-3.   Unit cells for three types of cubic lattice structures.

corner of the unit cell. The *body-centered cubic* (*bcc*) lattice has an additional atom at the center of the cube, and the *face-centered cubic* (*fcc*) unit cell has atoms at the eight corners and centered on the six faces.

As atoms are packed into the lattice in any of these arrangements, the distances between neighboring atoms will be determined by a balance between the forces which attract them together and other forces which hold them apart. We shall discuss the nature of these forces for particular solids in Section 3.1.1. For now, we can calculate the maximum fraction of the lattice volume which can be filled with atoms by approximating the atoms as hard spheres. For example, Fig. 1-4 illustrates the packing of spheres in a face-

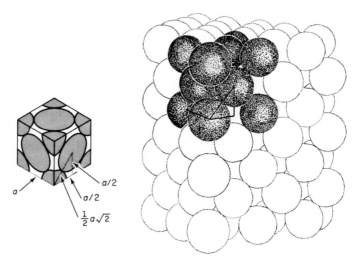

FIGURE 1-4.   Packing of hard spheres in an fcc lattice.

centered cubic cell of side a, such that the nearest neighbors touch. The dimension a for a cubic unit cell is called the *lattice constant*. For the fcc lattice the nearest neighbor distance is one-half the diagonal of a face, or $\frac{1}{2}(a\sqrt{2})$. Therefore, for the atom centered on the face to just touch the atoms at each corner of the face, the radius of the sphere must be one-half the nearest neighbor distance, or $\frac{1}{4}(a\sqrt{2})$.

Each corner atom in a cubic unit cell is shared with seven neighboring cells; thus each unit cell contains $\frac{1}{8}$ of a sphere at each of the eight corners for a total of one atom. Similarly the fcc cell contains half an atom at each of the six faces for a total of three. Thus we have

Atoms per cell $= 1$ (corners) $+ 3$ (faces) $= 4$
Nearest neighbor distance $= \frac{1}{2}(a\sqrt{2})$
Radius of each sphere $= \frac{1}{4}(a\sqrt{2})$
Volume of each sphere $= \frac{4}{3}\pi\left[\frac{1}{4}(a\sqrt{2})\right]^3 = \frac{\pi a^3\sqrt{2}}{24}$
Maximum fraction of cell filled

$$= \frac{\text{no. of spheres} \times \text{vol. of each sphere}}{\text{total vol. of cell}}$$

$$= \frac{4 \times (\pi a^3\sqrt{2})/24}{a^3}$$

$$= \frac{\pi\sqrt{2}}{6} = 74 \text{ per cent filled}$$

Therefore, if the atoms in an fcc lattice are packed as densely as possible, with no distance between the outer edges of nearest neighbors, 74 per cent of the volume is filled. This is a relatively high percentage compared with some other lattice structures (Prob. 1.1).

1.2.3 *Planes and Directions.*   In discussing crystals it is very help-ful to be able to refer to planes and directions within the lattice. The notation system generally adopted uses a set of three integers to describe the position of a plane or the direction of a vector within the lattice. The three integers describing a particular plane are found in the following way:

1. Find the intercepts of the plane with the crystal axes and express these intercepts as integral multiples of the basis vectors (the plane can be moved in and out from the origin, retaining its orientation, until such an integral intercept is discovered on each axis).
2. Take the reciprocals of the three integers found in step 1 and reduce these to a new set of integers h, k, and l, which have the same relationship to each other as the three reciprocals.
3. Label the plane (hkl).

Example:

The plane illustrated in Fig. 1-5 has intercepts at 2**a**, 4**b**, and 1**c** along the three crystal axes. Taking the reciprocals of these inter-cepts, we get $\frac{1}{2}$, $\frac{1}{4}$, and 1. These three fractions have the same relationship to each other as the integers 2, 1, and 4 (obtained by multiplying each fraction by 4). Thus the plane can be referred to as a (214) plane.

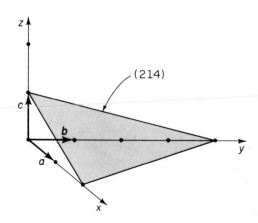

FIGURE 1-5. A (214) cry-stal plane.

The three integers h, k, and l are called the *Miller indices;* these three numbers define a set of parallel planes in the lattice. One advantage of taking the reciprocals of the intercepts is avoidance of infinities in the notation. One intercept is infinity for a plane parallel to an axis; however, the reciprocal of such an intercept is taken as zero. If a plane contains one of the axes, it is parallel to that axis and has a zero reciprocal intercept. If a plane passes through the origin, it can be translated to a parallel position for calculation of the Miller indices. If an intercept occurs on the negative branch of an axis, the minus sign is placed above the Miller index for convenience, such as $(h\bar{k}l)$.

From a crystallographic point of view, many planes in a lattice are equivalent; that is, a plane with given Miller indices can be shifted about in the lattice simply by choice of the position and orientation of the unit cell. The indices of such equivalent planes are enclosed in braces { } instead of parentheses. For example, in the cubic lattice of Fig. 1-6 all the cube faces

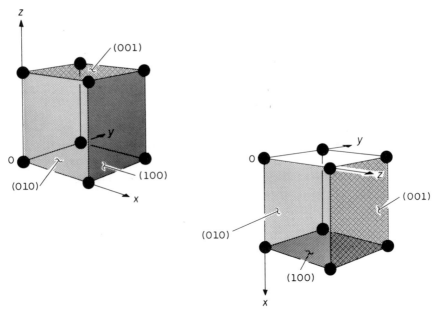

FIGURE 1-6. Equivalence of the cube faces ({100} planes) by rotation of the unit cell within the cubic lattice.

are crystallographically equivalent in that the unit cell can be rotated in various directions and still appear the same. The six equivalent faces are collectively designated as {100}.

A direction in a lattice is expressed as a set of three integers with the same relationship as the components of a vector in that direction. The three vector components are expressed in multiples of the basis vectors, and the three integers are reduced to their smallest values while retaining the relationship among them. For example, the body diagonal in the cubic lattice (Fig. 1-7a) is composed of the components 1**a**, 1**b**, and 1**c**; therefore this diagonal is the [111] direction. (Brackets are used for direction indices.) As in the case of planes, many directions in a lattice are equivalent, depending only on the

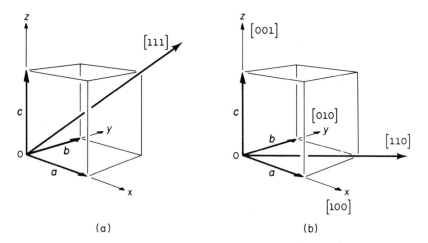

FIGURE 1-7.   Crystal directions in the cubic lattice.

arbitrary choice of orientation for the axes. Such equivalent direction indices are placed in angular brackets $\langle\ \rangle$. For example, the crystal axes in the cubic lattice [100], [010], and [001] are all equivalent and are called $\langle 100 \rangle$ directions (Fig. 1-7b).

   Comparing Figs. 1-6 and 1-7, we notice that in cubic lattices a direction [*hkl*] is perpendicular to the plane (*hkl*). This is convenient in analyzing lattices with cubic unit cells, but it should be remembered that it is not necessarily true in noncubic systems.

1.2.4 The Diamond Lattice.

The basic lattice structure for many important semiconductors is the diamond lattice, which is characteristic of Si and Ge. In many compound semiconductors, atoms are arranged in a basic diamond structure but are different on alternating sites. This is called a *zinc blende* lattice and is typical of the III–V compounds. One of the simplest ways of stating the construction of the diamond lattice is the following:

   The diamond lattice can be thought of as an fcc structure with an extra atom placed at $\mathbf{a}/4 + \mathbf{b}/4 + \mathbf{c}/4$ from each of the fcc atoms.

   Figure 1-8a illustrates the construction of a diamond lattice from an fcc unit cell. We notice that when the vectors are drawn with components one-fourth of the cube edge in each direction, only four additional points within the same unit cell are reached. Vectors drawn from any of the other fcc atoms simply determine corresponding points in adjacent unit cells. This method of constructing the diamond lattice implies that the original fcc has associated with it a second interpenetrating fcc displaced by $\frac{1}{4}, \frac{1}{4}, \frac{1}{4}$. The two interpenetrating fcc *sublattices* can be visualized by looking down on the

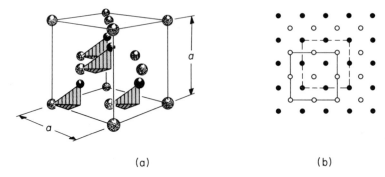

(a)   (b)

FIGURE 1-8. Diamond lattice structure: (a) a unit cell of the diamond lattice constructed by placing atoms $\frac{1}{4}, \frac{1}{4}, \frac{1}{4}$ from each atom in an fcc; (b) top view (along any $\langle 100 \rangle$ direction) of an extended diamond lattice. The open circles indicate one fcc sublattice and the solid circles indicate the interpenetrating fcc.

unit cell of Fig. 1-8a from the top (or along any $\langle 100 \rangle$ direction). In the top view of Fig. 1-8b, atoms belonging to the original fcc are represented by open circles, and the interpenetrating sublattice is shaded. If the atoms are all similar, we call this structure a diamond lattice; if the atoms differ on alternating sites, it is a zinc blende structure. For example, if one fcc sublattice is composed of Ga atoms and the interpenetrating sublattice is As, the zinc blende structure of GaAs results. Most of the compound semiconductors have this type of lattice, although some of the II–VI compounds are arranged in a slightly different structure called the *wurtzite* lattice. We shall restrict our discussions here to the diamond and zinc blende structures since they are typical of most of the commonly used semiconductors.

It is important from an electronic point of view to notice that each atom in the diamond and zinc blende structures is surrounded by four nearest neighbors (Fig. 1-9). The importance of this relationship of each atom to its neighbors will become evident in Section 3.1.1 when we discuss the bonding forces which hold the lattice together.

The fact that atoms in a crystal are arranged in certain planes is important to many of the mechanical, metallurgical, and chemical properties of the material. For example, crystals often can be cleaved along certain atomic planes, resulting in exceptionally planar surfaces. This is a familiar result in cleaved diamonds for jewelry; the facets of a diamond reveal clearly the triangular, hexagonal, and rectangular symmetries of intersecting planes in various crystallographic directions. Semiconductors with diamond and zinc blende lattices have similar cleavage planes. As an example, Fig. 1-10 shows a view along a $\langle 111 \rangle$ body diagonal of a small Si crystal which has been cleaved along intersecting {111} planes. This crystal clearly illustrates

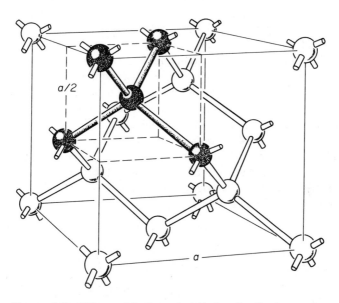

Figure 1-9.   Diamond lattice unit cell, showing the four nearest neighbor structure. From *Electrons and Holes in Semiconductors* by W. Shockley; © 1950 by Litton Educational Publishing Co., Inc.; by permission of Van Nostrand Reinhold Co., Inc.

Figure 1-10.   Silicon {111} surface with aluminum alloyed regions, revealing the symmetry of the diamond lattice.

the triangular symmetry of the {111} planes when viewed along the body diagonal. As an example of the influence of the lattice upon metallurgical properties, small regions of Al have been alloyed on the face of this crystal.†
We notice that the alloyed regions reflect the triangular and hexagonal symmetries of the atomic planes. Chemical reactions, such as etching of the crystal, often take place preferentially along certain crystal directions. These properties serve as interesting illustrations of crystal symmetry, but in addition, each plays an important role in fabrication processes for many semiconductor devices. We shall discuss these applications in later chapters. As a preview of later discussions, we can note a few applications here. Atomic cleavage planes are used to obtain flat and parallel faces in laser diodes (Section 7.4.2). Preferential alloying along crystal planes is used to obtain flat and parallel alloyed regions in alloy transistors (Section 8.3.3). Preferential etching is used to separate devices in the beam–lead method of integrated-circuit fabrication (Section 10.3.2).

1.3 Growth of Semiconductor Crystals

The progress of solid state device technology since the invention of the transistor has depended not only on the development of device concepts but also on the improvement of materials. For example, the fact that integrated circuits can be made today is the result of a considerable breakthrough in the growth of pure, single-crystal Si in the early and mid-1950's. The requirements on the growing of device-grade semiconductor crystals are more stringent than those for any other materials. Not only must semiconductors be available in large single crystals, but also the purity must be controlled within extremely close limits. For example, Si crystals now being used in devices are grown with densities of most impurities of less than one part in ten billion. Such purities require careful handling and treatment of the material at each step of the manufacturing process.

Elemental Si and Ge are obtained by chemical decomposition of compounds such as GeO_2, $SiCl_4$, and $SiHCl_3$, which are readily obtained from the mining, metallurgical, and chemical industries. Once the semiconductor material has been isolated and preliminary purification steps have been performed, it is melted and cast into ingots. Upon cooling from such a casting process, the Si or Ge is polycrystalline (Fig. 1-1c). The atoms are arranged in the diamond lattice over small regions of the ingot, since this is the natural structure for the material. However, unless some control is maintained over

---

†The process of alloying will be discussed in Section 5.1.2.

the cooling process, the crystalline regions occur with essentially random orientation. For the crystal to grow in a single orientation, it is necessary to maintain careful control over the boundary between the molten material and the solid during cooling.

In this section we shall discuss several methods for growing semiconductor crystals from molten material, from solutions of the semiconductor and certain metals, and from materials in vapor form. All these growth processes are useful for obtaining starting materials for semiconductor devices. Furthermore, several of the crystal growth processes play important roles in the fabrication of certain devices.

1.3.1 Growth from the Melt.
A common technique for growing single crystals involves selective cooling of the molten material so that solidification occurs along a particular crystal direction. For example, Fig. 1-11a

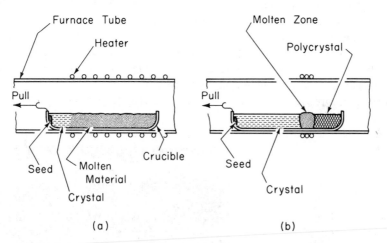

(a)   (b)

FIGURE 1-11.   Crystal growing from the melt in a crucible: (a) solidification from one end of the melt (horizontal Bridgman method); (b) melting and solidification in a moving zone.

shows a silica (quartz) crucible containing molten Ge, which may be pulled through a furnace such that solidification begins at one end and slowly proceeds down the length of the bar. Crystal growth can be enhanced if a small "seed" crystal is placed at the end which is cooled first. If the cooling rate is carefully controlled and the position of the interface between solid and melt is moved slowly along the crucible, the Ge atoms will be arranged in a diamond lattice as the crystal cools. The shape of the resulting crystal is determined by the crucible. Germanium, GaAs, and other semiconductor crystals are often grown by this technique, which is commonly called the horizontal *Bridgman* method. In a variation on this procedure, a small region of the polycrystalline material is melted, and the molten zone is moved down the crucible at such a rate that a crystal is formed behind the zone as it moves (Fig. 1-11b). There are several advantages of impurity control in this zone-growing process and its variations, as we shall see in Section 1.3.2. Historically, this method of crystal preparation led to material of the required purity to permit fabrication of the first transistors. The zone-growing process is one of the most important breakthroughs in modern materials technology. It is interesting that this advance in materials preparation owes its existence to the demands of a new electronic device—the transistor.

One disadvantage of growing the crystal in a containing crucible is that the molten material contacts the sides of the container; the resulting interference of the crucible wall introduces stresses during solidification and deviations from the perfect lattice structure. This is a particularly serious problem in Si, which has a high melting point and tends to adhere to crucible materials. An alternative method which eliminates this problem involves pulling the crystal from the melt as it grows (Fig. 1-12). In this method a seed crystal is lowered into the molten material and then is raised slowly, allowing the crystal to grow onto the seed. Generally, the crystal is rotated slowly as it grows to provide a slight stirring of the melt and to average out any temperature variations which would cause inhomogeneous solidification.

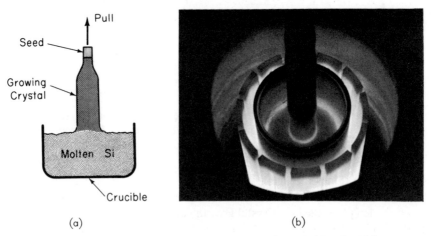

(a)   (b)

FIGURE 1-12. Pulling of a Si crystal from the melt (Czochralski method): (a) schematic diagram of the crystal growth process; (b) view through a port in the furnace, showing a Si crystal being pulled from the melt. (Photograph courtesy of Texas Instruments, Inc.)

This technique, commonly called the *Czochralski* method, is widely used in growing Si, Ge, and some of the compound semiconductors.

In Czochralski crystal growth, the shape of the ingot is determined by a combination of the tendency of the cross section to assume a polygonal shape due to the crystal structure and the influence of surface tension, which encourages a circular cross section. For Ge the cross section of the resulting ingot is polygonal with rounded corners, but the surface tension is so strong in Si that the cross section is almost circular (Fig. 1-13).

FIGURE 1-13. Silicon crystal grown by the Czochralski method. (Photograph courtesy of Monsanto Company.)

In each of these growth methods the important considerations are crystal perfection and freedom from impurities. Good crystals can be grown from molten material by these methods, but the presence of a crucible provides an important source of contamination. The primary impurity introduced by a silica (SiO_2) crucible is oxygen, and there are some semiconductor applications in which the oxygen content must be small. To eliminate contact of the molten material with a container, semiconductor crystals are often grown by a method known as *floating-zone* growth. In this technique the polycrystalline ingot is held vertically by clamps at both ends while a small region of the material is heated to the melting point (Fig. 1-14). Since it is

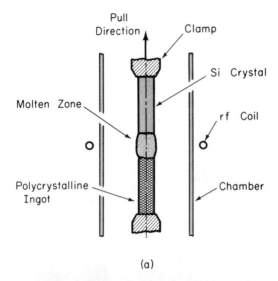

(a)

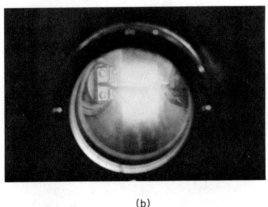

(b)

FIGURE 1-14. Floating-zone crystal growth: (a) schematic diagram of the growth process; (b) view through a furnace port of a Si crystal being grown by the floating-zone process. (Photograph courtesy of Monsanto Company.)

not necessary for the molten zone to be in contact with a container wall, impurity contamination can be greatly reduced. A seed crystal is included at one end of the ingot, and the molten zone is begun at this end. As the ingot is slowly moved relative to the heater, the molten zone travels along the bar, and crystal growth occurs behind the zone at the solid–liquid boundary. If the molten zone is short enough, surface tension will be sufficient to maintain the zone shape against the force of gravity. If the zone is made too long, it will collapse under gravitational force.

Common techniques for melting the ingot include radiation of heat from a resistance heater, induction heating, and heating by electron bombardment. In the induction heating process, electrical energy at radio frequencies (rf) is coupled to the sample by means of one or more turns of a water-cooled coil around the ingot. Large eddy currents are set up in the sample in the region of the coil, resulting in localized heating. This is a particularly convenient technique for the floating-zone growth process, since the heated region can be made small and carefully controlled. The electron bombardment technique involves the acceleration of electrons to very high energies by means of localized electric fields. Upon collision with the sample, the electrons give up their energy as heat. The electron bombardment technique is particularly useful for samples with high electrical resistance, in which it is difficult to induce large eddy currents by the application of rf power.

In all of these crystal growth techniques it is important that the atmosphere surrounding the sample be free of undesirable impurities. In some cases inert or reducing gas atmospheres may be used, and in others the crystals can be grown in a vacuum. The floating-zone method is particularly convenient for vacuum growth, since the sample can be held in an evacuated chamber while the rf coil is kept outside. As an alternative to vacuum growth, a hydrogen atmosphere often is used in cases which require a reducing atmosphere to further decrease the oxygen content of the crystal. When compound semiconductors are grown from the melt, evaporation of volatile materials must be controlled. This is an important problem for the arsenides and phosphides. For example, GaAs can be grown in a crucible or pulled from the melt if a positive pressure of arsenic is maintained over the molten material. This requires a sealed system in most cases, although one simple solution involves covering the melt with a viscous, nonreactive liquid to prevent evaporation.

1.3.2 Zone Refining. The use of a moving molten zone as in Fig. 1-11b and Fig. 1-14 can result in considerable purification of a crystal over the starting material, particularly when several passes are made along the ingot. At the solidifying interface between the melt and the crystal, there will be a certain distribution of impurities between the two phases. An important quantity which identifies this property is the *distribution coefficient* k_d, which is the ratio of the concentration of the impurity in the solid C_S to the concentration in the liquid C_L

(1-2)
$$k_d = \frac{C_S}{C_L}$$

The distribution coefficient is a function of the material, the impurity, the temperature of the solid–liquid interface, and the crystal growth rate. The implications of a distribution coefficient of one-half are illustrated in Fig. 1-15. As the molten zone is moved down the ingot, the relative concentration of the impurity in the molten liquid to that in the refreezing solid is two to one. Thus the concentration of impurities in that portion of material which solidifies first is one-half the original concentration C_o. As the zone

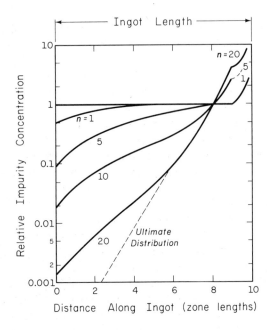

FIGURE **1-15.** Impurity distributions in a crystal after various numbers of passes (**n**) of a molten zone; the impurity distribution coefficient k_d is 0.5 in this example. From W. G. Pfann, "Zone Refining," *Scientific American*, vol. 217, no. 6, pp. 63–72, December 1967. © 1967 by Scientific American, Inc.

TABLE **1-2.**   Impurity distribution coefficients for some of the important impurities in Si and Ge. From F. A. Trumbore, "Solid Solubilities of Impurity Elements in Si and Ge," *Bell System Technical Journal*, vol. 39, pp. 205–233, January, 1960. Copyright 1960, The American Telephone and Telegraph Co., reprinted by permission.

Impurity	Ge	Si
boron	17	0.80
aluminum	0.073	0.0020
gallium	0.087	0.0080
indium	0.001	4×10^{-4}
phosphorus	0.080	0.35
arsenic	0.02	0.3
antimony	0.0030	0.023
lithium	0.002	0.01
copper	1.5×10^{-5}	4×10^{-4}
gold	1.3×10^{-5}	2.5×10^{-5}
manganese	$\sim 10^{-6}$	$\sim 10^{-5}$
iron	3×10^{-5}	8×10^{-6}
cobalt	$\sim 10^{-6}$	8×10^{-6}
zinc	4×10^{-4}	$\sim 1 \times 10^{-5}$

moves along the bar, however, impurities are driven along with the molten material until the concentration in the molten zone approaches C_o/k_d. At that point as many impurities enter the zone as leave it, and Eq. (1-2) is automatically satisfied with $C_S = C_o$ and $C_L = C_o/k_d$. After the first pass, a considerable region of the ingot exists with the original impurity concentration (curve $n = 1$ in Fig. 1-15). If another pass is made in the same direction, however, impurities are again swept with the molten zone until the condition $C_L = 2C_o$ is reached. On the second pass this point occurs farther along the bar, and the purified region is longer. If repeated passes are made, the bar can be purified over much of its length. After many passes, most of the impurities have moved to the end of the bar, which can then be cut away, leaving a highly purified crystal. Of course, if the impurity distribution coefficient k_d is smaller than the $k_d = 0.5$ case of Fig. 1-15, the condition $C_L = C_o/k_d$ is reached farther along the bar, and greater purification is obtained. As Table 1-2 indicates, many important impurities in Si and Ge have very small distribution coefficients; these impurities can be removed effectively with only a few passes of the molten zone. Because of its large distribution coefficient, boron cannot be removed from Si by the the usual zone-refining techniques. This impurity can be removed, however, by surrounding the hot molten zone with water vapor during growth; the boron reacts with the water vapor to form compounds at the surface of the zone, thereby removing boron from the interior of the crystal.

Zone refining is an extremely powerful technique, and it is applicable to materials other than semiconductors as well. Purification by zone refining has allowed studies of many important materials properties which would be completely masked by impurity effects without this technique. For example, copper prepared by zone refining makes possible high-power vacuum switches which could not be operated successfully without extremely pure copper. This application of zone refining illustrates the importance to other fields of the early transistor-motivated research in materials preparation and purification.

1.3.3 Solution Growth. It is possible to grow crystals of many semiconductors at temperatures well below their melting point. Since a mixture of the semiconductor with a second element may melt at a lower temperature than the semiconductor itself, it is often an advantage to grow the crystal from solution at the temperature of the mixture. For example, the melting point of GaAs is 1238°C, whereas a mixture of GaAs with Ga metal has a considerably lower melting point (depending on the proportions of the mixture). Thus, a GaAs seed crystal can be held in a Ga + GaAs solution, which is molten at a temperature low enough that the seed itself is not melted. If the solution is cooled slowly, a single-crystal GaAs layer grows on the seed. As the GaAs leaves the solution and grows on the parent crystal, the solution becomes richer in Ga and thus has a lower melting point. Upon further cooling, more GaAs leaves the solution, and the crystal continues to grow until finally the remaining solution is almost entirely Ga metal. By this technique single crystals can be grown at temperatures low enough to eliminate many problems of impurity introduction typical of growth at the crystal melting temperature. This method is particularly useful for III–V compounds in which Ga or In serves as the column III element, since these metals form solutions at conveniently low temperatures.

The most important application of solution techniques is in the growth of a thin crystalline layer on top of a second crystal, called the *substrate* (Fig. 1-16). The substrate crystal may be a wafer† of the same material as

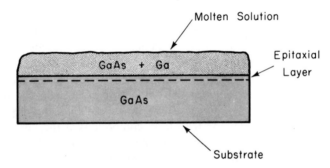

FIGURE 1-16. Solution growth of an epitaxial GaAs layer by slow cooling of a Ga + GaAs solution.

the grown layer or of a different material with a similar lattice structure. In this process the substrate serves as the seed crystal onto which the new crystalline material grows and also defines the shape of the growth. The growing crystal layer maintains the crystal structure and orientation of the substrate. The technique of growing an oriented single-crystal layer on a substrate is called *epitaxial growth*, or *epitaxy* (in this case *liquid epitaxy*). As we shall see in the following section, other methods of epitaxial growth make use of crystallization onto a substrate from a vapor (*vapor epitaxy*). In liquid epitaxy the solution can be placed onto the wafer in a number of ways; one of the most direct methods involves dipping the wafer into the solution and then withdrawing it after the solution has been cooled and growth has occurred on the substrate.

An obvious advantage of epitaxial growth is that pure material can be grown at temperatures well below the melting point of the semiconductor. Since the starting wafer serves mainly as a substrate, its impurity content

---

†*Wafers* are slices cut from a single-crystal ingot.

can differ from that of the epitaxial layer grown onto it. Another important advantage of epitaxy is that a crystal of one semiconductor can be grown on the surface of another. For example, we shall see in Chapter 7 that semiconductor diode lasers can be improved by the growth of the mixed compound AlGaAs onto a crystal of GaAs.

1.3.4 Vapor Growth. The advantages of low temperature and high purity which characterize solution growth can also be achieved by crystallization from the vapor phase. Crystalline layers can be grown onto a seed or substrate from a chemical vapor of the semiconductor material or from mixtures of chemical vapors containing the semiconductor. Epitaxial growth from a vapor phase is a particularly important source of semiconductor material for use in devices. Some compounds such as GaAs can be grown with better purity and crystal perfection by vapor epitaxy than by other methods. Furthermore, these techniques offer great flexibility in the actual fabrication of devices. When an epitaxial layer is grown on a substrate, it is relatively simple to obtain a sharp demarcation between the type of impurity doping in the substrate and in the grown layer. The advantages of this freedom to vary the impurity will be discussed in subsequent chapters. We point out here, however, that Si integrated-circuit devices (Chapter 10) are usually built on epitaxial layers grown by vapor epitaxy on Si wafers.

Epitaxial layers are grown on Si substrates by the controlled deposition of Si atoms onto the surface from a chemical vapor containing Si. Also, Si can be transferred directly from a source to the substrate by *vacuum deposition*, in which the Si source is evaporated onto the substrate in a vacuum chamber. However, this direct technique is not as suitable for mass production as are other methods which utilize the deposition of Si atoms by the chemical reaction of certain gases. In one method, a gas of silicon tetrachloride reacts with hydrogen gas to give Si and hydrochloric acid.†

$$(1\text{-}3) \qquad\qquad SiCl_4 + 2H_2 \longrightarrow Si + 4HCl$$

If this reaction occurs at the surface of a heated crystal, the Si atoms released in the reaction can be deposited as an epitaxial layer. The HCl remains gaseous at the reaction temperature and does not disturb the growing crystal.

The apparatus necessary for this vapor epitaxy technique is shown schematically in Fig. 1-17. The Si slice is heated in a silica tube into which the gases can be introduced. Since the chemical reactions take place in this chamber, it is called a *reaction chamber* or, more simply, a *reactor*. Hydrogen gas is passed through a heated chamber in which $SiCl_4$ is evaporated; then the two gases are introduced into the reactor over the substrate crystal, along with other gases containing the desired doping impurities. The Si slice is

---

†This reaction actually takes place in two steps; for simplicity, we have given only the overall reaction in Eq. (1-3).

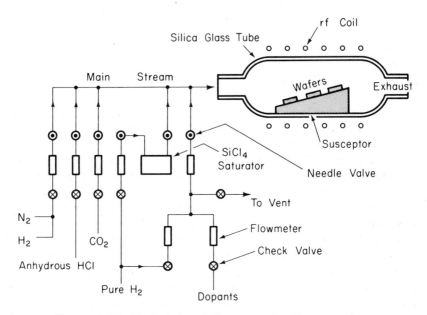

FIGURE 1-17. Typical deposition system for Si vapor epitaxy using a SiCl₄ source. After B. M. Berry, "Epitaxy," in *Fundamentals of Silicon Integrated Device Technology*, Vol. I, ed. R. M. Burger and R. P. Donovan. Englewood Cliffs, N. J.: Prentice-Hall, Inc., © 1967, p. 365.

placed on a graphite susceptor or some other material which can be heated to the reaction temperature with an rf heating coil. This method can be adapted to grow epitaxial layers of closely controlled impurity density on many Si slices simultaneously.

The reaction temperature for the hydrogen reduction of SiCl₄ is approximately 1250°C. Other reactions may be employed at somewhat lower temperatures, including the pyrolysis of silane (SiH₄) at 1000°C. Pyrolysis involves the breaking up of the silane at the reaction temperature

$$(1\text{-}4) \qquad\qquad SiH_4 \longrightarrow Si + 2H_2$$

There are several advantages of this technique, including the fact that the lower reaction temperature reduces migration of impurities from the substrate to the growing epitaxial layer. The hydrogen gas released in the reaction is useful in providing a reducing atmosphere to eliminate oxygen from the growing crystal.

We have discussed only a few of the many techniques of crystal growth which are commonly used in semiconductor device technology. The growth of crystals is an integral part of electronic device design and fabrication, and the problems and possibilities of the various growth techniques must be understood by the device designer. As we shall see in later chapters, crystal growth may be employed at several stages in the fabrication of devices.

READING LIST

N. Mott, "The Solid State," *Scientific American*, vol. 217, no. 3, pp. 62–72, September 1967.

A. H. Cottrell, "The Nature of Metals," *ibid.*, pp. 90–100.

W. G. Pfann, "Zone Refining," *Scientific American*, vol. 217, no. 6, pp. 62–72, December 1967.

M. Tanenbaum, "Semiconductor Crystal Growing," in *Semiconductors*, ed. N. B. Hannay. New York: Van Nostrand Reinhold Co., 1959, pp. 87–144.

B. M. Berry, "Epitaxy," in *Fundamentals of Silicon Integrated Device Technology, Vol. I: Oxidation, Diffusion, and Epitaxy*, eds. R. M. Burger and R. P. Donovan. Englewood Cliffs, N.J.: Prentice-Hall, Inc., 1967, pp. 349–470.

J. P. McKelvey, *Solid State and Semiconductor Physics*. New York: Harper & Row, Publishers, Inc., 1966, pp. 1–37 and 371–378.

R. A. Laudise, *Growth of Single Crystals*. Englewood Cliffs, N.J.: Prentice-Hall, Inc., 1970.

PROBLEMS

1.1 Show that the maximum fraction of the unit cell volume which can be filled by hard spheres in the sc, bcc, and diamond lattices are 0.52, 0.68, and 0.34, respectively.

1.2 Beginning with a sketch of an fcc lattice, add atoms at $(\frac{1}{4}, \frac{1}{4}, \frac{1}{4})$ from each fcc atom to obtain the diamond lattice. Show that only the four added atoms in Fig. 1-8a appear in the diamond unit cell.

1.3 Calculate the densities of Si and GaAs from the lattice constants (Appendix III), atomic weights, and Avogadro's number. Compare the results with densities given in Appendix III. The atomic weights of Si, Ga, and As are 28.1, 69.7, and 74.9 respectively.

1.4 Sodium chloride (NaCl) is a cubic crystal which differs from an sc in that alternating atoms are different; each Na is surrounded by six Cl nearest neighbors and vice versa in the three-dimensional lattice. Draw a two-dimensional NaCl lattice (e.g., looking down a $\langle 100 \rangle$ direction) and indicate a unit cell. Remember the unit cell must be repetitive upon displacement by the basis vectors.

1.5 (a) Label the plane illustrated in Fig. P1-5.
(b) Draw the equivalent $\langle 111 \rangle$, $\langle 100 \rangle$, and $\langle 110 \rangle$ directions in a cubic lattice; use a unit cube for illustrating each set of equivalent directions.

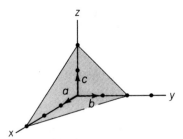

FIGURE **P1-5**

1.6 Show by a sketch that the bcc lattice can be represented by two inter-penetrating sc lattices. To simplify the sketch, show a $\langle 100 \rangle$ view of the lattice.

1.7 The atoms seen in Fig. 1-8b along a $\langle 100 \rangle$ direction of the diamond lattice are not all coplanar. Taking the top face of the cube in Fig. 1-8a to be (0), the parallel plane $a/4$ down to be ($\frac{1}{4}$), the plane through the center to be ($\frac{1}{2}$), and the second plane of shaded atoms to be ($\frac{3}{4}$), label the plane of each atom in Fig. 1-8b.

1.8 (a) Sketch an sc unit cell viewed along a $\langle 111 \rangle$ direction.
(b) Assume a slice of this material has $\{111\}$ surfaces top and bottom. If the crystal cleaves along $\{100\}$ planes, what sample shapes would you expect? What angles would the cleaved sides make with the top surface?

1.9 Sketch a view down a $\langle 110 \rangle$ direction of a diamond lattice. *Hint:* Fig. 1-9 can be mentally shifted so that the nearest vertical edge and its diagonally opposite counterpart are aligned. Include lines connecting nearest neighbors. What shape are the open channels in a $\langle 110 \rangle$ direction? For an interesting extended view of the diamond lattice in this direction, see W. Brandt, "Channeling in Crystals," *Scientific American*, vol. 218, no. 3, p. 94, March 1968.

1.10 A Si crystal is to be grown by the Czochralski method, and it is desired that the ingot contain 10^{16} phosphorus atoms/cm³.

(a) What concentration of phosphorus atoms should the melt contain to give this impurity density in the crystal during the initial growth?

(b) If the initial load of Si in the crucible is 500 g, how many grams of phosphorus should be added? The atomic weight of phosphorus is 31.

(c) Does the phosphorus concentration in the ingot remain at 10^{16} cm^{-3} throughout the growth? Discuss in qualitative terms the phosphorus distribution in the grown crystal. In practice, a more uniform impurity distribution is achieved by varying the growth rate (which varies k_d) appropriately during the crystal growth.

1.11 In this problem we wish to calculate the distribution $C_S(x)$ of a particular impurity in a solid after one pass of a molten zone of length *l*. The distribution coefficient is k_d, and Eq. (1-2) is valid at x, the boundary between the molten zone and the solidifying crystal. Assume the geometry shown in Fig. P1-11; the concentration of the impurity

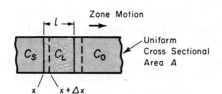

FIGURE P1-11

throughout the molten zone is $C_L(x)$, and the concentration to the right of the zone is the starting value C_0. Show that the distribution after one pass is

$$C_S(x) = C_0 - C_0(1 - k_d)e^{-k_d x/l}$$

Hint: Assume the zone in Fig. P1-11 moves a distance Δx to the right. Write the following balance: The quantity of impurity in the zone after the shift $lAC_L(x + \Delta x)$ equals that before the shift, minus the quantity solidified at the left, plus that melted into the zone at the right. Obtain the standard derivative form and let Δx approach zero. The resulting differential equation can be solved easily for $C_L(x)$, using the boundary condition that $C_L(x = 0) = C_0$.

1.12 We wish to zone refine an ingot of Ge which contains a uniform indium (In) concentration of 10^{16} cm^{-3}. One pass is made of a molten zone 1 cm long. Over what distance is the resulting In concentration below 5×10^{14} cm^{-3}? Assume the result of Prob. 1.11 as given.

ATOMS AND ELECTRONS $\mathbf{2}$

Since this book is primarily an introduction to solid state devices, it would be preferable not to delay this discussion with subjects such as atomic theory, quantum mechanics, and electron models. However, the behavior of solid state devices is directly related to these subjects. For example, it would be difficult to understand how an electron is transported through a semiconductor device without some knowledge of the electron and its interaction with the crystal lattice. Therefore, in this chapter we shall investigate some of the important properties of electrons, with special emphasis on two points: (1) the electronic structure of atoms and (2) the interaction of atoms and electrons with excitation, such as the absorption and emission of light. By studying electron energies in an atom, we lay the foundation for understanding the influence of the lattice on electrons participating in current flow through a solid. Our discussions concerning the interaction of light with electrons form the basis for later descriptions of changes in the conductivity of a semiconductor with optical excitation, properties of light-sensitive devices, and lasers.

First, we shall investigate some of the experimental observations which led to the modern concept of the atom, and then we shall give a brief introduction to the theory of quantum mechanics. Several important concepts will emerge from this introduction: The electrons in atoms are restricted to certain energy levels by quantum rules; the electronic structure of atoms is determined from these quantum conditions; and this "quantization" defines certain allowable transitions involving absorption and emission of energy by the electrons.

2.1 Introduction to Physical Models

The main effort of science is to describe what happens in nature, in as complete and concise a form as possible. In physics this effort involves observing natural phenomena, relating these observations to previously established theory, and finally establishing a physical model for the observations. The primary purpose of the model is to allow the information obtained in present observations to be used to understand new experiments. Therefore, the most useful models are expressed mathematically, so that quantitative explanations of new experiments can be made succinctly in terms of established principles. For example, we can explain the behavior of a spring-supported weight moving up and down periodically after an initial displacement, because the differential equations describing such a simple harmonic motion have been established and are understood by students of elementary physics. But the physical model upon which these equations of motion are based arises from serious study of natural phenomena such as gravitational force, the response of bodies to accelerating forces, the relationship of kinetic and potential energy, and the properties of springs. The mass and spring problem is relatively easy to solve because each of these properties of nature is well understood.

When a new physical phenomenon is observed, it is necessary to find out how it fits into the established models and "laws" of physics. In the vast majority of cases this involves a direct extension of the mathematics of well-established models to the particular conditions of the new problem. In fact, it is not uncommon for a scientist or engineer to predict that a new phenomenon should occur before it is actually observed, simply by a careful study and extension of existing models and laws. The beauty of science is that natural phenomena are not isolated events but are related to other events by a few analytically describable laws. However, it does happen occasionally that a set of observations cannot be described in terms of existing theories. In such cases it is necessary to develop models which are based as far as possible on existing laws, but which contain new aspects arising from the new phenomena. Postulating new physical principles is a serious business, and it is done only when there is no possibility of explaining the observations with established theory. When new assumptions and models are made, their justification lies in the following question: "Does the model describe precisely the observations, and can reliable predictions be made based on the model?" The model is good or poor depending on the answer to this question.

In the 1920's it became necessary to develop a new theory to describe phenomena on the atomic scale. A long series of careful observations had been made which clearly indicated that many events involving electrons and atoms did not obey the classical laws of mechanics. It was necessary, therefore, to develop a new kind of mechanics to describe the behavior of particles on this small scale. This new approach, called *quantum mechanics*, describes atomic phenomena very well and also properly predicts the way in which electrons behave in solids—our primary interest here. Through the years, quantum mechanics has been so successful that now it stands beside the classical laws as a valid description of nature.

A special problem often arises when students first encounter the theory of quantum mechanics. The problem is that quantum concepts are largely mathematical in nature and do not involve the "common sense" quality associated with classical mechanics. At first, many students find quantum concepts difficult, not so much because of the mathematics involved, but because they feel the concepts are somehow divorced from "reality." This is a reasonable reaction, since ideas which we consider to be real or intuitively satisfying are usually based on our own observation. Thus the classical laws of motion are easy to understand because we observe bodies in motion every day. On the other hand, we observe the effects of atoms and electrons only indirectly, and naturally we have very little feeling for what is happening on the atomic scale. It is necessary, therefore, to depend on the facility of the theory to predict experimental results rather than to attempt to force classical analogues onto the nonclassical phenomena of atoms and electrons.

Our approach in this chapter will be to investigate the important experimental observations which led to the quantum theory, and then to indicate how the theory accounts for these observations. Discussions of quantum theory must necessarily be largely qualitative in such a brief presentation, and those topics which are most important to solid state theory will be emphasized here. Several good references for further individual study are given at the end of this chapter.

2.2 Experimental Observations

The experiments which led to the development of quantum theory were concerned with the nature of light and the relation of optical energy to the energies of electrons within atoms. These experiments supplied only indirect evidence of the nature of phenomena on the atomic scale; however, the cumulative results of a number of careful experiments showed clearly that a new theory was needed.

2.2.1 The Photoelectric Effect. An important observation by
Planck indicated that radiation from a heated sample is emitted in discrete
units of energy, called *quanta;* the energy units were described by $h\nu$, where ν
is the frequency of the radiation, and h is a quantity now called Planck's
constant ($h = 6.63 \times 10^{-34}$ J-sec). Soon after Planck developed this hypothesis, Einstein interpreted an important experiment which clearly demonstrated the discrete nature (*quantization*) of light. This experiment involved
absorption of optical energy by the electrons in a metal and the relationship
between the amount of energy absorbed and the frequency of the light (Fig.
2-1). Let us suppose that monochromatic light is incident on the surface of

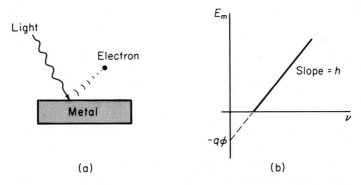

FIGURE 2-1.   The photoelectric effect: (a) electrons are ejected
from the surface of a metal when exposed to light of frequency
ν in a vacuum; (b) plot of the maximum kinetic energy of ejected
electrons vs. frequency of the incoming light.

a metal plate in a vacuum. The electrons in the metal absorb energy from
the light, and some of the electrons receive enough energy to be ejected from
the metal surface into the vacuum. This phenomenon is called the *photoelectric effect.* If the energy of the escaping electrons is measured, a plot can be
made of the maximum energy as a function of the frequency ν of the incident
light (Fig. 2-1b).

One simple way of finding the maximum energy of the ejected electrons
is to place another plate above the one shown in Fig. 2-1a and then create
an electric field between the two plates. The potential necessary to retard
all electron flow between the plates gives the energy E_m (Prob. 2.1). For a
particular frequency of light ν incident on the sample, a maximum energy
E_m is observed for the emitted electrons. The resulting plot of E_m vs. ν is
linear, with a slope equal to Planck's constant. The equation of the line
shown in Fig. 2-1b is

(2-1) $$E_m = h\nu - q\phi$$

where q is the magnitude of the electronic charge. The quantity ϕ (volts) is a characteristic of the particular metal used and is called the *work function* of the metal. When ϕ is multiplied by the electronic charge, an energy (joules) is obtained which represents the minimum energy required for an electron to escape from the metal into a vacuum. These results indicate that the electrons receive an energy $h\nu$ from the light and lose an amount of energy $q\phi$ in escaping from the surface of the metal.

This experiment demonstrates clearly that Planck's hypothesis was correct—light energy is contained in discrete units rather than in a continuous distribution of energies. Other experiments also indicate that, in addition to the wave nature of light, the quantized units of light energy can be considered as localized packets of energy, called *photons*. Some experiments emphasize the wave nature of light, while other experiments reveal the discrete nature of photons. This duality is fundamental to quantum processes and does not imply an ambiguity in the theory.

2.2.2 Atomic Spectra. One of the most valuable experiments of modern physics is the analysis of absorption and emission of light by atoms. For example, an electric discharge can be created in a gas, so that the atoms begin to emit light with wavelengths characteristic of the gas. We see this effect in a neon sign, which is typically a glass tube filled with neon or a gas mixture, with electrodes for creating a discharge. If the intensity of the emitted light is measured as a function of wavelength, one finds a series of sharp lines rather than a continuous distribution of wavelengths. By the early 1900's the characteristic spectra for several atoms were well known. A portion of the measured emission spectrum for hydrogen is shown in Fig. 2-2, in which the vertical lines represent the positions of observed emission peaks on the wavelength scale. Wavelength (λ) is usually measured in angstroms (1 Å = 10^{-10} m) and is related (in meters) to frequency by

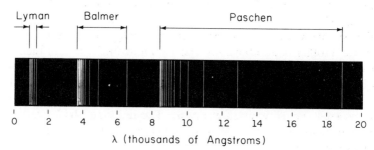

FIGURE 2-2. Some important lines in the emission spectrum of hydrogen.

(2-2)
$$\lambda = \frac{c}{v}$$

where c is the speed of light (3×10^8 m/sec).

The lines in Fig. 2-2 appear in several groups labeled the *Lyman, Balmer,* and *Paschen* series after their early investigators. Once the hydrogen spectrum was established, scientists noticed several interesting relationships among the lines. The various series in the spectrum were observed to follow certain empirical forms:

(2-3a)   Lyman: $v = cR\left(\dfrac{1}{1^2} - \dfrac{1}{n^2}\right),$ $n = 2, 3, 4, \ldots$

(2-3b)   Balmer: $v = cR\left(\dfrac{1}{2^2} - \dfrac{1}{n^2}\right),$ $n = 3, 4, 5, \ldots$

(2-3c)   Paschen: $v = cR\left(\dfrac{1}{3^2} - \dfrac{1}{n^2}\right),$ $n = 4, 5, 6, \ldots$

where R is a constant called the Rydberg constant ($R = 109{,}678$ cm^{-1}). If the photon energies hv are plotted for successive values of the integer n, we notice that each energy can be obtained by taking sums and differences of other photon energies in the spectrum (Fig. 2-3). For example, E_{42} in the

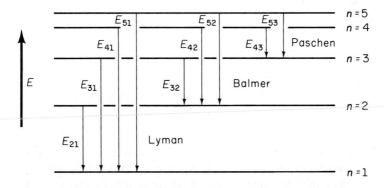

FIGURE 2-3. Relationships among photon energies in the hydrogen spectrum.

Balmer series is the difference between E_{41} and E_{21} in the Lyman series. This relationship among the various series is called the *Ritz Combination Principle*. Naturally, these empirical observations stirred a great deal of interest in constructing a comprehensive theory for the origin of the photons given off by atoms.

2.3 The Bohr Model

The results of emission spectra experiments led Niels Bohr to construct a model for the hydrogen atom, based on the mathematics of planetary systems. If the electron in the hydrogen atom has a series of planetary-type orbits available to it, it can be excited to an outer orbit and then can fall to any one of the inner orbits, giving off energy corresponding to one of the lines of Fig. 2-3. To develop the model, Bohr made several postulates:

1. Electrons exist in certain stable, circular orbits about the nucleus. This assumption implies that the orbiting electron does not give off radiation as classical electromagnetic theory would normally require of a charge experiencing angular acceleration; otherwise, the electron would not be stable in the orbit but would spiral into the nucleus as it lost energy by radiation.

2. The electron may shift to an orbit of higher or lower energy, thereby gaining or losing energy equal to the difference in the energy levels (by absorption or emission of a photon of energy $h\nu$).

(2-4) $$h\nu = E_2 - E_1$$

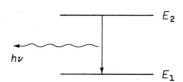

3. The angular momentum p_θ of the electron in an orbit is always an integral multiple of Planck's constant divided by 2π ($h/2\pi$ is often abbreviated $\hbar$ for convenience). This assumption

(2-5) $$p_\theta = n\hbar, \qquad n = 1, 2, 3, 4, \ldots$$

is necessary to obtain the observed results of Fig. 2-3.

If we visualize the electron in a stable orbit of radius r about the proton of the hydrogen atom, we can equate the electrostatic force between the charges to the centripetal force

(2-6)
$$-\frac{q^2}{Kr^2} = -\frac{mv^2}{r}$$

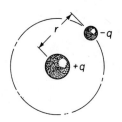

where $K = 4\pi\epsilon_0$ in MKS units, m is the mass of the electron, and v is its velocity. From assumption 3 we have

(2-7)
$$p_\theta = mvr = n\hbar$$

Since **n** takes on integral values, r should be denoted r_n to indicate the **n**'th orbit. Then Eq. (2-7) can be written

(2-8)
$$m^2v^2 = \frac{n^2\hbar^2}{r_n^2}$$

Substituting Eq. (2-8) into Eq. (2-6) we find

(2-9)
$$\frac{q^2}{Kr_n^2} = \frac{1}{mr_n} \cdot \frac{n^2\hbar^2}{r_n^2}$$

(2-10)
$$r_n = \frac{Kn^2\hbar^2}{mq^2}$$

for the radius of the **n**'th orbit of the electron. Now we must find the expression for the total energy of the electron in this orbit, so that we can calculate the energies involved in transitions between orbits.

From Eqs. (2-7) and (2-10) we have

(2-11)
$$v = \frac{n\hbar}{mr_n}$$

(2-12)
$$v = \frac{n\hbar q^2}{Kn^2\hbar^2} = \frac{q^2}{Kn\hbar}$$

Therefore, the kinetic energy of the electron is

(2-13)
$$\text{K.E.} = \frac{1}{2}mv^2 = \frac{mq^4}{2K^2n^2\hbar^2}$$

The potential energy is the product of the electrostatic force and the distance between the charges

(2-14)
$$\text{P.E.} = -\frac{q^2}{Kr_n} = -\frac{mq^4}{K^2\mathbf{n}^2\hbar^2}$$

Thus the total energy of the electron in the **n**'th orbit is

(2-15)
$$E_n = \text{K.E.} + \text{P.E.} = -\frac{mq^4}{2K^2\mathbf{n}^2\hbar^2}$$

The critical test of the model is whether energy differences between orbits correspond to the observed photon energies of the hydrogen spectrum. The transitions between orbits corresponding to the Lyman, Balmer, and Paschen series are illustrated in Fig. 2-4. The energy difference between orbits $\mathbf{n}_1$ and $\mathbf{n}_2$ is given by

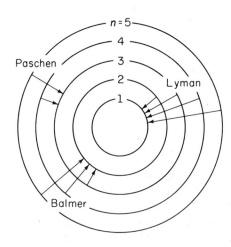

FIGURE 2-4. Electron orbits and transitions in the Bohr model of the hydrogen atom. Orbit spacing is not drawn to scale.

(2-16)
$$E_{n2} - E_{n1} = \frac{mq^4}{2K^2\hbar^2}\left(\frac{1}{\mathbf{n}_1{}^2} - \frac{1}{\mathbf{n}_2{}^2}\right)$$

The frequency of light given off by a transition between these two orbits is

(2-17)
$$\nu_{21} = \left[\frac{mq^4}{2K^2\hbar^2h}\right]\left(\frac{1}{\mathbf{n}_1{}^2} - \frac{1}{\mathbf{n}_2{}^2}\right)$$

The factor in brackets is essentially the Rydberg constant R times the speed of light c (Prob. 2.4). A comparison of Eq. (2-17) with the experimental results summed up by Eq. (2-3) indicates that the Bohr theory provides a good model for electronic transitions within the hydrogen atom, as far as the early experimental evidence is concerned.

Whereas the Bohr model accurately describes the gross features of the hydrogen spectrum, it does not include many fine points. For example, experimental evidence indicates some splitting of levels in addition to the levels predicted by the theory. Also, difficulties arise in extending the model to atoms more complicated than hydrogen. Attempts were made to modify the Bohr model for more general cases, but it soon became obvious that a more comprehensive theory was needed. However, the partial success of the Bohr model was an important step toward the eventual development of the quantum theory. The concept that electrons are quantized in certain allowed energy levels and the relationship of photon energy and transitions between levels had been established firmly by the Bohr theory.

2.4 Quantum Mechanics

The principles of quantum mechanics were developed from two different points of view at about the same time (the late 1920's). One approach, developed by Heisenberg, utilizes the mathematics of matrices and is called *matrix mechanics*. Independently, Schrödinger developed an approach utilizing a wave equation, now called *wave mechanics*. These two mathematical formulations appear to be quite different. However, closer examination reveals that beyond the formalism, the basic principles of the two approaches are the same. It is possible to show, for example, that the results of matrix mechanics reduce to those of wave mechanics after mathematical manipulation. We shall concentrate here on the wave mechanics approach, since solutions to a few simple problems can be obtained with it, involving less mathematical discussion.

2.4.1 Probability and the Uncertainty Principle. It is impossible to describe with absolute precision events involving individual particles on the atomic scale. Instead, we must speak of the average values (*expectation values*) of position, momentum, and energy of a particle such as an electron. It is important to note, however, that the uncertainties revealed in quantum calculations are not based on some shortcoming of the theory. In fact, a major strength of the theory is that it describes the probabilistic nature of events involving atoms and electrons. The fact is that such quantities as the position and momentum of an electron *do not exist* apart from a particular uncertainty. The magnitude of this inherent uncertainty is described by the *Heisenberg Uncertainty Principle:*†

> In any measurement of the position and momentum of a particle,
> the uncertainties in the two measured quantities will be related by

---

†This is often called the *Principle of Indeterminacy*.

(2-18)
$$(\Delta x)(\Delta p_x) \geq \hbar$$

Similarly, the uncertainties in an energy measurement will be related to the uncertainty in the time at which the measurement was made by

(2-19)
$$(\Delta E)(\Delta t) \geq \hbar$$

These limitations indicate that simultaneous measurement of position and momentum or of energy and time are inherently inaccurate to some degree. Of course, Planck's constant h is a rather small number (6.63×10^{-34} J-sec), and we are not concerned with this inaccuracy in the measurement of x and p_x for a truck, for example. On the other hand, measurements of the position of an electron and its speed are seriously limited by the Uncertainty Principle.

One implication of the Uncertainty Principle is that we cannot properly speak of *the* position of an electron, for example, but must look for the *probability* of finding an electron at a certain position. Thus one of the important results of quantum mechanics is that a *probability density function* can be obtained for a particle in a certain environment, and this function can be used to find the expectation value of important quantities such as position, momentum, and energy. We are familiar with the methods for calculating discrete (single-valued) probabilities from common experience. For example, it is clear that the probability of drawing a particular card out of a random deck is $\frac{1}{52}$, and the probability that a tossed coin will come up heads is $\frac{1}{2}$. The techniques for making predictions when the probability varies are less familiar, however. In such cases it is common to define a probability density function which describes, for example, the probability of finding a particle within a certain volume. Given a probability density function $P(x)$ for a one-dimensional problem, the probability of finding the particle in a range from x to $x + dx$ is $P(x)\,dx$. Since the particle will be *somewhere*, this definition implies that

(2-20)
$$\int_{-\infty}^{\infty} P(x)\,dx = 1$$

if the function $P(x)$ is properly chosen. Equation (2-20) is implied by stating that the function $P(x)$ is *normalized* (i.e., the integral equals unity).

To find the average value of a function of x, we need only multiply the value of that function in each increment dx by the probability of finding the particle in that dx and sum over all x. Thus the average value of $f(x)$ is

(2-21a)
$$\langle f(x) \rangle = \int_{-\infty}^{\infty} f(x)\,P(x)\,dx$$

If the probability density function is not normalized, this equation should be written

(2-21b)
$$\langle f(x) \rangle = \frac{\int_{-\infty}^{\infty} f(x)\, P(x)\, dx}{\int_{-\infty}^{\infty} P(x)\, dx}$$

2.4.2 The Schrödinger Wave Equation. There are several ways to develop the wave equation by applying quantum concepts to various classical equations of mechanics. One of the simplest approaches is to consider a few basic postulates, develop the wave equation from them, and rely on the accuracy of the results to serve as a justification of the postulates. In more advanced texts these assumptions are dealt with in more convincing detail.

Basic Postulates:

1. Each particle in a physical system is described by a wave function $\Psi(x, y, z, t)$. This function and its space derivative $(\partial\Psi/\partial x + \partial\Psi/\partial y + \partial\Psi/\partial z)$ are continuous, finite, and single valued.

2. In dealing with classical quantities such as energy E and momentum p, we must relate these quantities with abstract quantum mechanical operators defined in the following way:

Classical variable	Quantum operator
x	x
$f(x)$	$f(x)$
p(x)	$\dfrac{\hbar}{j}\dfrac{\partial}{\partial x}$
E	$-\dfrac{\hbar}{j}\dfrac{\partial}{\partial t}$

and similarly for the other two directions.

3. The probability of finding a particle with wave function Ψ in the volume $dx\, dy\, dz$ is $\Psi^*\Psi\, dx\, dy\, dz$.† The product $\Psi^*\Psi$ is normalized according to Eq. (2-20) so that

$$\int_{-\infty}^{\infty} \Psi^*\Psi\, dx\, dy\, dz = 1$$

---

†Ψ^* is the *complex conjugate* of Ψ obtained by reversing the sign on each j. Thus $(e^{jx})^* = e^{-jx}$.

and the average value $\langle Q \rangle$ of any variable Q is calculated from the wave function by using the operator form Q_{op} defined in postulate 2

$$\langle Q \rangle = \int_{-\infty}^{\infty} \Psi^* Q_{op} \Psi \, dx \, dy \, dz$$

Once we find the wave function Ψ for a particle, we can calculate its average position, energy, and momentum, within the limits of the Uncertainty Principle. Thus a major part of the effort in quantum calculations involves solving for Ψ within the conditions imposed by a particular physical system. We notice from assumption 3 that the probability density function is $\Psi^*\Psi$, or $|\Psi|^2$.

The classical equation for the energy of a particle can be written

$$\text{Kinetic energy} + \text{potential energy} = \text{total energy}$$

(2-22)
$$\frac{1}{2m}p^2 \quad + \quad V \quad = \quad E$$

In quantum mechanics we use the operator form for these variables (postulate 2); the operators are allowed to operate on the wave function Ψ. For a one-dimensional problem Eq. (2-22) becomes†

(2-23)
$$-\frac{\hbar^2}{2m}\frac{\partial^2 \Psi(x, t)}{\partial x^2} + V(x)\,\Psi(x, t) = -\frac{\hbar}{j}\frac{\partial \Psi(x, t)}{\partial t}$$

which is the Schrödinger wave equation. In three dimensions the equation is

(2-24)
$$-\frac{\hbar^2}{2m}\nabla^2\Psi + V\Psi = -\frac{\hbar}{j}\frac{\partial \Psi}{\partial t}$$

where $\nabla^2\Psi$ is

$$\frac{\partial^2 \Psi}{\partial x^2} + \frac{\partial^2 \Psi}{\partial y^2} + \frac{\partial^2 \Psi}{\partial z^2}$$

The wave function Ψ in Eqs. (2-23) and (2-24) includes both space and time dependencies. It is common to calculate these dependencies separately and combine them later. Furthermore, many problems are time independent, and only the space variables are necessary. Thus we try to solve the wave equation by breaking it into two equations by the technique of separation of variables. Let $\Psi(x, t)$ be represented by the product $\psi(x)\,\phi(t)$. Using this product in Eq. (2-23) we obtain

---

†The operational interpretation of $(\partial/\partial x)^2$ is the second derivative form $\partial^2/\partial x^2$; the square of j is -1.

$$(2\text{-}25) \qquad -\frac{\hbar^2}{2m}\frac{\partial^2\psi(x)}{\partial x^2}\,\phi(t) + V(x)\,\psi(x)\,\phi(t) = -\frac{\hbar}{j}\,\psi(x)\,\frac{\partial\phi}{\partial t}$$

Now the variables can be separated (Prob. 2.8) to obtain the time-dependent equation in one dimension

$$(2\text{-}26) \qquad \frac{d\phi(t)}{dt} + \frac{jE}{\hbar}\,\phi(t) = 0$$

and the time-independent equation

$$(2\text{-}27) \qquad \frac{d^2\psi(x)}{dx^2} + \frac{2m}{\hbar^2}[E - V(x)]\,\psi(x) = 0$$

We can show that the separation constant E corresponds to the energy of the particle when particular solutions are obtained, such that a wave function ψ_n corresponds to a particle energy E_n.

These equations are the basis of wave mechanics. From them we can determine the wave functions for particles in various simple systems. For calculations involving electrons, the potential term $V(x)$ usually results from an electrostatic or magnetic field.

2.4.3 Potential Well Problem.

It is quite difficult to find solutions to the Schrödinger equation for most realistic potential fields. One can solve the problem with some effort for the hydrogen atom, for example, but solutions for more complicated atoms are hard to obtain. There are several important problems, however, which illustrate the theory without complicated manipulation. The simplest problem is the potential energy well with infinite boundaries. Let us assume a particle is trapped in a potential well with $V(x)$ zero except at the boundaries $x = 0$ and L, where it is infinitely large (Fig. 2-5a)

$$(2\text{-}28) \qquad \begin{aligned} V(x) &= 0, & 0 < x < L \\ V(x) &= \infty, & x = 0, L \end{aligned}$$

Inside the well we set $V(x) = 0$ in Eq. (2-27)

$$(2\text{-}29) \qquad \frac{d^2\psi(x)}{dx^2} + \frac{2m}{\hbar^2}E\,\psi(x) = 0, \qquad 0 < x < L$$

This is the wave equation for a free particle; it applies to the potential well problem in the region with no potential $V(x)$.

Possible solutions to Eq. (2-29) are $\sin kx$ and $\cos kx$, where k is $\sqrt{2mE}/\hbar$. In choosing a solution, however, we must examine the boundary conditions. The only allowable value of ψ at the walls is zero. Otherwise, there would

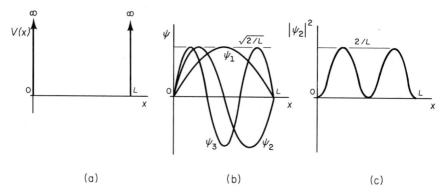

FIGURE 2-5. The problem of a particle in a potential well: (a) potential energy diagram; (b) wave functions in the first three quantum states; (c) probability density distribution for the second state.

be a nonzero $|\psi|^2$ outside the potential well, which is impossible because a particle cannot penetrate an infinite barrier. Therefore, we must choose only the sine solution and define k such that sin kx goes to zero at $x = L$.

$$(2\text{-}30) \qquad\qquad \psi = A \sin kx, \qquad k = \frac{\sqrt{2mE}}{\hbar}$$

The constant A is the amplitude of the wave function and will be evaluated from the normalization condition (postulate 3). If ψ is to be zero at $x = L$, then k must be some integral multiple of π/L

$$(2\text{-}31) \qquad\qquad k = \frac{n\pi}{L}, \qquad n = 1, 2, 3, \ldots$$

From Eqs. (2-30) and (2-31) we can solve for the total energy E_n for each value of the integer **n**

$$(2\text{-}32) \qquad\qquad \frac{\sqrt{2mE_n}}{\hbar} = \frac{n\pi}{L}$$

$$(2\text{-}33) \qquad\qquad E_n = \frac{n^2\pi^2\hbar^2}{2mL^2}$$

Thus for each allowable value of **n** the particle energy is described by Eq. (2-33). We notice that *the energy is quantized.* Only certain values of energy are allowed. The integer **n** is called a *quantum number;* the particular wave function ψ_n and corresponding energy state E_n describe the *quantum state* of the particle.

The constant A is found from postulate 3

(2-34) $$\int_{-\infty}^{\infty} \psi^*\psi \, dx = \int_0^L A^2 \left(\sin \frac{n\pi}{L}x\right)^2 dx = A^2 \frac{L}{2}$$

Setting this equal to unity we obtain

(2-35) $$A = \sqrt{\frac{2}{L}}, \qquad \psi_n = \sqrt{\frac{2}{L}} \sin \frac{n\pi}{L}x$$

The first three wave functions ψ_1, ψ_2, ψ_3 are sketched in Fig. 2-5b. The probability density function $\psi^*\psi$, or in this case $|\psi|^2$, is sketched for ψ_2 in Fig. 2-5c.

2.4.4 Tunneling. The wave functions are relatively easy to obtain for the potential well with infinite walls, since the boundary conditions force ψ to zero at the walls. A slight modification of this problem illustrates a principle which is very important in some solid state devices—the quantum mechanical *tunneling* of an electron through a barrier of finite height and thickness. Let us consider the potential barrier of Fig. 2-6. If the barrier is not infinite,

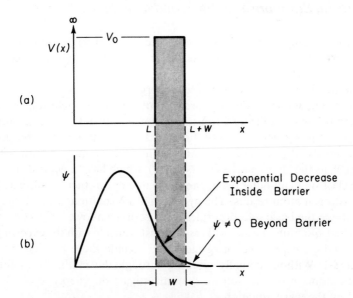

FIGURE 2-6. Quantum mechanical tunneling: (a) potential barrier of height V_0 and thickness W; (b) wave function ψ for an electron with energy $E < V_0$, indicating a non-zero value of the wave function beyond the barrier.

the boundary conditions do not force ψ to zero at the barrier. Instead, we must use the condition that ψ and its slope $d\psi/dx$ are continuous at each boundary of the barrier (postulate 1). Thus ψ must have a finite value within the barrier and also on the other side. Since ψ has a value to the right of the barrier, $\psi^*\psi$ exists there also, implying that there is some probability of finding the particle beyond the barrier. We notice that the particle does not go over the barrier; its total energy is assumed to be less than the barrier height V_0. The mechanism by which the particle "penetrates" the barrier is called tunneling. However, no classical analogue, including classical descriptions of tunneling through barriers, is appropriate for this effect. Quantum mechanical tunneling is intimately bound to the Uncertainty Principle. If the barrier is sufficiently thin, we cannot say with certainty that the particle exists only on one side. However, the wave function amplitude for the particle is reduced by the barrier as Fig. 2-6 indicates, so that by making the thickness W greater, we can reduce ψ on the right-hand side to the point that negligible tunneling occurs. Tunneling is important only over very small dimensions, but it can be of great importance in the conduction of electrons in solids, as we shall see in Chapters 5 and 6.

2.5 Atomic Structure and the Periodic Table

The Schrödinger equation describes accurately the interactions of particles with potential fields, such as electrons within atoms. Indeed, the modern understanding of atomic theory (the modern atomic *models*) comes from the wave equation and from Heisenberg's matrix mechanics. It should be pointed out, however, that the problem of solving the Schrödinger equation directly for complicated atoms is extremely difficult. In fact, only the hydrogen atom is generally solved directly; atoms of atomic number greater than one are usually handled by techniques involving approximations. Many atoms such as the alkali metals (Li, Na, etc.), which have a neutral core with a single electron in an outer orbit, can be treated by a rather simple extension of the hydrogen atom results. The hydrogen atom solution is also important in identifying the basic selection rules for describing allowed electron energy levels. These quantum mechanical results must coincide with the experimental spectra, and we expect the energy levels to include those predicted by the Bohr model. Without actually working through the mathematics for the hydrogen atom, in this section we shall investigate the energy level schemes dictated by the wave equation.

2.5.1 The Hydrogen Atom. Finding the wave functions for the hydrogen atom requires a solution of the Schrödinger equation in three dimensions for a coulombic potential field. Since the problem is spherically

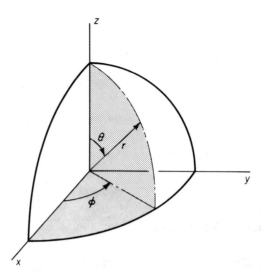

FIGURE 2-7. The spherical
coordinate system.

symmetric, the spherical coordinate system is used in the calculation (Fig. 2-7). The term $V(x, y, z)$ in Eq. (2-24) must be replaced by $V(r, \theta, \phi)$, representing the Coulomb potential which the electron experiences in the vicinity of the proton. The Coulomb potential varies only with r in spherical coordinates

$$(2\text{-}36) \qquad V(r, \theta, \phi) = V(r) = -(4\pi\epsilon_0)^{-1}\frac{q^2}{r}$$

as in Eq. (2-14).

When the separation of variables is made, the time-independent equation can be written as

$$(2\text{-}37) \qquad \psi(r, \theta, \phi) = R(r)\,\Theta(\theta)\,\Phi(\phi)$$

Thus the wave functions are found in three parts. Separate solutions must be obtained for the r-dependent equation, the θ-dependent equation, and the ϕ-dependent equation. After these three equations are solved, the total wave function ψ is obtained from the product.

As in the simple potential well problem, each of the three hydrogen atom equations gives a solution which is quantized. Thus we would expect a quantum number to be associated with *each* of the three parts of the wave equation. As an illustration, the ϕ-dependent equation obtained after separation of variables is

$$(2\text{-}38) \qquad \frac{d^2\Phi}{d\phi^2} + \mathbf{m}^2\Phi = 0$$

where $\mathbf{m}$ is a constant. The solution to this equation is

(2-39) $$\Phi_m(\phi) = Ae^{jm\phi}$$

where A can be evaluated by the normalization condition as before

(2-40) $$\int_0^{2\pi} \Phi_m^*(\phi)\,\Phi_m(\phi)\,d\phi = 1$$

(2-41) $$A^2 \int_0^{2\pi} e^{-jm\phi}e^{jm\phi}\,d\phi = A^2 \int_0^{2\pi} d\phi = 2\pi A^2$$

Thus the value of A is

(2-42) $$A = \frac{1}{\sqrt{2\pi}}$$

and the ϕ-dependent wave function is

(2-43) $$\Phi_m(\phi) = \frac{1}{\sqrt{2\pi}} e^{jm\phi}$$

Since values of ϕ repeat every 2π radians, Φ should repeat also. This occurs if **m** is an integer, including negative integers and zero (Prob. 2.9). Thus the wave functions for the ϕ-dependent equation are quantized with the following selection rule for the quantum numbers:

(2-44) $$\mathbf{m} = \ldots, -3, -2, -1, 0, +1, +2, +3, \ldots$$

By similar treatments, the functions $R(r)$ and $\Theta(\theta)$ can be obtained, each being quantized by its own selection rule. For the r-dependent equation, the quantum number **n** can be any positive integer (not zero), and for the θ-dependent equation the quantum number l can be zero or a positive integer. However, there are interrelationships among the equations which restrict the various quantum numbers used with a single wave function ψ_{nlm}

(2-45) $$\psi_{nlm}(r, \theta, \phi) = R_n(r)\,\Theta_l(\theta)\,\Phi_m(\phi)$$

These restrictions are summarized as follows:

(2-46a) $\mathbf{n} = 1, 2, 3, \ldots$

(2-46b) $l = 0, 1, 2, \ldots, (\mathbf{n}-1)$

(2-46c) $\mathbf{m} = -l, \ldots, -2, -1, 0, +1, +2, \ldots, +l$

In addition to the three quantum numbers arising from the three parts of the wave equation, there is an important quantization condition on the "spin" of the electron. Investigations of electron spin employ the theory of relativity as well as quantum mechanics; therefore, we shall simply state here that the spin condition **s** of an electron with ψ_{nlm} specified is

(2-47) $$\mathbf{s} = \pm\tfrac{1}{2}$$

That is, the spin of the particle can assume only the half-integer values $+\tfrac{1}{2}$ and $-\tfrac{1}{2}$.

The fact that the spin quantum number is given as $\pm\tfrac{1}{2}$ while the others are integers is not important to our limited discussion. It is only necessary for us to note that for any ψ_{nlm} the electron can have one of two spin conditions. We could refer to these two conditions as "spin up" and "spin down". The important point for our discussion is that each allowed energy state of the electron in the hydrogen atom is uniquely described by four quantum numbers: **n, l, m,** and **s**.†

Using these four quantum numbers, we can identify the various states which the electron can occupy in a hydrogen atom. The number **n**, called the *principal* quantum number, specifies the "orbit" of the electron in Bohr terminology. Of course, the concept of orbit is replaced by probability density functions in quantum mechanical calculations. It is common to refer to states with a given principal quantum number as belonging to a *shell* rather than an orbit.

There is considerable fine structure in the energy levels about the Bohr orbits, due to the dictates of the other three quantum conditions. For example, an electron with $\mathbf{n} = 1$ (the first Bohr orbit) can have only $l = 0$ and $\mathbf{m} = 0$ according to Eq. (2-46), but there are two spin states allowed from Eq. (2-47). For $\mathbf{n} = 2$, l can be 0 or 1, and $\mathbf{m}$ can be -1, 0, or $+1$. The various allowed combinations of quantum numbers appear in the first four columns of Table 2-1. From these combinations it is apparent that the electron in a hydrogen atom can occupy any one of a large number of excited states in addition to the lowest (*ground*) state ψ_{100}. Energy differences between the various states properly account for the observed lines in the hydrogen spectrum.

2.5.2 The Periodic Table.
The quantum numbers discussed in the previous section arise from solutions to the hydrogen atom problem. Thus the energies obtainable from the wave functions are unique to the hydrogen atom and cannot be extended to more complicated atoms without appropriate alterations. However, the quantum number selection rules are valid for more complicated structures, and we can use these rules to gain an understanding of the arrangement of atoms in the periodic table of chemical elements. Without these selection rules, it is difficult to understand why only two electrons fit into the first Bohr orbit of an atom, whereas eight electrons are allowed in the second orbit. After even the brief discussion of quantum

---

†In many texts the numbers we have called **m** and **s** are referred to as $\mathbf{m}_l$ and $\mathbf{m}_s$, respectively.

TABLE 2-1.   Quantum numbers to $n = 3$ and allowable states for the electron in a hydrogen atom: The first four columns show the various combinations of quantum numbers allowed by the selection rules of Eq. (2-46); the last two columns indicate the number of allowed states (combinations of **n**, *l*, **m**, and **s**) for each *l* (*subshell*) and **n** (shell, or Bohr orbit).

n	*l*	m	s	Allowable states in subshell	Allowable states in completed shell
1	0	0	$\pm\frac{1}{2}$	2	2
2	0	0	$\pm\frac{1}{2}$	2	8
	1	−1	$\pm\frac{1}{2}$		
		0	$\pm\frac{1}{2}$	6	
		1	$\pm\frac{1}{2}$		
3	0	0	$\pm\frac{1}{2}$	2	18
	1	−1	$\pm\frac{1}{2}$		
		0	$\pm\frac{1}{2}$	6	
		1	$\pm\frac{1}{2}$		
	2	−2	$\pm\frac{1}{2}$		
		−1	$\pm\frac{1}{2}$		
		0	$\pm\frac{1}{2}$	10	
		1	$\pm\frac{1}{2}$		
		2	$\pm\frac{1}{2}$		

numbers given above, we should be able to answer these questions with more insight.

Before discussing the periodic table, we must be aware of an important principle of quantum theory, the *Pauli Exclusion Principle*. This rule states that no two electrons in an interacting system† can have the same set of quantum numbers **n**, *l*, **m**, **s**. In other words, only two electrons can have the same three quantum numbers **n**, *l*, **m**, and those two must have opposite spin. The importance of this principle cannot be overemphasized; it is basic to the electronic structure of all atoms in the periodic table. One implication of this principle is that by listing the various combinations of quantum numbers, we can determine into which shell each electron of a complicated atom fits, and how many electrons are allowed per shell. The quantum states summarized in Table 2-1 can be used to indicate the electronic configurations for atoms in the lowest energy state.

---

†An interacting system is one in which electron wave functions overlap—in this case an atom with two or more electrons.

In the first electronic shell ($n = 1$), l can be only zero since the maximum value of l is always $n - 1$. Similarly, m can be only zero since m runs from the negative value of l to the positive value of l. Two electrons with opposite spin can fit in this ψ_{100} state; therefore, the first shell can have at most two electrons. For the helium atom (atomic number $Z = 2$) in the ground state, both electrons will be in the first Bohr orbit ($n = 1$), both will have $l = 0$ and $m = 0$, and they will have opposite spin. Of course, one or both of the He atom electrons can be excited to one of the higher energy states of Table 2-1 and subsequently relax to the ground state, giving off a photon characteristic of the He spectrum.

As Table 2-1 indicates, there can be two electrons in the $l = 0$ subshell, six electrons when $l = 1$, and ten electrons for $l = 2$. The electronic configurations of various atoms in the periodic table can be deduced from this list of allowed states. The ground state electron structures for a number of atoms are listed in Table 2-2. There is a simple shorthand notation for electronic structures which is commonly used instead of such a table. The only new convention to remember in this notation is the naming of the l values:

$$l = 0, 1, 2, 3, 4, \ldots$$

$$s, p, d, f, g, \ldots$$

This convention was created by early spectroscopists who referred to the first four spectral groups as *sharp*, *principal*, *diffuse*, and *fundamental*. Alphabetical order is used beyond f. With this convention for l, we can write an electron state as follows:

┌6 electrons in the 3p subshell

$3p^6$

$(n = 3)$───┘ └($l = 1$)

For example, the total electronic configuration for Si ($Z = 14$) in the ground state is

$$1s^2\ 2s^2\ 2p^6\ 3s^2\ 3p^2$$

We notice that Si has a closed Ne configuration (see Table 2-2) plus four electrons in an outer $n = 3$ orbit ($3s^2 3p^2$). These are the four valence electrons of Si; two valence electrons are in an s state and two are in a p state. The Si electronic configuration can be written [Ne] $3s^2 3p^2$ for convenience, since the Ne configuration $1s^2 2s^2 2p^6$ forms a closed shell (typical of the inert elements). The column IV semiconductor Ge ($Z = 32$) has an electronic structure similar to Si, except that the four valence electrons are outside a closed $n = 3$ shell. Thus the Ge configuration is [Ar] $3d^{10}4s^24p^2$. In Chapter 3 we

TABLE 2-2. Electronic configurations for atoms in the ground state.

Atomic number (Z)	Element	n = 1, l = 0 — 1s	n = 2, l = 0 1 — 2s 2p	n = 3, l = 0 1 2 — 3s 3p 3d	n = 4, l = 0 1 — 4s 4p	Shorthand notation
			Number of electrons			
1	H	1				$1s^1$
2	He	2				$1s^2$
3	Li		1			$1s^2\ 2s^1$
4	Be	helium core, 2 electrons	2			$1s^2\ 2s^2$
5	B		2 1			$1s^2\ 2s^2\ 2p^1$
6	C		2 2			$1s^2\ 2s^2\ 2p^2$
7	N		2 3			$1s^2\ 2s^2\ 2p^3$
8	O		2 4			$1s^2\ 2s^2\ 2p^4$
9	F		2 5			$1s^2\ 2s^2\ 2p^5$
10	Ne		2 6			$1s^2\ 2s^2\ 2p^6$
11	Na			1		[Ne] $3s^1$
12	Mg			2		$3s^2$
13	Al		neon core, 10 electrons	2 1		$3s^2\ 3p^1$
14	Si			2 2		$3s^2\ 3p^2$
15	P			2 3		$3s^2\ 3p^3$
16	S			2 4		$3s^2\ 3p^4$
17	Cl			2 5		$3s^2\ 3p^5$
18	Ar			2 6		$3s^2\ 3p^6$
19	K				1	[Ar] $4s^1$
20	Ca				2	$4s^2$
21	Sc			1	2	$3d^1\ 4s^2$
22	Ti			2	2	$3d^2\ 4s^2$
23	V			3	2	$3d^3\ 4s^2$
24	Cr			5	1	$3d^5\ 4s^1$
25	Mn			5	2	$3d^5\ 4s^2$
26	Fe	argon core, 18 electrons		6	2	$3d^6\ 4s^2$
27	Co			7	2	$3d^7\ 4s^2$
28	Ni			8	2	$3d^8\ 4s^2$
29	Cu			10	1	$3d^{10}\ 4s^1$
30	Zn			10	2	$3d^{10}\ 4s^2$
31	Ga			10	2 1	$3d^{10}\ 4s^2\ 4p^1$
32	Ge			10	2 2	$3d^{10}\ 4s^2\ 4p^2$
33	As			10	2 3	$3d^{10}\ 4s^2\ 4p^3$
34	Se			10	2 4	$3d^{10}\ 4s^2\ 4p^4$
35	Br			10	2 5	$3d^{10}\ 4s^2\ 4p^5$
36	Kr			10	2 6	$3d^{10}\ 4s^2\ 4p^6$

shall see that the four valence electrons of Ge and Si play very important roles in the bonding forces of the diamond lattice and in the conduction of charges through these semiconductors.

There are several cases in Table 2-2 which do not follow the most straightforward choice of quantum numbers. For example, we notice that in K $(Z = 19)$ and Ca $(Z = 20)$ the $4s$ state is filled before the $3d$ state; in Cr $(Z = 24)$ and Cu $(Z = 29)$ there is a transfer of an electron back to the $3d$ state. These exceptions, required by minimum energy considerations, are discussed more fully in most atomic physics texts.

READING LIST

G. FEINBERG, "Light," *Scientific American*, vol. 219, no. 3, pp. 50–59, September 1968.

A. H. WAHL, "Chemistry by Computer," *Scientific American*, vol. 222, no. 4, pp. 54–70, April 1970.

A. D'ABRO, *The Rise of the New Physics, Vol. II*. New York: Dover Publications, Inc., 1951.

H. A. POHL, *Quantum Mechanics for Science and Engineering*. Englewood Cliffs, N.J.: Prentice-Hall, Inc., 1967.

R. P. FEYNMAN, R. B. LEIGHTON, and M. SANDS, *The Feynman Lectures on Physics, Vol. III: Quantum Mechanics*. Reading, Mass: Addison–Wesley Publishing Company, Inc., 1965.

R. L. SPROULL, *Modern Physics*. New York: John Wiley & Sons, Inc., 1963.

C. W. SHERWIN, *Introduction to Quantum Mechanics*. New York: Holt, Rinehart & Winston, Inc., 1960.

T. L. MARTIN, Jr., and W. F. LEONARD, *Electrons and Crystals*. Belmont, Calif.: Brooks/Cole Publishing Company, 1970.

PROBLEMS

2.1 (a) Sketch a simple vacuum tube device and the associated circuitry for measuring E_m in the photoelectric effect experiment. The electrodes can be placed in a sealed glass envelope.

(b) Sketch the photocurrent I vs. retarding voltage V that you would expect to measure for a given electrode material and configuration. Make the sketch for several intensities of light at a given wavelength. (Continued on page 50.)

2.1 (continued 2.1)

   (c) The work function of platinum is 4.09 eV. What retarding potential will be required to reduce the photocurrent to zero in a photoelectric experiment with Pt electrodes if the wavelength of incident light is 2440 Å? Remember that an energy of $q\phi$ is lost by each electron in escaping the surface.

2.2 (a) Show that the various lines in the hydrogen spectrum can be expressed in angstroms as

$$\lambda(\text{Å}) = \frac{911 n_1{}^2 n^2}{n^2 - n_1{}^2}$$

   where $n_1 = 1$ for the Lyman series, 2 for the Balmer series, and 3 for the Paschen series. The integer n is larger than n_1.

   (b) Calculate λ for the Lyman series to $n = 5$, the Balmer series to $n = 7$, and the Paschen series to $n = 10$. Plot the results as in Fig. 2-2. What are the wavelength limits for each of the three series?

2.3 Calculate the Bohr radius and energy for $n = 1, 2$, and 3.

2.4 Show that the calculated Bohr expression for frequency of emitted light in the hydrogen spectrum, Eq. (2-17), corresponds to the experimental expressions, Eq. (2-3).

2.5 Because of the Uncertainty Principle, it is improper to consider a photon as being localized in space more precisely than about one wavelength. Assume a photon with $\lambda = 1 \, \mu\text{m}$ is detected and its position is measured to an accuracy of one wavelength. What is the uncertainty in the photon's energy?

2.6 A golfer wishing to calibrate his nine-iron shot hits 100 balls and measures the distance of each to the nearest 10-yard interval. He finds 40 balls near the 100-yard mark, 20 near 90 yards and 20 more near 110 yards, 10 near 80 yards and the last 10 balls near the 120-yard mark. He concludes that his average distance is 100 yards.

   (a) Use Eq. (2-21) to show that this rather obvious conclusion is really correct.

   (b) Calculate his average two-iron distance from the following data: 7 balls at 180 yards, 15 at 190 yards, 40 at 200 yards, 30 at 210 yards, and 8 balls at 220 yards.

2.7 A sample of radioactive material undergoes decay such that the number of atoms $N(t)$ remaining in the unstable state at time t is related to the number N_0 at $t = 0$ by the relation $N(t) = N_0 \exp(-t/\tau)$. Show that τ is the average lifetime of an atom in the unstable state before it spontaneously decays. Equation (2-21b) can be used with t substituted for x.

2.8 Separate the variables in Eq. (2-25) to obtain Eqs. (2-26) and (2-27). What is the solution to Eq. (2-26)? Assuming $\psi(x)$ has a particular time-independent value $\psi_n(x)$, use postulate 3 to show that the corresponding energy equals the separation constant E_n.

2.9 Show that Eq. (2-44) is a proper choice for **m**.

2.10 What do Li, Na, and K have in common? What do F, Cl, and Br have in common? What are the electron configurations for ionized Na and Cl?

ENERGY BANDS AND CHARGE
CARRIERS IN SEMICONDUCTORS   **3**

In this chapter we begin to discuss the specific mechanisms by which current flows in a solid. In examining these mechanisms we shall learn why some materials are good conductors of electric current, whereas others are poor conductors. We shall see how the conductivity of a semiconductor can be varied by changing the temperature or the number of impurities. These fundamental concepts of charge transport form the basis for later discussions of solid state device behavior.

3.1 Bonding Forces and Energy Bands in Solids

In Chapter 2 we found that electrons are restricted to sets of discrete energy levels within atoms. Large gaps exist in the energy scale in which no energy states are available. In a similar fashion, electrons in solids are restricted to certain energies and are not allowed at other energies. The basic difference between the case of an electron in a solid and that of an electron in an isolated atom is that in the solid the electron has a *range*, or *band*, of available energies. The discrete energy levels of the isolated atom spread into bands of energies in the solid because in the solid the wave functions of electrons in neighboring atoms overlap, and an electron is not necessarily localized at a particular atom. Thus, for example, an electron in the outer orbit of one atom feels the influence of neighboring atoms, and its overall wave function is altered. Naturally, this influence affects the potential energy term and the boundary conditions in the Schrödinger equation, and we would expect to obtain different energies in the solution. Usually, the influence of neighboring atoms on the energy levels of a particular atom can be treated as a small perturbation, giving rise to shifting and splitting of energy states into energy bands.

3.1.1 Bonding Forces in Solids.

The interaction of electrons in neighboring atoms of a solid serves the very important function of holding the crystal together. For example, alkali halides such as NaCl are typified by *ionic bonding*. In the NaCl lattice, each Na atom is surrounded by six nearest neighbor Cl atoms, and vice versa. Four of the nearest neighbors are evident in the two-dimensional representation shown in Fig. 3-1. The electronic

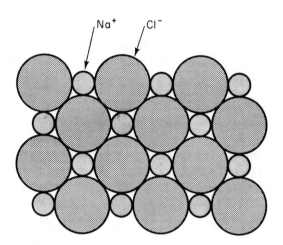

FIGURE 3-1. The NaCl lattice: an example of ionic bonding.

structure of Na $(Z = 11)$ is [Ne]$3s^1$, and Cl $(Z = 17)$ has the structure [Ne] $3s^2 3p^5$. In the lattice each Na atom gives up its outer $3s$ electron to a Cl atom, so that the crystal is made up of ions with the electronic structures of the inert atoms Ne and Ar (Ar has the electronic structure [Ne] $3s^2 3p^6$). However, the ions have net electric charges after the electron exchange. The Na$^+$ ion has a net positive charge, having lost an electron, and the Cl$^-$ ion has a net negative charge, having gained an electron.

Each Na$^+$ ion exerts an electrostatic attractive force upon its six Cl$^-$ neighbors, and vice versa. These coulombic forces pull the lattice together until the wave functions for neighboring atoms begin to overlap. Then repulsive restraints prevent the atoms from moving closer together. A reasonably accurate calculation of the atomic spacing can be made by considering the ions as hard spheres being attracted together (Prob. 3.1).

An important observation in the NaCl structure is that all electrons are tightly bound to atoms. Once the electron exchanges have been made between the Na and Cl atoms to form the Na$^+$ and Cl$^-$ ions, the outer orbits of all atoms are completely filled. Since the ions have the closed-shell configurations

of the inert atoms Ne and Ar, there are no loosely bound electrons to participate in current flow; as a result, NaCl is a good insulator.

In a metal atom the outer electronic shell is only partially filled, usually
by no more than three electrons. We have already noted that the alkali metals
(e.g., Na) have only one electron in the outer orbit. This electron is loosely
bound and is given up easily in ion formation. This accounts for the great
chemical activity of the alkali metals, as well as for their high electrical conductivity. In the lattice the outer electron of each alkali atom is contributed
to the lattice as a whole, so that the crystal is made up of ions with closed
shells immersed in a sea of free electrons. The forces holding the lattice
together arise from an interaction between the positive ion cores and the
surrounding free electrons. This is one type of *metallic bonding*. Obviously,
there are complicated differences in the bonding forces for various metals,
as evidenced by the wide range of melting temperatures (234°K for Hg,
3643°K for W). However, the metals have the sea of electrons in common,
and these electrons are free to move about the lattice under the influence of
an electric field.

A third type of bonding is exhibited by the diamond lattice semiconductors. We recall that each atom in the Ge, Si, or C diamond lattice is
surrounded by four nearest neighbors, each with four electrons in the outer
orbit. In these crystals each atom shares its valence electrons with its four
neighbors (Fig. 3-2). Bonding between nearest neighbor atoms is illustrated

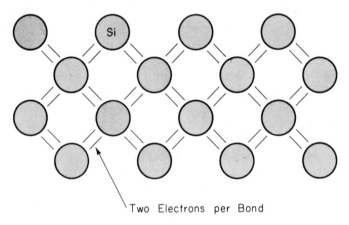

Two Electrons per Bond

FIGURE 3-2. Covalent bonding in the Si lattice, viewed along a
⟨100⟩ direction (see also Figs. 1-8 and 1-9).

in the diamond lattice diagram of Fig. 1-9. The bonding forces arise from a
quantum mechanical interaction between the shared electrons. This is known
as *covalent bonding;* each electron pair constitutes a covalent bond. In the

sharing process it is no longer relevant to ask which electron belongs to a particular atom—both belong to the bond. The two electrons are indistinguishable, except that they must have opposite spin to satisfy the Pauli Exclusion Principle.

As in the case of the ionic crystals, no free electrons are available to the lattice in the covalent diamond structure of Fig. 3-2. By this reasoning Ge and Si should also be insulators. However, we have pictured an idealized lattice at $0°K$ in this figure. As we shall see in subsequent sections, an electron can be thermally or optically excited out of a covalent bond and thereby become free to participate in conduction.

Compound semiconductors such as GaAs have mixed bonding, in which both ionic and covalent bonding forces participate. Some ionic bonding is to be expected in a crystal such as GaAs because of the difference in placement of the Ga and As atoms in the periodic table. The ionic character of the bonding becomes more important as the atoms of the compound become further separated in the periodic table, as in the II–VI compounds. The degree of ionic and covalent bonding is determined by the difference in *electronegativity* of the atoms. This concept is discussed in most physical chemistry texts.

3.1.2 Energy Bands.

As isolated atoms are brought together to form a solid, various interactions occur between neighboring atoms, including those described in the preceding section. The forces of attraction and repulsion between atoms will find a balance at the proper interatomic spacing for the crystal. In the process, important changes occur in the electron energy level configurations of the atoms, and these changes result in the varied electrical properties of solids.

Qualitatively, we can see that as atoms are brought together, the application of the Pauli Exclusion Principle becomes important. When two atoms are completely isolated from each other so that there is no interaction of electron wave functions between them, they can have identical electronic structures. As the spacing between the two atoms becomes smaller, however, electron wave functions begin to overlap. The Exclusion Principle dictates that no two electrons in a given interacting system may have the same quantum state; thus there must be a splitting of the discrete energy levels of the isolated atoms into new levels belonging to the pair rather than to individual atoms. This splitting of levels is important in the formation of molecules, such as H_2 (Fig. 3-3).

In the hydrogen molecule, the two electrons can have opposite spins in an energy level lower than the atomic $1s$ level. This is the *bonding state;* a slightly higher level accounts for the *antibonding state.* The atomic $1s$ levels in effect split into two "$1s$" levels in the molecule. The quotation

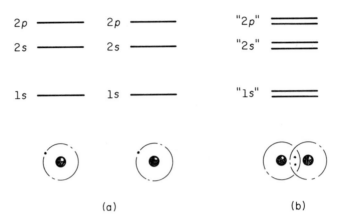

2p ———— 2p ———— "2p" ≡≡≡≡

2s ———— 2s ———— "2s" ≡≡≡≡

1s ———— 1s ———— "1s" ≡≡≡≡

(a)   (b)

FIGURE 3-3. Energy level splitting in the diatomic hydrogen
molecule: (a) isolated hydrogen atoms; (b) H_2 molecule.

marks around "1s" indicate that the two split levels are no longer really 1s
in nature; each differs slightly from 1s to accommodate the two levels in the
same region of space.

In a solid, many atoms are brought together, so that the split energy
levels form essentially continuous *bands* of energies. As an example, Fig.
3-4 illustrates the imaginary formation of a diamond crystal from isolated
carbon atoms. Each isolated carbon atom has an electronic structure
$1s^2 2s^2 2p^2$ in the ground state. Each atom has available two 1s states, two 2s
states, six 2p states, and higher states (see Tables 2-1 and 2-2). If we consider
N atoms, there will be 2N, 2N, and 6N states of type 1s, 2s, and 2p, respectively. As the interatomic spacing decreases, these energy levels split into
bands, beginning with the outer (**n** = 2) shell. As the "2s" and "2p" bands
grow, they merge into a single band composed of a mixture of energy levels.
This band of "2s–2p" levels contains 8N available states. As the distance
between atoms approaches the equilibrium interatomic spacing of diamond,
this band splits into two bands separated by an *energy gap* E_g. The upper
band (called the *conduction band*) contains 4N states, as does the lower
(*valence*) band. Thus, apart from the low-lying and tightly bound "1s" levels,
the diamond crystal has two bands of available energy levels separated by
an energy gap E_g wide, which contains no allowed energy levels for electrons
to occupy. This gap is sometimes called a "forbidden band," since in a perfect
crystal it contains no electron energy states.

We should pause at this point and count electrons. The lower "1s" band
is filled with the 2N electrons which originally resided in the collective 1s
states of the isolated atoms. However, there were 4N electrons in the original
isolated **n** = 2 shells (2N in 2s states and 2N in 2p states). These 4N electrons
must occupy states in the valence band or the conduction band in the crystal.

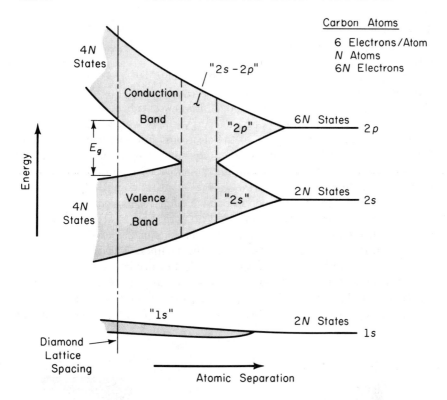

FIGURE 3-4. Formation of energy bands as a diamond crystal is formed by bringing together isolated carbon atoms. From *Electrons and Holes in Semiconductors*, by W. Shockley; © 1950 by Litton Educational Publishing Co., Inc.; by permission of Van Nostrand Reinhold, Co., Inc.

At 0°K the electrons will occupy the lowest energy states available to them. In the case of the diamond crystal, there are exactly $4N$ states in the valence band available to the $4N$ electrons. Thus at 0°K, *every* state in the valence band will be filled, while the conduction band will be completely empty of electrons. As we shall see, this arrangement of completely filled and empty energy bands has an important effect on the electrical conductivity of the solid.

3.1.3 Metals, Semiconductors, and Insulators.
Every solid has its own characteristic energy band structure. This variation in band structure is responsible for the wide range of electrical characteristics observed in various materials. The diamond band structure of Fig. 3-4, for example, can give a good picture of why carbon in the diamond lattice is a

good insulator. To reach such a conclusion, we must consider the properties of completely filled and completely empty energy bands in the current conduction process.

Before discussing the mechanisms of current flow in solids further, we can observe here that for electrons to experience acceleration in an applied electric field, they must be able to move into new energy states. This implies there must be empty states (allowed energy states which are not already occupied by electrons) available to the electrons. For example, if relatively few electrons reside in an otherwise empty band, ample unoccupied states are available into which the electrons can move. On the other hand, the diamond structure is such that the valence band is completely filled with electrons at 0°K and the conduction band is empty. There can be no charge transport within the valence band, since no empty states are available into which electrons can move. There are no electrons in the conduction band, so no charge transport can take place there either. Thus carbon in the diamond structure has a high resistivity typical of insulators.

Semiconductor materials at 0°K have basically the same structure as insulators—a filled valence band separated from an empty conduction band by a band gap containing no allowed energy states (Fig. 3-5). The difference

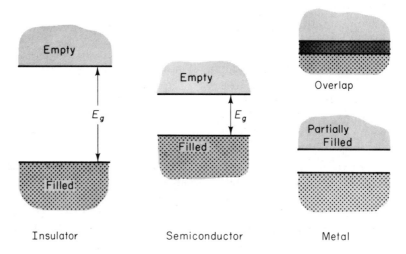

FIGURE 3-5. Typical band structures at 0°K.

lies in the size of the band gap E_g, which is much smaller in semiconductors than in insulators. For example, the semiconductor Si has a band gap of about 1.1 eV compared with 5 eV for diamond. The relatively small band gaps of semiconductors (Appendix III) allow for excitation of electrons from the lower (valence) band to the upper (conduction) band by reasonable amounts of thermal or optical energy. For example, at room temperature a semiconductor with a 1-eV band gap will have a significant number of electrons excited thermally across the energy gap into the conduction band, whereas an insulator with $E_g = 10$ eV will have a negligible number of such excitations. Thus an important difference between semiconductors and insulators is that the number of electrons available for conduction can be increased greatly in semiconductors by thermal or optical energy.

In metals the bands either overlap or are only partially filled. Thus electrons and empty energy states are intermixed within the bands so that electrons can move freely under the influence of an electric field. As expected, the metallic band structures suggest metals have a high electrical conductivity.

3.1.4 Direct and Indirect Semiconductors.
The "thought experiment" of Section 3.1.2, in which isolated atoms were brought together to form a solid, is useful in pointing out the existence of energy bands and some of their properties. Other techniques are generally used, however, when quantitative calculations are made of band structures. In a typical calculation, a single electron is assumed to travel through a perfectly periodic lattice. The wave function of the electron is assumed to be in the form of a plane wave† moving, for example, in the x-direction with propagation constant **k**. The space-dependent wave function for the electron is

$$(3\text{-}1) \qquad \psi_k(x) = U(\mathbf{k}_x, x)\, e^{j\mathbf{k}_x x}$$

where the function $U(\mathbf{k}_x, x)$ modulates the wave function according to the periodicity of the lattice.

In such a calculation, allowed values of energy can be plotted vs. the propagation constant **k**. Since the periodicity of most lattices is different in the various directions, the $(E, \mathbf{k})$ diagram must be plotted for the various crystal directions. Figure 3-6 shows general features of the $(E, \mathbf{k})$ diagrams for Si and GaAs. In this diagram we have indicated two of the crystal directions for **k**; actually, the full relationship between E and **k** is a complex surface which should be visualized in three dimensions.

There are a number of complicated features in Fig. 3-6, which could be examined in a more advanced text. However, we can make one observation here which will be particularly useful to us in later chapters. We notice that the band structure of GaAs has a minimum in the conduction band and a maximum in the valence band for the same **k** value (**k** = 0). On the other hand, Si has its valence band maximum at a different value of **k** than its conduction band minimum. Thus an electron making a smallest-energy

---

†Discussions of plane waves are available in most sophomore physics texts or in introductory electromagnetics texts.

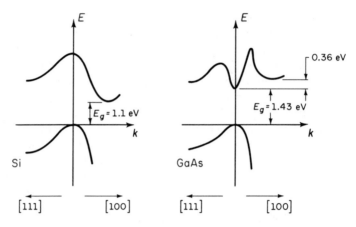

FIGURE 3-6. General features of the energy band structures for Si and GaAs,† showing allowed electron energies versus electron propagation vector **k**.

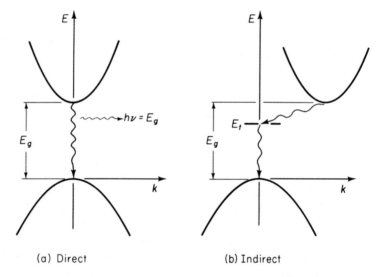

(a) Direct   (b) Indirect

FIGURE 3-7. Direct and indirect electron transitions in semi-conductors: (a) direct transition with accompanying photon emission; (b) indirect transition via a defect level.

---

†For more complete band diagrams, see M. L. Cohen and T. K. Bergstresser, "Band Structures and Pseudopotential Form Factors for Fourteen Semiconductors of the Diamond and Zincblende Structures," *Physical Review*, vol. 141, no. 2, pp. 789–796, January 1966.

transition from the conduction band to the valence band in GaAs can do so without a change in **k** value; on the other hand, a transition from the minimum point in the Si conduction band to the maximum point of the valence band requires some change in **k**. Thus there are two classes of semiconductor energy bands: *direct* and *indirect* (Fig. 3-7).

In a direct semiconductor, an electron in the conduction band can fall to an empty state in the valence band, giving off the energy difference E_g as a photon of light. On the other hand, an electron in the conduction band minimum of an indirect semiconductor cannot fall directly to the valence band maximum but must go through some defect state (E_t) within the band gap. We shall discuss such defect states in Sections 4.2.1 and 4.3.2. In an indirect transition which involves a change in **k**, the energy is given up largely as heat to the lattice rather than as an emitted photon.† This difference between direct and indirect band structures is very important for deciding which semiconductors can be used in devices requiring light output. For example, semiconductor light emitters (Section 6.4) and lasers (Chapter 7) generally must be made of materials capable of direct band-to-band transitions or of indirect materials with vertical transitions between defect states.

Band diagrams such as those shown in Fig. 3-6 are cumbersome to draw in analyzing devices. Therefore, in most discussions we shall use the simple band pictures of Section 3.1.3, remembering that electron transitions across the band gap may be direct or may occur via defect levels.

3.2 Charge Carriers in Semiconductors

The mechanism of current conduction is relatively easy to visualize in the case of a metal; the metal atoms are imbedded in a "sea" of relatively free electrons, and these electrons can move as a group under the influence of an electric field. This free electron view is oversimplified, but many important conduction properties of metals can be derived from just such a model. However, we cannot account for all of the electrical properties of semiconductors in this way. Since the semiconductor has a filled valence band and an empty conduction band at 0°K, we must consider the increase in conduction band electrons by thermal excitations across the band gap as the temperature is raised. In addition, after electrons are excited to the conduction band, the empty states left in the valence band can contribute to the conduction process. The introduction of impurities has an important effect on the

---

†Here we are discussing the *most probable* events, which dominate typical indirect semiconductors. Many possible secondary effects are neglected.

energy band structure and on the availability of charge carriers. Thus there is considerable flexibility in controlling the electrical properties of semiconductors.

3.2.1 Electrons and Holes.

As the temperature of a semiconductor is raised from $0°K$, some electrons in the valence band receive enough thermal energy to be excited across the band gap to the conduction band. The result is a material with some electrons in an otherwise empty conduction band and some unoccupied states in an otherwise filled valence band (Fig. 3-8).† For

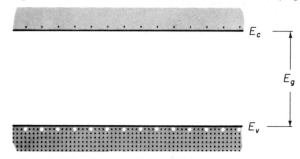

FIGURE 3-8.   Electron-hole pairs in a semiconductor.

convenience, an empty state left in the valence band is referred to as a *hole*. If the conduction band electron and the hole in the valence band are created during a single event (the excitation of a valence band electron to the conduction band), they are called an *electron-hole pair* (abbreviated EHP).

After excitation to the conduction band, an electron is surrounded by a large number of unoccupied energy states (approximately $4N$ states for diamond-like semiconductors, according to Fig. 3-4). For example, the equilibrium number of electron-hole pairs in Si at room temperature is only about 10^{10} EHP/cm^3, compared to the Si atom density N of more than 10^{22} atoms/cm^3. Thus the few electrons in the conduction band are free to move about via the many available empty states.

The corresponding problem of charge transport in the valence band is somewhat more complicated. However, it is possible to show that the effects of valence band electrons moving via holes can be accounted for by simply keeping track of the holes themselves. For example, if a valence band electron moves into an empty state, the hole in effect moves to a new position. If we focus our attention on the motion of the hole itself, we can account for the charge transport which has taken place. Hole motion can be visualized as in Fig. 3-9, which is a one-dimensional figure for convenience. If we consider one small segment of the valence band, as defined by the dashed lines,

---

†In Fig. 3-8 and in subsequent discussions we refer to the bottom of the conduction band as E_c and the top of the valence band as E_v.

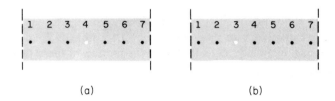

FIGURE 3-9. Motion of valence band electrons and a hole in a one-dimensional lattice: (a) original positions; (b) positions after a movement of the electrons to the left.

we can number the spatial positions of six electrons and one hole. Now if the group of electrons moves to the left under the influence of an electric field, we can observe the new positions after one step. The electron at position 1 moves to the left and out of our picture. A new electron moves from the right of the dashed line into position 7, while each of the remaining electrons in the group moves one step to the left. The hole now appears at position 3. It has moved along with the adjacent valence band electrons.

Whereas the equations governing the motion of valence band electrons are rather complicated, it turns out that *all of the charge transport within the valence band can be accounted for by treating the holes as positive charge carriers.* The usual equations of motion can be applied as if the hole were a particle with charge $+q$ (where q is the magnitude of the electronic charge) and with some effective mass. We shall discuss the concept of effective mass for both electrons and holes in the following section.

A careful study of Fig. 3-9 raises an interesting question. The valence band electrons and the hole move in the same direction during a net movement to the left of the group of electrons. But if the hole is a positive charge carrier, should it not move in the *opposite* direction to the electrons in an electric field? The explanation of this dilemma is that, unlike holes or conduction band electrons, electrons near the top of the valence band do not follow the usual equations of motion. It is necessary to treat these electrons by examining the influence of an electric field on their position in **k**-space (Fig. 3-6). The equations of motion can be applied to these electrons only by considering their mass as a negative number. This complication illustrates why the concept of a hole is valuable to us in analyzing conduction processes in semiconductors.

We should remember that the valence band contains a large number of energy states and is three dimensional in **k**-space. This implies that propagation vectors **k** for various directions are associated with the valence band states. In a completely filled band, all states are occupied and there is no *net* motion of electrons in any direction. There can be a net charge transport only when a few states are empty (holes). Therefore, when the group of valence band electrons in Fig. 3-9 moves in a field, only the hole motion is counted in summing up the net charge transport.

In all the following discussions we shall concentrate on the electrons in the conduction band and on the holes in the valence band. We can account for the current flow in a semiconductor by the motion of these two types of charge carriers.

3.2.2 Effective Mass.

3.2.2 Effective Mass. Since the electrons in a crystal are not completely free but interact with the periodic potential of the lattice, their "wave-particle" motion cannot be expected to be the same as for electrons in free space. Thus, in applying the usual equations of electrodynamics to charge carriers in a solid, we must use altered values of particle mass. In doing so, we account for most of the influences of the lattice, so that the electrons and holes can be treated as "almost free" carriers in most computations. Calculating the appropriate values of effective mass to use in the equations of motion is an involved theoretical problem. The calculation of effective mass must take into account the shape of the energy bands in three-dimensional **k**-space (see Fig. 3-6), which is difficult for complicated band structures. However, an average value of the effective mass m^* can be measured experimentally in semiconductors by the use of an effect called *cyclotron resonance*. Basically, this is an experiment in which the charge carrier is induced to move along a circular path in a magnetic field in resonance with an applied micro-wave signal. The resonant frequency is governed by the particle effective mass and the magnetic field.

Assume an electron moves in a magnetic field as in Fig. 3-10. From sophomore physics we know that the resulting path of the electron is circular and that the magnetic field exerts a force on the electron of $q\mathrm{v}\mathscr{B}$, where **v** is the electron's velocity and $\mathscr{B}$ is the magnetic flux density. Equating this force to the centripetal force, we obtain

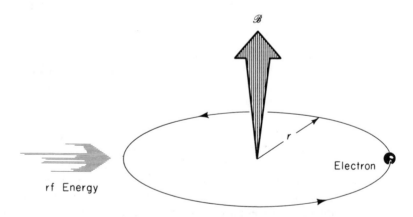

FIGURE 3-10. Cyclotron resonance experiment.

(3-2)
$$\frac{m^* v^2}{r} = q v \mathscr{B}$$

(3-3)
$$v = \frac{q \mathscr{B} r}{m^*}$$

But the angular frequency of rotation is

(3-4)
$$\omega_c = \frac{v}{r} = \frac{q \mathscr{B}}{m^*}$$

The electron revolves in the circular path with an angular frequency ω_c. Now, if an alternating electromagnetic field is applied to the sample, the electrons can absorb energy from the field. The absorption is small unless the frequency of the alternating field just matches the angular frequency of the electrons, at which point a resonance absorption of energy takes place. Thus by measuring the frequency of the absorbed signal, usually in the rf or microwave range, we can deduce the value of ω_c and thereby evaluate the effective mass m^*. Of course, the same type of experiment can be performed to measure the hole effective mass.

The results of cyclotron resonance experiments provide values of effective mass for electrons and holes for a particular semiconductor material. In any calculation involving the mass of the charge carriers, we must use effective mass values for the particular material involved. Table 3-1 lists the

TABLE 3-1. Effective mass values for Ge and Si.† The free electron rest mass is m_0. From E. M. Conwell, "Properties of Silicon and Germanium: II," *Proc. IRE*, vol. 46, no. 6, pp. 1281–1300, June 1958.

	Ge	Si
m_n^*	$0.55 m_0$	$1.1 m_0$
m_p^*	$0.37 m_0$	$0.59 m_0$

†These are average values for effective mass, appropriate for use in the density of states calculation of Section 3.3.2; they are computed at 4°K. A different average is used for effective mass values in conductivity calculations.

effective masses for Ge and Si. In this table and in all subsequent discussions, the electron effective mass is denoted by m_n^* and the hole effective mass by m_p^*. The n subscript indicates the electron as a negative charge carrier, and the p subscript indicates the hole as a positive charge carrier.

3.2.3 Intrinsic Material. A perfect semiconductor crystal with no impurities or lattice defects is called an *intrinsic* semiconductor. In such material there are no charge carriers at 0°K, since the valence band is filled with electrons and the conduction band is empty. At higher temperatures electron-hole pairs are generated as valence band electrons are excited thermally across the band gap to the conduction band. These EHP's are the only charge carriers in intrinsic material.

The generation of EHP's can be visualized in a qualitative way by considering the breaking of covalent bonds in the lattice (Fig. 3-11). If one

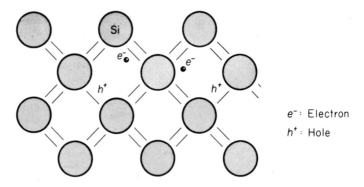

FIGURE 3-11. Electron-hole pairs in the covalent bonding model.

of the Si valence electrons is broken away from its position in the bonding structure such that it becomes free to move about in the lattice, a conduction electron is created and a broken bond (hole) is left behind. The energy required to break the bond is the band gap energy E_g. This model helps in visualizing the physical mechanism of EHP creation, but the energy band model is more productive for purposes of quantitative calculation. One important difficulty in the "broken bond" model is that the free electron and the hole seem deceptively localized in the lattice. Actually, the positions of the free electron and the hole are spread out over several lattice spacings and should be considered quantum mechanically by probability distributions (see Section 2.4).

Since the electrons and holes are created in pairs, the conduction band electron density n (electrons per cm³) is equal to the density of holes in the valence band p (holes per cm³). Each of these intrinsic carrier densities is commonly referred to as n_i. Thus *for intrinsic material*

$$(3-5) \qquad\qquad n = p = n_i$$

At a given temperature there is a certain density of electron-hole pairs n_i. Obviously, if a steady state carrier density is maintained, there must be *recombination* of EHP's at the same rate at which they are generated. Recombination occurs when an electron in the conduction band makes a transition (direct or indirect) to an empty state (hole) in the valence band, thus annihilating the pair. If we denote the generation rate of EHP's as g_i (EHP/cm^3-sec) and the recombination rate as r_i, equilibrium requires that

$$(3\text{-}6) \qquad\qquad r_i = g_i$$

Each of these rates is temperature dependent. For example, $g_i(T)$ increases when the temperature is raised, and a new carrier density n_i is established such that the higher recombination rate $r_i(T)$ just balances generation. At any temperature, we can predict that the rate of recombination of electrons and holes r_i is proportional to the equilibrium density of electrons n_0 and the density of holes p_0

$$(3\text{-}7) \qquad\qquad r_i = \alpha_r n_0 p_0 = \alpha_r n_i^2 = g_i$$

where α_r is a constant of proportionality which depends on the particular mechanism by which recombination takes place. We shall discuss the calculation of n_i as a function of temperature in Section 3.3.3; recombination processes will be discussed in Chapter 4.

3.2.4 Extrinsic Material.

In addition to the intrinsic carriers generated thermally, it is possible to create carriers in semiconductors by purposely introducing impurities into the crystal. This process, called *doping*, is the most common technique for varying the conductivity of semiconductors. By doping, a crystal can be altered so that it has a predominance of either electrons or holes. Thus there are two types of doped semiconductors, n-type (mostly electrons) and p-type (mostly holes). When a crystal is doped such that the equilibrium carrier densities n_0 and p_0 are different from the intrinsic carrier density n_i, the material is said to be *extrinsic*.

When impurities or lattice defects are introduced into an otherwise perfect crystal, additional levels are created in the energy band structure, usually within the band gap. For example, an impurity from column V of the periodic table (P, As, and Sb) introduces an energy level very near the conduction band in Ge or Si. This level is filled with electrons at 0°K, and very little thermal energy is required to excite these electrons to the conduction band (Fig. 3-12). Thus at about 50–100°K virtually all of the electrons in the impurity level are "donated" to the conduction band. Such an impurity level is called a *donor* level, and the column V impurities in Ge or Si are called donor impurities. From Fig. 3-12 we note that material doped with donor impurities can have a considerable density of electrons in the conduction band, even when the temperature is too low for the intrinsic EHP density

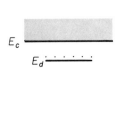

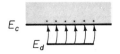

$T = 0$ $T \simeq 50\,^\circ K$

FIGURE 3-12. Donation of electrons from a donor level to the conduction band.

to be appreciable. Thus semiconductors doped with a significant number of donor atoms will have $n \gg (n_i, p)$ at room temperature. This is n-type material.

Atoms from column III (B, Al, Ga, and In) introduce impurity levels in Ge or Si near the valence band. These levels are empty of electrons at $0\,^\circ K$ (Fig. 3-13). At low temperatures, enough thermal energy is available

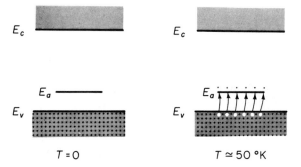

$T = 0$ $T \simeq 50\,^\circ K$

FIGURE 3-13. Acceptance of valence band electrons by an acceptor level, and the resulting creation of holes.

to excite electrons from the valence band into the impurity level, leaving behind holes in the valence band. Since this type of impurity level "accepts" electrons from the valence band, it is called an *acceptor* level, and the column III impurities are acceptor impurities in Ge and Si. As Fig. 3-13 indicates, doping with acceptor impurities can create a semiconductor with a hole density p much greater than the conduction band electron density (p-type material).

In the covalent bonding model, donor and acceptor atoms can be visualized as shown in Fig. 3-14. An Sb atom (column V) in the Si lattice has the four necessary valence electrons to complete the covalent bonds with the

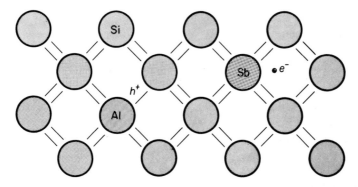

FIGURE 3-14. Donor and acceptor atoms in the covalent bonding model of a Si lattice.

neighboring Si atoms, plus one extra electron. This fifth electron does not fit into the bonding structure of the lattice and is therefore loosely bound to the Sb atom. A small amount of thermal energy enables this extra electron to overcome its coulombic binding to the impurity atom and be donated to the lattice as a whole. Thus it is free to participate in current conduction. This process is a qualitative model of the excitation of electrons out of a donor level and into the conduction band (Fig. 3-12). Similarly, the column III impurity Al has only three valence electrons to contribute to the covalent bonding (Fig. 3-14), thereby leaving one bond incomplete. With a small amount of thermal energy, this incomplete bond can be transferred to other atoms as the bonding electrons exchange positions. Again, the idea of an electron "hopping" from an adjacent bond into the incomplete bond at the Al site provides some physical insight into the behavior of an acceptor, but the model of Fig. 3-13 is preferable for most discussions.

We can calculate rather simply the approximate energy required to excite the fifth electron of a donor atom into the conduction band. Let us assume for rough calculations that the Sb atom of Fig. 3-14 has its four covalent bonding electrons rather tightly bound and the fifth "extra" electron loosely bound to the atom. We can approximate this situation by using the Bohr model results, considering the loosely bound electron as ranging about the tightly bound "core" electrons in a hydrogen-like orbit. From Eq. (2-15) the magnitude of the ground-state energy ($n = 1$) of such an electron is

(3-8)
$$E = \frac{mq^4}{2K^2\hbar^2}$$

The value of K must be modified from the free-space value $4\pi\epsilon_0$ used in the hydrogen atom problem to $4\pi\epsilon_0\epsilon_r$, where ϵ_r is the relative dielectric constant of the semiconductor material. In addition, we must use the effective mass m_n^* typical of the semiconductor.† For example, in Ge ($\epsilon_r = 16$, $m_n^* = 0.25m_0$) we calculate

(3-9) $$E = \frac{m_n^* q^4}{8(\epsilon_0 \epsilon_r)^2 h^2} = 2.1 \times 10^{-21}\,\text{J} = 0.013\,\text{eV}$$

Thus the energy required to excite the donor electron from the **n** = 1 state to the free state (**n** = ∞) is ≃ 0.01 eV. This corresponds to the energy difference $E_c - E_d$ in Fig. 3-12 and is in very close agreement with actual measured values. Generally, the column V donor levels lie approximately 0.01 eV below the conduction band in Ge, and the column III acceptor levels lie about 0.01 eV above the valence band. In Si the usual donor and acceptor levels lie about 0.03–0.06 eV from a band edge.

The importance of doping will become obvious when we discuss electronic devices made from junctions between p-type and n-type semiconductor material. The extent to which doping controls the electronic properties of semiconductors can be illustrated here by considering changes in the sample resistance which occur with doping. In Si, for example, the intrinsic carrier density n_i is about 10^{10} cm^{-3} at room temperature. If we dope Si with 10^{15} Sb atoms/cm^3, the conduction electron density changes by five orders of magnitude. The resistivity of Si changes from about 2×10^5 Ω-cm to 5 Ω-cm with this doping.

When a semiconductor is doped n-type or p-type, one type of carrier dominates. In the example given above, the conduction band electrons outnumber the holes in the valence band by many orders of magnitude. We refer to the small number of holes in n-type material as *minority carriers* and the relatively large number of conduction electrons as *majority carriers*. Similarly, electrons are the minority carriers in p-type material, and holes are the majority carriers.

3.3 Carrier Densities

In calculating semiconductor electrical properties and analyzing device behavior, it is often necessary to know the density of charge carriers in the material. The majority carrier density is usually obvious in heavily doped material, since one majority carrier is obtained for each impurity atom (for the standard doping impurities). The density of minority carriers is not obvious, however, nor is the temperature dependence of the carrier densities.

---

†The "density of states" effective mass listed in Table 3-1 is not appropriate for this calculation. The value used here is an average of the effective mass in different crystallographic directions, called the "conductivity effective mass."

To obtain equations for the carrier densities we must investigate the distribution of carriers over the available energy states. This type of distribution is not difficult to calculate, but the derivation requires some background in statistical methods. Since we are primarily concerned here with the application of these results to semiconductor materials and devices, we shall accept the distribution function as given.

3.3.1 The Fermi Level.

Electrons in solids obey *Fermi–Dirac* statistics.[†] In the development of this type of statistics, one must consider the indistinguishability of the electrons, their wave nature, and the Pauli Exclusion Principle. The rather simple result of these statistical arguments is that the distribution of electrons over a range of allowed energy levels at thermal equilibrium is

(3-10)
$$f(E) = \frac{1}{1 + e^{(E-E_F)/kT}}$$

where k is Boltzmann's constant ($k = 8.62 \times 10^{-5}$ eV/°K $= 1.38 \times 10^{-23}$ J/°K). The function $f(E)$, the *Fermi–Dirac distribution function*, gives the probability that an available energy state at E will be occupied by an electron at absolute temperature T. The quantity E_F is called the *Fermi level*, and it represents an important quantity in the analysis of semiconductor behavior. We notice that, for an energy E equal to the Fermi level energy E_F, the occupation probability is

(3-11)
$$f(E_F) = [1 + e^{(E_F - E_F)/kT}]^{-1} = \frac{1}{1+1} = \frac{1}{2}$$

Thus an energy state at the Fermi level has a probability of $\frac{1}{2}$ of being occupied by an electron.

A closer examination of $f(E)$ indicates that at 0°K the distribution takes the simple rectangular form shown in Fig. 3-15. With $T = 0$ in the denominator of the exponent, $f(E)$ is $1/(1 + 0) = 1$ when the exponent is negative

---

[†]Examples of other types of statistics are *Maxwell–Boltzmann* for classical particles (e.g., gas) and *Bose–Einstein* for photons. For two discrete energy levels E_2 and E_1 (with $E_2 > E_1$), classical gas atoms follow a Boltzmann distribution; the number n_2 of atoms in state E_2 is related to the number n_1 in E_1 at thermal equilibrium by

$$\frac{n_2}{n_1} = \frac{e^{-E_2/kT}}{e^{-E_1/kT}} = e^{-(E_2 - E_1)/kT}$$

assuming the two levels have an equal number of states. The exponential term $\exp(-\Delta E/kT)$ is commonly called the *Boltzmann factor*. It appears also in the denominator of the Fermi–Dirac distribution function. We shall return to the Boltzmann distribution in Chapter 7 in discussions of the properties of lasers.

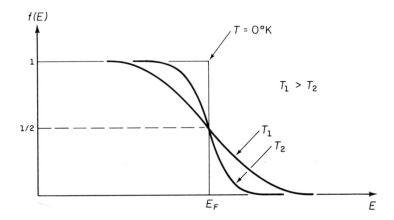

FIGURE 3-15.   The Fermi-Dirac distribution function.

$(E < E_F)$, and is $1/(1 + \infty) = 0$ when the exponent is positive $(E > E_F)$. This rectangular distribution implies that at 0°K every available energy state up to E_F is filled with electrons, and all states above E_F are empty.

At temperatures higher than 0°K, some probability exists for states above the Fermi level to be filled. For example, at $T = T_1$ in Fig. 3-15 there is some probability $f(E)$ that states above E_F are filled, and there is a corresponding probability $[1 - f(E)]$ that states below E_F are empty. The Fermi function is symmetrical about E_F for all temperatures; that is, the probability $f(E_F + \Delta E)$ that a state ΔE above E_F is filled is the same as the probability $[1 - f(E_F - \Delta E)]$ that a state ΔE below E_F is empty (Prob. 3.3). The symmetry of the distribution of empty and filled states about E_F makes the Fermi level a natural reference point in calculations of electron and hole densities in semiconductors.

In applying the Fermi–Dirac distribution to semiconductors, we must recall that $f(E)$ is the probability of occupancy of an *available* state at E. Thus if there is no available state at E (e.g., in the band gap of a semiconductor), there is no possibility of finding an electron there. We can best visualize the relation between $f(E)$ and the band structure by turning the $f(E)$ vs. E diagram on its side so that the E scale corresponds to the energies of the band diagram (Fig. 3-16). For intrinsic material we know that the density of holes in the valence band is equal to the density of electrons in the conduction band. Therefore, the Fermi level E_F must lie at the middle of the band gap in intrinsic material.† Since $f(E)$ is symmetrical about E_F, the

---

†Actually the intrinsic E_F is displaced slightly from the middle of the gap, since the densities of available states in the valence and conduction bands are not equal (Section 3.3.2).

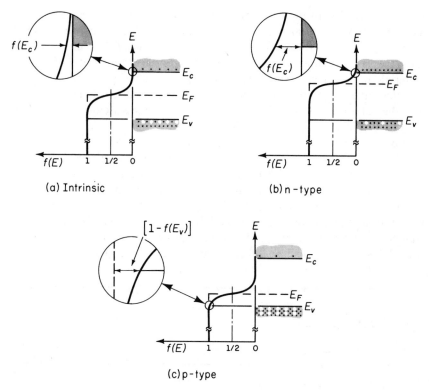

(a) Intrinsic   (b) n-type

(c) p-type

FIGURE 3-16. The Fermi distribution function applied to semiconductors: (a) intrinsic material; (b) n-type material; (c) p-type material.

electron probability "tail" of $f(E)$ extending into the conduction band of Fig. 3-16a is symmetrical with the hole probability tail $[1 - f(E)]$ in the valence band. The distribution function has values within the band gap between E_v and E_c, but there are no energy states available, and no electron occupancy results from $f(E)$ in this range.

The tails in $f(E)$ are exaggerated in Fig. 3-16 for illustrative purposes. Actually, the probability values at E_v and E_c are quite small for intrinsic material at reasonable temperatures. For example, in Si at 300°K, $n_i = p_i \simeq 10^{10}$ cm^{-3}, whereas the densities of available states at E_v and E_c are on the order of 10^{19} cm^{-3}. Thus the probability of occupancy $f(E)$ for an individual state in the conduction band and the hole probability $[1 - f(E)]$ for a state in the valence band are quite small. Because of the relatively large density of states in each band, small changes in $f(E)$ can result in significant changes in carrier concentration.

In n-type material there is a high density of electrons in the conduction band compared with the hole density in the valence band (recall Fig. 3-12). Thus in n-type material the distribution function $f(E)$ must lie above its intrinsic position on the energy scale (Fig. 3-16b). Since $f(E)$ retains its shape for a particular temperature, the larger density of electrons at E_c in n-type material implies a correspondingly smaller hole density at E_v. We notice that the value of $f(E)$ for each energy level in the conduction band (and therefore the total electron density n) increases as E_F moves closer to E_c. Thus the energy difference $(E_c - E_F)$ gives a measure of n; we shall express this relation mathematically in the following section.

For p-type material the Fermi level lies near the valence band (Fig. 3-16c) such that the $[1 - f(E)]$ tail below E_v is larger than the $f(E)$ tail above E_c. The value of $(E_F - E_v)$ indicates how strongly p-type the material is.

It is usually inconvenient to draw $f(E)$ vs. E on every energy band diagram to indicate the electron and hole distributions. Therefore, it is common practice merely to indicate the position of E_F in band diagrams (Fig. 3-17). This is sufficient information, since for a particular temperature

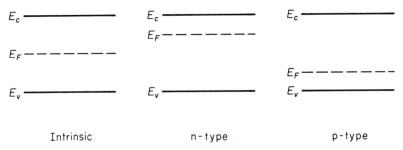

FIGURE 3-17. Simplified energy band diagrams for intrinsic, n-type, and p-type semiconductors.

the position of E_F implies the distributions in Fig. 3-16. In most discussions we shall use simplified band diagrams such as those in Fig. 3-17.

3.3.2 Electron and Hole Densities at Equilibrium. The Fermi distribution function can be used to calculate the densities of electrons and holes in a semiconductor, if the densities of available states in the valence and conduction bands are known. For example, the density of electrons in the conduction band is

(3-12)
$$ n = \int_{E_c}^{\infty} f(E)\, N(E)\, dE $$

where $N(E)dE$ is the density of states (cm^{-3}) in the energy range dE. The number of electrons per unit volume in the energy range dE is the product of the density of states and the probability of occupancy $f(E)$. Thus the total electron density is the integral over the entire conduction band, as in Eq. (3-12).† The function $N(E)$ can be calculated by using quantum mechanics and the Pauli Exclusion Principle (Appendix IV). The result of the integration of Eq. (3-12) is the same as that obtained if we represent all of the distributed electron states in the conduction band by an *effective density of states* N_c located at the conduction band edge E_c. Therefore, the conduction band electron density is simply the effective density of states at E_c times the probability of occupancy at E_c.‡

$$(3\text{-}13) \qquad\qquad n = N_c f(E_c)$$

In this expression we assume the Fermi level E_F lies at least several kT below the conduction band. Then the exponential term is large compared with unity, and the Fermi function $f(E_c)$ can be simplified as

$$(3\text{-}14) \qquad\qquad f(E_c) = \frac{1}{1 + e^{(E_c - E_F)/kT}} \simeq e^{-(E_c - E_F)/kT}$$

Since kT at room temperature is only 0.026 eV, this is generally a good approximation. For this condition the density of electrons in the conduction band is

$$(3\text{-}15) \qquad\qquad n = N_c e^{-(E_c - E_F)/kT}$$

The effective density of states N_c is shown in Appendix IV to be

$$(3\text{-}16) \qquad\qquad N_c = 2\left(\frac{2\pi m_n^* kT}{h^2}\right)^{3/2}$$

Since the quantities in Eq. (3-16) are known, values of N_c can be tabulated as a function of temperature. As Eq. (3-15) indicates, the electron density increases as E_F moves closer to the conduction band. This is the result we would predict from Fig. 3-16b. By similar arguments, the density of holes in the valence band is

$$(3\text{-}17) \qquad\qquad p = N_v[1 - f(E_v)]$$

---

†The upper limit is actually improper in Eq. (3-12), since the conduction band does not extend to infinite energy. This is unimportant in the calculation of n, however, since $f(E)$ becomes negligibly small for large values of E. Most electrons occupy states near the bottom of the conduction band at equilibrium.

‡The simple expression for n obtained in Eq. (3-13) is the direct result of integrating Eq. (3-12), as in Appendix IV. Equations (3-15) and (3-19) properly include the effects of the conduction and valence bands through the density of states terms.

where N_v is the effective density of states in the valence band. The probability of finding an empty state at E_v is

(3-18) $$1 - f(E_v) = 1 - \frac{1}{1 + e^{(E_v - E_F)/kT}} \simeq e^{-(E_F - E_v)/kT}$$

for E_F larger than E_v by several kT. From these equations, the density of holes in the valence band is

(3-19) $$p = N_v e^{-(E_F - E_v)/kT}$$

The effective density of states in the valence band reduced to the band edge is

(3-20) $$N_v = 2\left(\frac{2\pi m_p^* kT}{h^2}\right)^{3/2}$$

As expected from Fig. 3-16c, Eq. (3-19) predicts that the hole density increases as E_F moves closer to the valence band.

The electron and hole densities predicted by Eqs. (3-15) and (3-19) are valid whether the material is intrinsic or doped, provided thermal equilibrium is maintained. Thus *for intrinsic material, E_F lies at some intrinsic level E_i* near the middle of the band gap (Fig. 3-16a), and the intrinsic electron and hole concentrations are

(3-21) $$n_i = N_c e^{-(E_c - E_i)/kT}, \qquad p_i = N_v e^{-(E_i - E_v)/kT}$$

The product of n and p at equilibrium is a constant for a particular material and temperature, even if the doping is varied:

(3-22a) $$np = (N_c e^{-(E_c - E_F)/kT})(N_v e^{-(E_F - E_v)/kT}) = N_c N_v e^{-(E_c - E_v)/kT}$$
$$= N_c N_v e^{-E_g/kT}$$

(3-22b) $$n_i p_i = (N_c e^{-(E_c - E_i)/kT})(N_v e^{-(E_i - E_v)/kT}) = N_c N_v e^{-E_g/kT}$$

The intrinsic electron and hole densities are equal (since the carriers are created in pairs), $n_i = p_i$; thus the intrinsic density is

(3-23) $$n_i = \sqrt{N_c N_v}\, e^{-E_g/2kT}$$

The constant product of electron and hole densities in Eq. (3-22) can be written conveniently as

(3-24) $$np = n_i^2$$

This is an important relation, and we shall use it extensively in later calculations.

Comparing Eqs. (3-21) and (3-23), we note that the intrinsic level E_i is the middle of the band gap ($E_c - E_i = E_g/2$), if the effective densities of states N_c and N_v are equal. There is usually some difference in effective mass for electrons and holes, however, and N_c and N_v are slightly different as Eqs. (3-16) and (3-20) indicate. Thus the intrinsic level E_i is displaced from the middle of the band gap, although the deviation is usually slight (Prob. 3.6).

Another convenient way of writing Eqs. (3-15) and (3-19) is

(3-25a) $$n = n_i e^{(E_F - E_i)/kT}$$

(3-25b) $$p = n_i e^{(E_i - E_F)/kT}$$

obtained by the application of Eq. (3-21). This form of the equations indicates directly that the electron density is n_i when E_F is at the intrinsic level E_i, and that n increases exponentially as the Fermi level moves away from E_i toward the conduction band. Similarly, the hole density p varies from n_i to larger values as E_F moves from E_i toward the valence band. Since these equations reveal the qualitative features of carrier density so directly, they are particularly convenient to remember.

3.3.3 Temperature Dependence of Carrier Densities. The variation of carrier density with temperature is indicated by Eq. (3-25). Initially, the variation of n and p with T seems relatively straightforward in these relations. The problem is complicated, however, by the fact that n_i has a strong temperature dependence [Eq. (3-23)] and that E_F can also vary with temperature. Let us begin by examining the intrinsic carrier density. By combining Eqs. (3-23), (3-16), and (3-20) we obtain

(3-26) $$n_i(T) = 2\left(\frac{2\pi kT}{h^2}\right)^{3/2}(m_n^* m_p^*)^{3/4} e^{-E_g/2kT}$$

The exponential temperature dependence dominates $n_i(T)$, and a plot of $\ln n_i$ vs. $10^3/T$ appears almost linear (Fig. 3-18).† In this figure we neglect variations due to the $T^{3/2}$ dependence of the density of states function and the fact that E_g varies somewhat with temperature.‡ The value of n_i at any temperature is a definite number for a given semiconductor, and its value is

---

†When plotting quantities such as carrier density, which involve a Boltzmann factor, it is common to use an inverse temperature scale. This allows terms which are exponential in $1/T$ to appear linear in the semilogarithmic plot. When reading such graphs, remember that temperature increases from right to left.

‡For Si the band gap E_g varies from about 1.11 eV at 300°K to about 1.16 eV at 0°K.

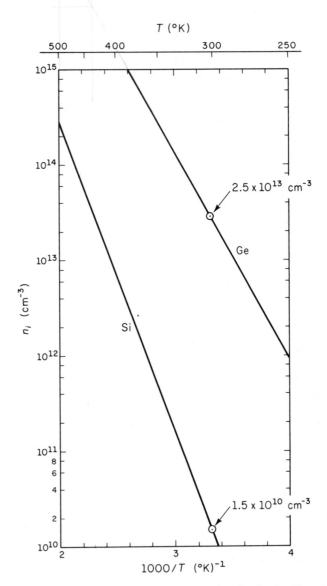

FIGURE 3-18. Approximate intrinsic carrier density for Si and Ge as a function of inverse temperature. The room temperature values are marked for reference.

known for most materials. Thus we can take n_i as given in calculating n or p from Eq. (3-25).†

With n_i and T given, the unknowns in Eq. (3-25) are the carrier densities and the Fermi level position relative to E_i. One of these two quantities must be given if the other is to be found. If the carrier density is held at a certain value, as in heavily doped extrinsic material, E_F can be obtained from Eq. (3-25). The temperature dependence of electron concentration in a doped semiconductor can be visualized as shown in Fig. 3-19. In this example,

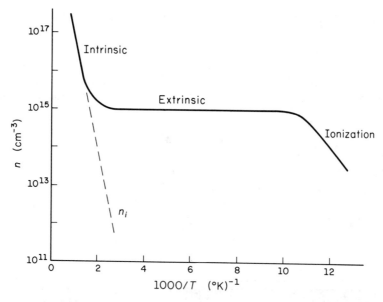

FIGURE 3-19.   Carrier concentration vs. inverse temperature for Si doped with 10^{15} donors/cm³.

Si is doped n-type with a donor density N_d of 10^{15} cm⁻³. At very low temperatures (large $1/T$), negligible intrinsic EHP's exist, and the donor electrons are bound to the donor atoms. As the temperature is raised, these electrons are donated to the conduction band, and at about 100°K ($1000/T = 10$) all the donor atoms are ionized. This temperature range is called the *ionization* region. Once the donors are ionized, the conduction band electron density is $n \simeq N_d = 10^{15}$ cm⁻³, since one electron is obtained for each donor atom.

---

†Care must be taken to use consistent units in these calculations. For example, if an energy such as E_g is expressed in electron volts (eV), it should be multiplied by q (1.6 × 10⁻¹⁹ C) to convert to joules if k is in J/°K; alternatively, E_g can be kept in eV and the value of k in eV/°K can be used.

When every available extrinsic electron has been transferred to the conduction band, n is virtually constant with temperature until the concentration of intrinsic carriers n_i becomes comparable to the extrinsic density N_d. Finally, at higher temperatures n_i is much greater than N_d, and the intrinsic carriers dominate. In most devices it is desirable to control the carrier density by doping rather than by thermal EHP generation. Thus one usually dopes the material such that the extrinsic range extends beyond the highest temperature at which the device is to be used.

3.3.4 Compensation and Space Charge Neutrality.

When the concept of doping was introduced, we assumed the material contained either N_d donors or N_a acceptors, so that the extrinsic majority carrier concentrations were $n \simeq N_d$ or $p \simeq N_a$, respectively, for the n-type or p-type material. It often happens, however, that a semiconductor contains both donors and acceptors. For example, Fig. 3-20 illustrates a semiconductor for which both

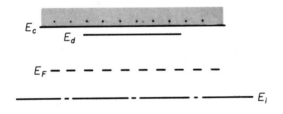

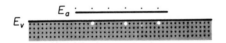

FIGURE 3-20. Compensation in an n-type semiconductor ($N_d > N_a$).

donors and acceptors are present, but $N_d > N_a$. The predominance of donors makes the material n-type, and the Fermi level is therefore in the upper part of the band gap. Since E_F is well above the acceptor level E_a, this level is essentially filled with electrons. However, with E_F above E_i, we cannot expect a hole density in the valence band commensurate with the acceptor density. In fact, the filling of the E_a states occurs at the expense of the donated conduction band electrons. The mechanism can be visualized as follows: Assume an acceptor state is filled with a valence band electron as described in Fig. 3-13, with a hole resulting in the valence band. This hole is then filled by recombination with one of the conduction band electrons. Extending this logic to all the acceptor atoms, we expect the resultant density of electrons in the conduction band to be $N_d - N_a$ instead of the total N_d. This process is called *compensation*. By this process it is possible to begin with an n-type semiconductor and add acceptors until $N_a = N_d$ and no donated electrons remain in the conduction band. In such compensated material, $n = n_i = p$ and intrinsic conduction is obtained. With further acceptor doping the semiconductor becomes p-type with a hole density of essentially $N_a - N_d$.

The exact relationship among the electron, hole, donor, and acceptor densities can be obtained by considering the requirements for *space charge neutrality*. If the material is to remain electrostatically neutral, the sum of the positive charges (holes and ionized donor atoms) must balance the sum of the negative charges (electrons and ionized acceptor atoms)

(3-27) $$p + N_d^+ = n + N_a^-$$

Thus in Fig. 3-20 the net electron density in the conduction band is

(3-28) $$n = p + (N_d^+ - N_a^-)$$

If the material is doped n-type ($n \gg p$) and all the impurities are ionized, we can approximate Eq. (3-28) by $n \simeq N_d - N_a$.

Since the intrinsic semiconductor itself is electrostatically neutral and the doping atoms we add are also neutral, the requirement of Eq. (3-27) must be maintained at equilibrium. The electron and hole densities and the Fermi level adjust such that Eq. (3-27) and Eq. (3-25) are satisfied.

3.4 Drift of Carriers in Electric and Magnetic Fields

Knowledge of carrier densities in a solid is necessary for calculating current flow in the presence of electric or magnetic fields. In addition to the values of n and p, we must be able to take into account the collisions of the charge carriers with the lattice and with the impurities. These processes will affect the ease with which electrons and holes can flow through the crystal, that is, their *mobility* within the solid. As should be expected, these collision and scattering processes depend on temperature, which affects the thermal motion of the lattice atoms and the velocity of the carriers.

3.4.1 Conductivity and Mobility.
The charge carriers in a solid are in constant motion, even at thermal equilibrium. At room temperature, for example, the thermal motion of an individual electron may be visualized as random scattering from lattice atoms, impurities, other electrons, and defects (Fig. 3-21). Since the scattering is random, there is no net motion of the group of n electrons/cm^3 over any period of time. This is not true of an individual electron, of course. The probability of the electron in Fig. 3-21

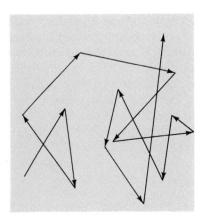

FIGURE 3-21. Thermal motion of an electron in a solid.

returning to its starting point after some time t is negligibly small. However, if a large number of electrons is considered (e.g., 10^{16} cm^{-3} in an n-type semiconductor), there will be no preferred direction of motion for the group of electrons and no net current flow.

If an electric field $\mathscr{E}_x$ is applied in the x-direction, each electron experiences a net force $-q\mathscr{E}_x$ from the field. This force may be insufficient to alter appreciably the random path of an individual electron; the effect when averaged over all the electrons, however, is a net motion of the group in the $-x$-direction. If $\mathbf{p}_x$ is the x-component of the total momentum of the group, the force of the field on the n electrons/cm^3 is

$$(3\text{-}29) \qquad -nq\mathscr{E}_x = \frac{d\mathbf{p}_x}{dt}\bigg|_{\text{field}}$$

Initially, Eq. (3-29) seems to indicate a continuous acceleration of the electrons in the $-x$-direction. This is not the case, however, because the net acceleration of Eq. (3-29) is just balanced in steady state by the decelerations of the collision processes. Thus while the steady field $\mathscr{E}_x$ does produce a net momentum $\mathbf{p}_{-x}$, the net rate of change of momentum when collisions are included must be zero in the case of steady state current flow.

To find the total rate of momentum change from collisions, we must investigate the collision probabilities more closely. If the collisions are truly random, there will be a constant probability of collision at any time for each electron. Let us consider a group of N_0 electrons at time $t = 0$ and define $N(t)$ as the number of electrons which *have not* undergone a collision by time t. The rate of decrease in $N(t)$ at any time t is proportional to the number left unscattered at t,

$$(3\text{-}30) \qquad -\frac{dN(t)}{dt} = \frac{1}{\bar{t}}N(t)$$

where $\bar{t}^{-1}$ is a constant of proportionality.

The solution to Eq. (3-30) is an exponential function

(3-31) $$N(t) = N_0 e^{-t/\bar{t}}$$

and we see that $\bar{t}$ represents the mean time between scattering events† (Prob. 3.10), called the *mean free time*. The probability that any electron has a collision in the time interval dt is $dt/\bar{t}$. Thus the differential change in p_x due to collisions in time dt is

(3-32) $$dp_x = -p_x \frac{dt}{\bar{t}}$$

The rate of change of p_x due to the decelerating effect of collisions is

(3-33) $$\left. \frac{dp_x}{dt} \right|_{\text{collisions}} = -\frac{p_x}{\bar{t}}$$

The sum of acceleration and deceleration effects must be zero for steady state. Taking the sum of Eqs. (3-29) and (3-33), we have

(3-34) $$-\frac{p_x}{\bar{t}} - nq\mathscr{E}_x = 0$$

The average momentum per electron is

(3-35) $$\langle p_x \rangle = \frac{p_x}{n} = -q\bar{t}\mathscr{E}_x$$

where the angular brackets indicate an average over the entire group of electrons. As expected for steady state, Eq. (3-35) indicates that the electrons have *on the average* a constant net velocity in the negative x-direction

(3-36) $$\langle v_x \rangle = \frac{\langle p_x \rangle}{m_n^*} = -\frac{q\bar{t}}{m_n^*}\mathscr{E}_x$$

Actually, the individual electrons move in many directions by thermal motion during a given time period, but Eq. (3-36) tells us the *net drift* of an average electron in response to the electric field. The drift speed described by

---

†Equations (3-30) and (3-31) are typical of events dominated by random processes, and the forms of these equations occur often in many branches of physics and engineering. For example, in the radioactive decay of unstable nuclear isotopes, N_0 nuclides decay exponentially with a mean lifetime $\bar{t}$. Other examples will be found in this text, including the absorption of light in a semiconductor and the recombination of excess EHP's.

Eq. (3-36) is usually much smaller than the random speed due to thermal motion (Prob. 3.12).

The current density resulting from this net drift is just the number of electrons crossing a unit area per unit time ($n\langle v_x \rangle$) multiplied by the charge on the electron ($-q$)

$$(3\text{-}37) \qquad\qquad J_x = -qn\langle v_x \rangle$$

$$\frac{\text{ampere}}{\text{cm}^2} = \frac{\text{coulomb}}{\text{electron}} \cdot \frac{\text{electrons}}{\text{cm}^3} \cdot \frac{\text{cm}}{\text{sec}}$$

Using Eq. (3-36) for the average velocity, we obtain

$$(3\text{-}38) \qquad\qquad J_x = \frac{nq^2\bar{t}}{m_n^*}\mathscr{E}_x$$

Thus the current density is proportional to the electric field, as we expect from Ohm's law

$$(3\text{-}39) \qquad J_x = \sigma\mathscr{E}_x, \qquad \text{where } \sigma \equiv \frac{nq^2\bar{t}}{m_n^*}$$

The conductivity σ (Ω-cm)$^{-1}$ can be written

$$(3\text{-}40) \qquad\qquad \sigma = qn\mu_n, \qquad \text{where } \mu_n \equiv \frac{q\bar{t}}{m_n^*}$$

The quantity μ_n, called the *electron mobility*, describes the ease with which electrons drift in the material. Mobility is a very important quantity in characterizing semiconductor materials and in device development.

The mobility defined in Eq. (3-40) can be expressed as the average particle drift velocity per unit electric field. Comparing Eqs. (3-36) and (3-40), we have

$$(3\text{-}41) \qquad\qquad \mu_n = -\frac{\langle v_x \rangle}{\mathscr{E}_x}$$

The units of mobility are (cm/sec)/(V/cm) = cm^2/V-sec, as Eq. (3-41) suggests. The minus sign in the definition results in a positive value of mobility, since electrons drift opposite to the field.

The current density can be written in terms of mobility as

$$(3\text{-}42) \qquad\qquad J_x = qn\mu_n\mathscr{E}_x$$

This derivation has been based on the assumption that the current is carried primarily by electrons. For hole conduction we change n to p, $-q$

to $+q$, and μ_n to μ_p, where $\mu_p = +\langle v_x\rangle/\mathscr{E}_x$ is the mobility for holes. If both electrons and holes participate, we must modify Eq. (3-42) to

$$(3\text{-}43) \qquad J_x = q(n\mu_n + p\mu_p)\mathscr{E}_x = \sigma\mathscr{E}_x$$

Values of μ_n and μ_p are given for many of the common semiconductor materials in Appendix III.

Let us look more closely at the drift of electrons and holes. If the semiconductor bar of Fig. 3-22 contains both types of carriers, Eq. (3-43) gives the conductivity of the material. The resistance of the bar is then

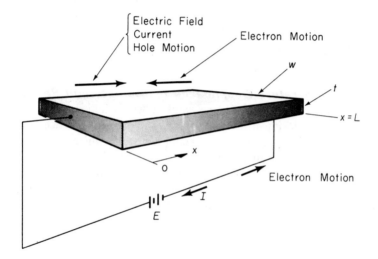

FIGURE 3-22.   Drift of electrons and holes in a semiconductor bar.

$$(3\text{-}44) \qquad R = \frac{\rho L}{wt} = \frac{L}{wt}\frac{1}{\sigma}$$

where ρ is the resistivity (Ω-cm). The physical mechanism of carrier drift requires that the holes in the bar move as a group in the direction of the electric field and that the electrons move as a group in the opposite direction. Both the electron and the hole components of current are in the direction of the $\mathscr{E}$ field, since conventional current is positive in the direction of hole flow and opposite to the direction of electron flow. The drift current described by Eq. (3-43) is constant throughout the bar. A valid question arises, therefore, concerning the nature of the electron and hole flow at the contacts and in the external circuit. We should specify that the contacts to the bar of Fig. 3-22 are *ohmic*, meaning that they are perfect sources and sinks of both carrier types and have no special tendency to inject or collect either electrons or holes.

If we consider that current is carried around the external circuit by electrons, there is no problem in visualizing electrons flowing into the bar at one end and out at the other (always opposite to I). Thus for every electron leaving the left end ($x = 0$) of the bar in Fig. 3-22, there is a corresponding electron entering at $x = L$, so that the electron concentration in the bar remains constant at n. But what happens to the holes at the contacts? As a hole reaches the ohmic contact at $x = L$, it recombines with an electron which must be supplied through the external circuit. As this hole disappears, a corresponding hole must appear at $x = 0$ to maintain space charge neutrality. It is reasonable to consider the source of this hole as the generation of an EHP at $x = 0$, with the hole flowing into the bar and the electron flowing into the external circuit.

3.4.2 Temperature Dependence of Mobility and Resistivity.

The two basic types of scattering mechanisms which influence electron and hole mobility are *lattice scattering* and *impurity scattering*. In lattice scattering a carrier moving through the lattice encounters atoms which are out of their normal lattice positions due to thermal vibration. The frequency of such scattering events increases as the lattice temperature increases, since the thermal agitation of the lattice becomes greater. Therefore, we should expect the mobility to decrease as the sample is heated (Fig. 3-23). On the other

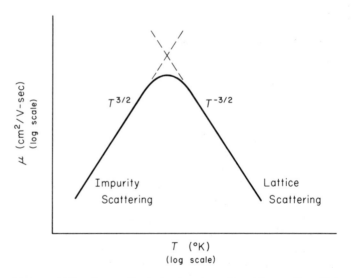

FIGURE 3-23. Approximate temperature dependence of mobility with both lattice and impurity scattering.

hand, scattering from lattice defects such as ionized impurities becomes the dominant mechanism at low temperatures. Since the atoms of the cooler lattice are less agitated, lattice scattering is less important; however, the thermal motion of the carriers is also slower. Since a slowly moving carrier is likely to be scattered more strongly by an interaction with a charged ion than is a carrier with greater momentum, impurity scattering events cause a decrease in mobility with decreasing temperature. As Fig. 3-23 indicates, the approximate temperature dependencies are $T^{-3/2}$ for lattice scattering and $T^{3/2}$ for impurity scattering. Since the scattering probability of Eq. (3-32) is inversely proportional to the mean free time and therefore to mobility, the mobilities due to two or more scattering mechanisms add inversely

$$(3\text{-}45) \qquad \frac{1}{\mu} = \frac{1}{\mu_1} + \frac{1}{\mu_2}$$

As a result, the mechanism causing the lowest mobility value dominates, as shown in Fig. 3-23.

As the density of impurities increases, the effects of impurity scattering are felt at higher temperatures. For example, the electron mobility μ_n of intrinsic silicon at 300°K is 1350 cm²/(V-sec). With a donor doping density of 10^{17} cm^{-3}, however, μ_n is 700 cm²/(V-sec). Thus the presence of the 10^{17} ionized donors/cm³ introduces a significant amount of impurity scattering.

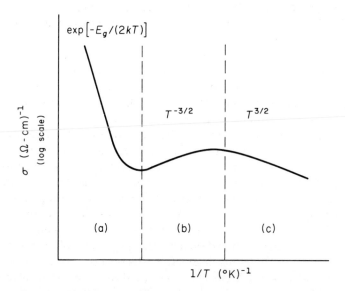

FIGURE 3-24. Temperature dependence of conductivity for a semiconductor: (a) intrinsic region; (b) extrinsic region with lattice scattering dominant; (c) extrinsic region with impurity scattering dominant.

The temperature dependence of mobility plays an important role in the extrinsic range of conductivity. Figure 3-24 is a typical plot of σ vs. $1/T$. Since for n-type material σ is proportional to the product of electron concentration and mobility [Eq. (3-40)], the temperature variation of μ_n (Fig. 3-23) must be combined with that of n (Fig. 3-19). At high temperatures the exponential variation of the intrinsic carrier density dominates, and in the extrinsic range (for which n is constant at N_d) the temperature variation of μ_n dominates.

The strong temperature dependence of conductivity for a semiconductor is used in a device called the *thermistor*. One important use of this device is in the control of temperature sensitive systems. For example, an intrinsic semiconductor sample can be placed in a chamber which is to be heated to a specific temperature, and a voltage is placed across the sample. When the temperature varies from the prescribed value, σ changes for the thermistor, and therefore the current through the device changes. This variation in current can be used to initiate corrections in the temperature control circuit. Since σ varies as $\exp(-E_g/2kT)$, the thermistor is very sensitive to temperature (see Prob. 3.13). With proper calibration, a thermistor can be used as a "thermometer"—its resistance can be measured and the corresponding temperature can be obtained from a predetermined calibration chart.

3.4.3 Hot Carrier Effects.

One assumption implied in the derivation of Eq. (3-39) was that Ohm's law is valid in the carrier drift processes. That is, it was assumed that the drift current is proportional to the electric field and that the proportionality constant (σ) is not a function of field $\mathscr{E}$. This assumption is valid over a wide range of $\mathscr{E}$. However, large electric fields ($> 10^3$ V/cm) can cause the current $J = -qn\mathrm{v}$ to exhibit a sublinear dependence on the electric field (Fig. 3-25). This dependence of σ upon $\mathscr{E}$ for large fields is an example of the *hot carrier effect*.

In the linear region of Fig. 3-25, the mean velocity of the electrons (or holes) is dictated by the temperature of the lattice, and the added drift velocity is a comparatively small perturbation. Thus, if the electrons are considered as a collection of particles similar to the molecules of a gas, the "temperature" of the particles can be related to their mean velocity. This electron temperature will correspond to the lattice temperature at equilibrium. Since energy is imparted to the drifting electrons by the field, the lattice temperature will increase as this energy is transferred to the atoms by collision processes. This is the familiar "Joule heating" effect which causes the temperature of a resistor to increase when current is passed through it. At relatively low values of field, the transfer of energy from the electrons to the lattice is efficient, and a steady state temperature balance can be maintained. At high fields, however, the electrons cannot transfer their kinetic energy to the atoms

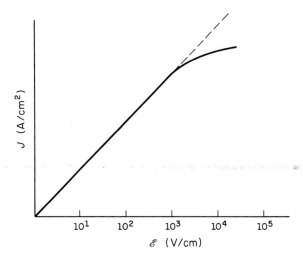

Figure 3-25. Deviation from Ohm's law behavior at high electric fields.

fast enough to maintain thermal balance. The electrons then become "hotter" in terms of their kinetic energy than the lattice temperature. These "hot" electrons suffer a decrease in mobility within the lattice, and the result is a sublinear dependence of current on the field as Fig. 3-25 indicates.

In many cases an upper limit is reached for the carrier velocity in a high field. This limit occurs near the mean thermal velocity ($\simeq 10^7$ cm/sec) and represents the point at which added energy imparted by the field is transferred to the lattice rather than increasing the carrier velocity. The result of this *scattering limited velocity* is a fairly constant current at high fields, as shown in Fig. 3-25. This behavior is typical of Si, Ge, and some other semiconductors. However, there are other important effects in some materials; for example, in Chapter 12 we shall discuss a decrease in electron velocity at high fields for GaAs and certain other materials, which results in negative conductivity and current instabilities in the sample. Another high-field effect is avalanche multiplication, which we shall discuss in Section 5.4.2.

3.4.4 The Hall Effect. If a magnetic field is applied perpendicular to the direction at which holes drift in a p-type bar, the path of the holes tends to be deflected (Fig. 3-26). Using vector notation, the total force on a single hole due to the electric and magnetic fields is

$$(3\text{-}46) \qquad\qquad \mathbf{F} = q(\mathscr{E} + \mathbf{v} \times \mathscr{B})$$

In the *y*-direction the force is

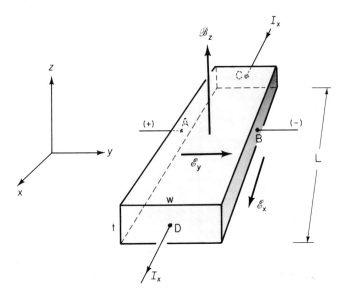

FIGURE 3-26. The Hall effect.

$$(3\text{-}47) \qquad\qquad F_y = q(\mathscr{E}_y - \mathsf{v}_x \mathscr{B}_z)$$

The important result of Eq. (3-47) is that unless an electric field $\mathscr{E}_y$ is established along the width of the bar, each hole will experience a net force (and therefore an acceleration) in the $-y$-direction due to the $q\mathsf{v}_x\mathscr{B}_z$ product. Therefore, to maintain a steady state flow of holes down the length of the bar, the electric field $\mathscr{E}_y$ must just balance the product $\mathsf{v}_x\mathscr{B}_z$

$$(3\text{-}48) \qquad\qquad \mathscr{E}_y = \mathsf{v}_x \mathscr{B}_z$$

so that the net force F_y is zero. Physically, this electric field is set up when the magnetic field shifts the hole distribution slightly in the $-y$-direction. Once the electric field $\mathscr{E}_y$ becomes as large as $\mathsf{v}_x\mathscr{B}_z$, no net lateral force is experienced by the holes as they drift along the bar. The establishment of the electric field $\mathscr{E}_y$ is known as the *Hall effect*, and the resulting voltage $V_{AB} = \mathscr{E}_y w$ is called the *Hall voltage*. If we use the expression derived in Eq. (3-37) for the drift velocity (using $+q$ and p for holes), the field $\mathscr{E}_y$ becomes

$$(3\text{-}49) \qquad\qquad \mathscr{E}_y = \frac{J_x}{qp}\mathscr{B}_z = R_H J_x \mathscr{B}_z, \qquad R_H \equiv \frac{1}{qp}$$

Thus the Hall field is proportional to the product of the current density and the magnetic flux density. The proportionality constant $R_H = (qp)^{-1}$ is called the *Hall coefficient*. A measurement of the Hall voltage for a known current and magnetic field yields a value for the hole density p

$$(3\text{-}50) \qquad p = \frac{1}{qR_H} = \frac{J_x \mathscr{B}_z}{q\mathscr{E}_y} = \frac{(I_x/wt)\mathscr{B}_z}{q(V_{AB}/w)} = \frac{I_x \mathscr{B}_z}{qtV_{AB}}$$

Since all of the quantities in the right-hand side of Eq. (3-50) can be measured, the Hall effect can be used to give quite accurate values for carrier density. For example, it is possible to measure a hole density of less than 10^{12} cm^{-3} in a lightly doped Si sample, thus detecting an ionized acceptor density of about one part in 10^{10} (considering $\sim 10^{22}$ Si atoms/cm^3). This sensitivity is several orders of magnitude greater than that achieved by the best chemical analysis.

If a measurement of resistance R is made, the sample resistivity ρ can be calculated

$$(3\text{-}51) \qquad \rho(\Omega\text{-cm}) = \frac{Rwt}{L} = \frac{V_{CD}/I_x}{L/wt}$$

Since the conductivity $\sigma = 1/\rho$ is given by $q\mu_p p$, the mobility is simply the ratio of the Hall coefficient and the resistivity

$$(3\text{-}52) \qquad \mu_p = \frac{\sigma}{qp} = \frac{1/\rho}{q(1/qR_H)} = \frac{R_H}{\rho}$$

Measurements of the Hall coefficient and the resistivity over a range of temperatures yield plots of carrier density and mobility vs. temperature, such as shown in Figs. 3-19 and 3-23. Such measurements are extremely useful in the analysis of semiconductor materials. Although the discussion here has been related to p-type material, similar results are obtained for n-type material. A negative value of q is used for electrons, and the Hall voltage V_{AB} and Hall coefficient R_H are negative. In fact, a measurement of the sign of the Hall voltage is a common technique for determining if an unknown sample is p-type or n-type.

A convenient sample geometry for Hall effect and resistivity measurements is shown in Fig. 3-27. Samples of this shape can be cut from a wafer of semiconductor material by an ultrasonic cutting tool.† The main advantage of this geometry is that wires can be soldered to the large contact areas without significantly disturbing the field distribution along the main body of

---

†The cutting tool is machined such that the desired sample shape is removed from the cutting face. Thus the tool touches the sample everywhere except those areas shown in Fig. 3-27. While this tool is vibrated vertically at high frequency and low amplitude, a fine, abrasive slurry is passed between the sample and the tool. The brittle semiconductor material is abraded away where the tool touches it, leaving the desired sample shape.

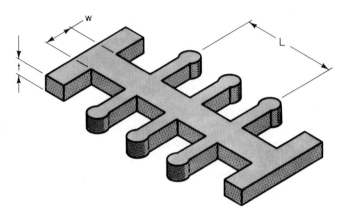

FIGURE 3-27. Sample geometry for Hall effect and resistivity measurements.

the bar. In the sample shown in Fig. 3-27, three Hall voltages can be measured along the bar (to check for uniformity of doping), and resistance voltages can be measured between several combinations of contacts.

There are several applications of the Hall effect in addition to its important uses in measuring material properties. For example, a bar of known geometry and Hall coefficient can be used as a gaussmeter to measure magnetic flux density. From Eq. (3-49) we note that measurement of the Hall voltage for a given sample current allows the calculation of $\mathscr{B}_z$. The Hall effect bar may be very small and often can be mounted conveniently on a pole face of the magnet for continuous measurement of the field. Another interesting example of Hall effect applications is the electrical performance of product operations. For example, if $\mathscr{B}_z$ is made proportional to one current (e.g., driving an electromagnet in the linear range) and J_x is made proportional to another current, $\mathscr{E}_y$ gives a product of the two currents. As a result, V_{AB} becomes a measure of the product of two quantities.

READING LIST

J. C. Phillips, "Bonds and Bands in Semiconductors," *Science*, vol. 169, pp. 1035–1042, 11 September 1970.

H. Ehrenreich, "The Electrical Properties of Materials," *Scientific American*, vol. 217, no. 3, pp. 195–204, September 1967.

A. Nussbaum, *Electronic and Magnetic Behavior of Materials.* Englewood Cliffs, N.J.: Prentice-Hall, Inc., 1967.

W. Shockley, *Electrons and Holes in Semiconductors*. New York: Van Nostrand Reinhold Co., 1950.

T. L. Martin, Jr., and W. F. Leonard, *Electrons and Crystals*. Belmont, Calif.: Brooks/Cole Publishing Company, 1970.

J. P. McKelvey, *Solid State and Semiconductor Physics*. New York: Harper & Row, Publishers, 1966.

A. van der Ziel, *Solid State Physical Electronics* (2nd ed.). Englewood Cliffs, N.J.: Prentice-Hall, Inc., 1968.

PROBLEMS

3.1 The ionic radii of Na^+ (atomic weight 23) and Cl^- (atomic weight 35.5) are 1.0 and 1.8 Å, respectively. Treating the ions as hard spheres, calculate the density of NaCl. Compare this with the measured density of 2.17 g/cm³.

3.2 Assuming U is constant in Eq. (3-1) for an essentially free electron, show that the x-component of the electron momentum in the crystal is given by $\langle p_x \rangle = \hbar k_x$. This result implies that $(E, \mathbf{k})$ diagrams such as shown in Figs. 3-6 and 3-7 can be considered plots of electron energy vs. momentum, with a scaling factor $\hbar$.

3.3 Show the probability that a state ΔE above the Fermi level E_F is filled equals the probability that a state ΔE below E_F is empty.

3.4 Given m_n^* and m_p^* from Table 3-1, calculate the effective densities of states N_c and N_v for Ge at 300°K (assume m_n^* and m_p^* do not vary with temperature). Calculate the intrinsic carrier concentration and compare this with the value given in Fig. 3-18. Expect some error to arise from the simplifying assumptions.

3.5 Show that Eqs. (3-25) result from Eqs. (3-15) and (3-19). If $n = 10^{17}$ cm⁻³, where is the Fermi level relative to E_i in Ge at 300°K?

3.6 Derive an expression relating the intrinsic level E_i to the center of the band gap $E_g/2$. Calculate the displacement of E_i from $E_g/2$ for Si at 300°K, assuming the effective mass values in Table 3-1 are valid at this temperature.

3.7 Calculate the band gap of Si from Eq. (3-23) and the plot of n_i vs. $1000/T$ (Fig. 3-18). *Hint*: the slope cannot be measured directly from a semilogarithmic plot; read the values from two points on the plot and take the natural logarithm with slide rule or tables as needed for the solution.

3.8   A semiconductor device requires n-type material; it is to be operated at 100°C. Would Si doped with 10^{14} atoms/cm^3 of phosphorus be useful in this application? Could Ge doped with 10^{14} cm^{-3} As be used?

3.9   (a) A Si sample is doped with 10^{17} As atoms/cm^3. What is the hole density p at 300°K?

(b) A Ge sample is doped with 10^{14} Sb atoms/cm^3. Using the requirements of space charge neutrality, calculate the electron density n at 300°K.

3.10   Show that $\bar{t}$ in Eq. (3-31) represents the mean time between scattering events.

3.11   (a) Show that the minimum conductivity of a semiconductor sample occurs when $n = n_i\sqrt{\mu_p/\mu_n}$. *Hint*: begin with Eq. (3-43) and apply Eq. (3-24).

(b) What is the expression for the minimum conductivity σ_{min}?

(c) Calculate σ_{min} for Ge at 300°K and compare with the intrinsic conductivity.

3.12   Assume a conduction electron in Ge ($\mu_n = 3900$ cm^2/V-sec) has a thermal energy of kT, related to its mean thermal velocity by $E_{th} = (m_0 v_{th}^2)/2$. This electron is placed in an electric field of 10 V/cm. Show that the drift velocity of the electron in this case is small compared to its thermal velocity. Repeat for a field of 10^4 V/cm, using the same value of μ_n. Comment on the actual mobility effects at this higher value of field.

3.13   An intrinsic sample of Ge is used as a thermistor. Calculate the ratio of the conductivity of this sample at 100°C to its value at 27°C. Assume the mobilities vary as $T^{-3/2}$, where T is in °K.

3.14   In soldering wires to a sample such as that shown in Fig. 3-26, it is difficult to align the Hall probes A and B precisely. If B is displaced slightly down the length of the bar from A, an erroneous Hall voltage results. Show that the true Hall voltage V_H can be obtained from two measurements of V_{AB}, with the magnetic field first in the $+z$-direction and then in the $-z$-direction.

3.15   A Ge sample is properly contacted and oriented in a 5-kGauss magnetic field as in Fig. 3-26. The current is 1 mA. The sample dimensions are $w = 10$ mils, $t = 2$ mils, and $L = 100$ mils. The following data are taken: $V_{AB} = -6.15$ mV and $V_{CD} = 400$ mV. Find the type and concentration of the majority carrier and its mobility. *Note*: 1 kGauss $= 10^{-5}$ Wb/cm^2. Comment on the mobility value.

EXCESS CARRIERS
IN SEMICONDUCTORS **4**

Most semiconductor devices operate by the creation of charge carriers in excess of the thermal equilibrium values. These excess carriers can be created by optical excitation or electron bombardment, or as we shall see in Chapter 5, they can be injected across a forward-biased p-n junction. However the excess carriers arise, they can dominate the conduction processes in the semiconductor material. In this chapter we shall investigate the creation of excess carriers by optical absorption and the resulting properties of photoluminescence and photoconductivity. We shall investigate more closely the mechanism of electron-hole pair recombination and the effects of carrier trapping. Finally, we shall discuss the diffusion of excess carriers due to a carrier gradient, which serves as a basic mechanism of current conduction along with the mechanism of drift in an electric field.

4.1 Optical Absorption†

An important technique for measuring the band gap energy of a semiconductor is the absorption of incident photons by the material. In this experiment photons of selected wavelength are directed at the sample, and the relative transmission of the various photons is observed. Since photons with energies greater than the band gap energy are absorbed while photons with energies less than band gap are transmitted, this experiment gives an accurate measure of the band gap energy.

---

†In this context the word "optical" does not necessarily imply that the photons absorbed are in the visible part of the spectrum. Many semiconductors absorb photons in the infrared region, but this is included in the term "optical absorption."

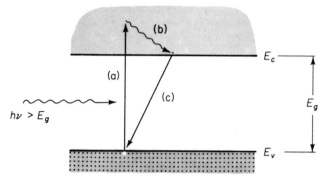

FIGURE 4-1.   Optical absorption of a photon with $hv > E_g$: (a) an EHP is created during photon absorption; (b) the excited electron gives up energy to the lattice by scattering events; (c) the electron recombines with a hole in the valence band.

4.1.1 Transmission and Absorption.   It is apparent that a photon with energy $hv \geq E_g$ can be absorbed in a semiconductor (Fig. 4-1). Since the valence band contains many electrons and the conduction band has many empty states into which the electrons may be excited, the probability of photon absorption is high. As Fig. 4-1 indicates, an electron excited to the conduction band by optical absorption may have more energy than is common for conduction band electrons (almost all electrons are near E_c unless the sample is very heavily doped). Thus the excited electron loses energy to the lattice in scattering events until its velocity reaches the thermal equilibrium velocity of other conduction band electrons. The electron and hole created by this absorption process are *excess carriers*; since they are out of balance with their environment, they must eventually recombine. While the excess carriers exist in their respective bands, however, they are free to contribute to the conductivity of the material.

A photon with energy less than E_g is unable to excite an electron from the valence band to the conduction band. Thus in a pure semiconductor, there is negligible absorption of photons with $hv < E_g$. One exception to this rule is that a small amount of absorption can occur within a given band (*free carrier absorption*); for example, a low-energy photon can excite a conduction band electron temporarily to a higher state within the conduction band. No excess carriers are created in this process. This component of absorption is usually negligible compared with band-to-band excitation (intrinsic absorption) by photons of higher energy, and most photons with $hv < E_g$ are transmitted through the material. This explains why some materials are transparent in certain wavelength ranges. We are able to "see through" certain insulators, such as a good NaCl crystal, because a large energy gap containing no electron states exists in the material. If the band gap is about 2 eV wide, only long wavelengths (infrared) and the red part of the visible spectrum are transmitted; on the other hand, a band gap of about 3 eV allows infrared and the entire visible spectrum to be transmitted.†

4.1.2 Absorption Constant and Band Gap. If a beam of photons with $hv > E_g$ falls on a semiconductor, there will be some predictable amount of absorption, determined by the properties of the material. We would expect the ratio of transmitted to incident light intensity to depend on the photon wavelength and the thickness of the sample. To calculate this dependence, let us assume that a photon beam of intensity I_0 (photons/cm²-sec) is directed at a sample of thickness l (Fig. 4-2). The beam contains only photons of

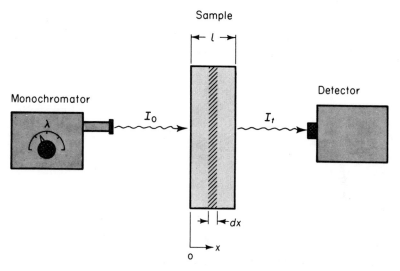

FIGURE 4-2.   Optical absorption experiment.

wavelength λ, selected by a monochromator. As the beam passes through the sample, its intensity at a distance x from the surface can be calculated by considering the probability of absorption within any increment dx. Since a photon which has survived to x without absorption has no memory of how far it has traveled, its probability of absorption in any dx is constant.‡ Thus

†Transmission of photons through the band gap does not explain the color of all materials; for example, many colored glasses contain impurity states within the band gap which can cause absorption of certain photons. Other materials have characteristic colors because of selective reflection rather than transmission. For an interesting discussion of these topics, see the article by Weisskopf in the reading list for this chapter.

‡This is analogous to the problem of carrier scattering discussed in Section 3.4.1, in which the probability of scattering in any dt for a given carrier is constant.

the degradation of the intensity $-d\mathbf{I}(x)/dx$ is proportional to the intensity remaining at x

(4-1) $$-\frac{d\mathbf{I}(x)}{dx} = \alpha\mathbf{I}(x)$$

The solution to this equation is

(4-2) $$\mathbf{I}(x) = \mathbf{I}_0 e^{-\alpha x}$$

and the intensity of light transmitted through the sample thickness l is

(4-3) $$\mathbf{I}_t = \mathbf{I}_0 e^{-\alpha l}$$

The coefficient α is called the *absorption constant* and has units of cm^{-1}. This coefficient will of course vary with the photon wavelength and with the material. In a typical plot of α vs. wavelength (Fig. 4-3), there is negligible

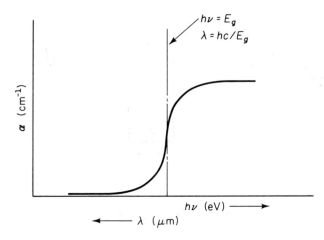

FIGURE 4-3. Dependence of optical absorption constant α for a semiconductor on the wavelength of incident light.

absorption at long wavelengths ($h\nu$ small) and considerable absorption of wavelengths shorter than hc/E_g (where c is the speed of light).

In most cases the onset of absorption at the band gap is sharp, and a reasonable measurement of E_g can be obtained for wide band gap semiconductors by using the eye as a detector in Fig. 4-2. As the monochromator is varied from long to short wavelengths, light will be transmitted through the semiconductor until absorption begins at a fairly well-defined wavelength. Of course, for more accurate measurement a detector must be used which provides an output proportional to light intensity.

As an alternative to a transmission experiment, it is possible to monitor the resistance of the sample and observe the photon energy at which absorption takes place. Since absorbed photons create EHP's and these carriers are able to participate in conduction, the conductivity of the sample will increase when photons are absorbed. This change in the conductivity with optical absorption, called *photoconductivity*, will be discussed further in Section 4.3.4.

Figure 4-4 indicates the band gap energies of some of the common

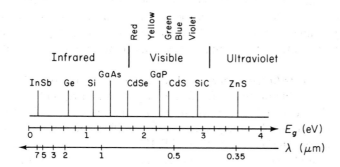

FIGURE 4-4.   Band gaps of some common semiconductors relative to the optical spectrum.

semiconductors, relative to the visible, infrared, and ultraviolet portions of the spectrum. We observe that GaAs, Si, Ge, and InSb lie outside the visible region, so that if the eye is to be used as a detector in the transmission experiment, an infrared converter tube must be used.† Other semiconductors, such as GaP and CdS, have band gaps wide enough to pass photons in the visible range. It is important to note here that a semiconductor absorbs photons with energies equal to the band gap, *or larger*. Thus Si absorbs not only band gap light (~ 1 μm) but also shorter wavelengths, including those in the visible part of the spectrum.

4.2 Luminescence

When electron-hole pairs are generated in a semiconductor, or when carriers are excited into higher impurity levels from which they fall to their

---

†Many devices for "converting" infrared into visible light have been developed, largely for military uses in night vision. A typical converter tube receives infrared photons, which create electrons at a metal plate by the photoelectric effect (Section 2.2.1). These electrons are accelerated to a luminescent screen similar to that of a television tube, where they are absorbed, with visible photons given off (the mechanism of luminescence will be discussed in Section 4.2).

equilibrium states, light can be given off by the material. Many of the semiconductors are well suited for light emission, particularly the compound semiconductors with direct band gaps. The general property of light emission is called *luminescence*.† This overall category can be subdivided according to the excitation mechanism: If carriers are excited by photon absorption, the radiation resulting from the recombination of the excited carriers is called *photoluminescence*; if the excited carriers are created by high-energy electron bombardment of the material, the mechanism is called *cathodoluminescence*; if the excitation occurs by the introduction of current into the sample, the resulting luminescence is called *electroluminescence*. Other types of excitation are possible, but these three are the most important for device applications.

4.2.1 Photoluminescence. The simplest example of light emission from a semiconductor occurs for direct excitation and recombination of an EHP, as depicted in Fig. 4-1. If the recombination occurs directly rather than via a defect level, band gap light is given off in the process. For steady state excitation, the recombination of EHP's occurs at the same rate as the generation, and one photon is emitted for each photon absorbed. Direct recombination is a fast process; the mean lifetime of the EHP is usually on the order of 10^{-8} sec or less. Thus the emission of photons stops within approximately 10^{-8} sec after the excitation is turned off. Such fast luminescent processes are often referred to as *fluorescence*. In some materials, however, emission continues for periods up to seconds or minutes after the excitation is removed. These slow processes are called *phosphorescence*, and the materials are called *phosphors*. An example of a slow process is shown in Fig. 4-5. This material contains a defect level (perhaps due to an impurity) in the band gap which has a strong tendency to temporarily capture (*trap*) electrons from the conduction band. The events depicted in the figure are as follows: (a) An incoming photon with $hv_1 > E_g$ is absorbed, creating an EHP; (b) the excited electron gives up energy to the lattice by scattering until it nears the bottom of the conduction band; (c) the electron is *trapped* by the impurity level E_t and remains trapped until it can be thermally reexcited to the conduction band (d); (e) finally direct recombination occurs as the electron falls to an empty state in the valence band, giving off a photon (hv_2) of approximately the band gap energy. The delay time between excitation and recombination can be relatively long if the probability of thermal reexcitation from the trap (d) is small. Even longer delay times result if the electron is retrapped several times before recombination. If the trapping probability is

---

†The emission processes considered here should not be confused with radiation due to incandescence which occurs in heated materials. The various luminescent mechanisms can be considered "cold" processes as compared to the "hot" process of incandescence, which increases with temperature. In fact, most luminescent processes become more efficient as the temperature is lowered.

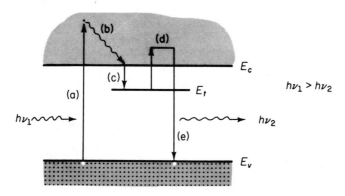

FIGURE 4-5. Excitation and recombination mechanisms in photoluminescence with a trapping level for electrons.

greater than the probability of recombination, an electron may make several trips between the trap and the conduction band before recombination finally occurs. In such a material the emission of phosphorescent light persists for a relatively long time after the excitation is removed.

To understand the trapping process, we must discuss briefly the mechanism of recombination via a defect level. After a conduction band electron falls to a level in the band gap, it can either be reexcited to the conduction band or can fall to the valence band, annihilating a hole. This latter process is equivalent to a hole being "captured" at the recombination level. An example of a defect center which temporarily traps one type of carrier (as in Fig. 4-5) is the highly charged impurity. An impurity with a double positive charge at equilibrium is still singly positive after an electron is captured from the conduction band. Subsequent capture of a hole (completing the recombination process) requires that the hole overcome a repulsive coulombic barrier at the positive center. The probability of this occurring may be small compared with the probability of reexcitation of the electron to the conduction band. If reexcitation dominates, the defect center serves only to trap the electron temporarily, thereby delaying the recombination process. We shall discuss these recombination and trapping events in more detail in Section 4.3.2.

Many photoluminescent materials utilize electron transitions between defect energy levels rather than band-to-band recombination. For example, most luminescent phosphors are either wide band gap semiconductors or insulators, containing doping impurities with multiple levels in the band gap of the host material. In many phosphors the transitions of interest occur via impurity levels (Fig. 4-6). The impurities responsible for these levels are often called *activators*. In many cases a second type of impurity (*coactivator*) must be added to achieve the desired luminescence, or simply to maintain charge

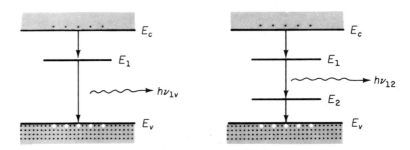

FIGURE 4-6. Luminescent phosphors with transitions via impurity levels.

balance in the ions which make up the crystal. For example, a common activator in ZnS is Cu, which replaces a Zn^{++} ion in the host lattice with a singly charged Cu^+ ion. The resulting reduction of positive charge can be compensated by the introduction of a coactivator such as Al^{3+} on a Zn^{++} site or Cl^- on a $S^=$ site. In either case the charge imbalance introduced by the activator is compensated by the coactivator. If no coactivator is introduced, the crystal may automatically compensate for the missing positive charge by forming lattice defects (*self-compensation*). For example, one missing $S^=$ ion (a *vacancy* at one S site) will compensate two Cu^+ activator atoms, as will one extra Zn^{++} ion (*interstitial*) in the lattice. This tendency to form compensating lattice defects in the presence of doping impurities is common in semiconductors which have considerable ionic bonding (e.g., most of the II–VI compounds). Thus it is difficult or impossible to dope many II–VI materials n-type or p-type at will (Appendix III), because the crystal automatically compensates for the dopants with lattice defects. In the III–V compounds the bonding is more covalent than ionic, and the column IV crystals Si and Ge are essentially covalent semiconductors. Thus these materials are relatively easy to dope n-type or p-type. We shall return to the discussion of self-compensation of II–VI compounds in Chapter 6 in reference to materials for light-emitting diodes.

Depending on the type and amount of coactivator, Cu activation of ZnS can result in blue, green, or red luminescence. By varying the type of host lattice, activator, and coactivator, phosphors can be obtained with single-color light emission or "white" light emission from many transitions. The single-color (monochromatic) phosphors are useful in applications such as color television screens, while white light is used in other applications such as the fluorescent lamp.

The advantages of a fluorescent lamp over an incandescent lamp can be understood by considering the spectral output of an incandescent filament.

Much of the emitted radiation lies in the ultraviolet (high-energy) or the infrared (low-energy) regions of the spectrum, rather than in the visible range. If the lamp is intended for optical lighting, all photons outside the visible range are wasted. The ultraviolet photons need not be lost, however, since a coating of appropriate luminescent material on the surface of the lamp will absorb the high-energy photons and emit visible photons. In most cases a fluorescent lamp is composed of a glass tube filled with gas (e.g., a mixture of argon and mercury), with a fluorescent coating on the inside of the tube. When an electric discharge is induced between electrodes in the tube, the excited atoms of the gas emit photons, largely in the visible and ultraviolet regions of the spectrum. This light is absorbed by the luminescent coating, and visible photons are emitted. The efficiency of such a lamp is considerably better than that of an incandescent bulb, and the wavelength mixture in light given off can be adjusted by proper selection of the fluorescent material.

4.2.2 Cathodoluminescence; The Cathode-Ray Tube.

The most common example of the excitation of luminescent materials by energetic electrons is the cathode-ray tube (CRT). This light-emitting tube is the basis of the oscilloscope, television set, and other visual-display systems. The basic principle of the CRT is the selective excitation of a phosphorescent screen by a beam of energetic electrons within a vacuum tube (Fig. 4-7). Electrons are emitted from a heated cathode and are accelerated by an electric field to a positively charged anode. A beam of electrons passes through a hole in the anode and proceeds to the deflection system where the path of the electrons is bent by an electric or a magnetic field. By varying the deflection, the electron beam can be directed to any spot on the screen. As the

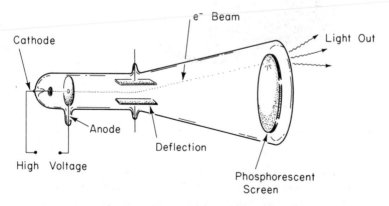

FIGURE 4-7.   The cathode-ray tube.

energetic electrons collide with the phosphor coating, energy is given up to the phosphor as electrons are excited to higher states within the material. When these electrons recombine, the screen gives off light at the desired spot. In a television picture tube, the intensity and position of the beam is varied to give a visible replica of the image being received.

It is important to choose an appropriate phosphor which gives not only the desired color of emitted light but also the proper *persistence*. Persistence refers to the length of time over which the luminescence persists after the excitation is turned off. Since the entire pattern is traced out by a single electron beam in the CRT, it is necessary for the individual spots to persist while the rest of the pattern is being traced (there is also some persistence of an image in the eye of the viewer, and this is taken into account in choosing a CRT phosphor).

In color television the screen is coated with a pattern of phosphor dots which are doped to emit the primary additive colors—red, green, and blue. To excite the proper dots, three electron beams are swept together through the pattern. Each beam controls the excitation of one type of phosphor dot (Fig. 4-8). By varying the relative intensity of light emitted by each of the dots, a broad range of colors can be synthesized.

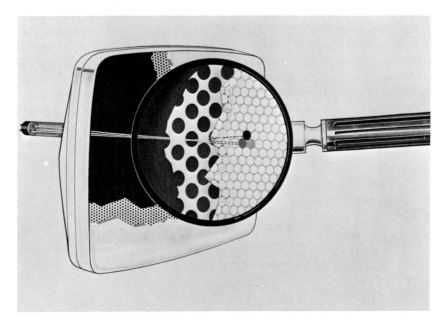

FIGURE 4-8. Cutaway view of a color television screen. The three electron beams converge at holes in a metal mask and illuminate red, green, and blue phosphor dots on the screen. (Illustration courtesy of Zenith Radio Corporation.)

4.2.3 *Electroluminescence.* There are many ways by which electrical energy can be used to generate photon emission in a solid. In some cases an electric current causes the injection of minority carriers into regions of the crystal where they can recombine with majority carriers, resulting in the emission of recombination radiation. This important effect (*injection electroluminescence*) will be discussed in Chapters 6 and 7 in terms of p-n junction theory.

The first electroluminescent effect to be observed was the emission of photons by certain phosphors in an alternating electric field (the Destriau effect). In this device, a phosphor powder such as Cu-activated ZnS is held in a binder material (often a plastic) of a high dielectric constant. When an a-c electric field is applied, light is given off by the phosphor (Fig. 4-9). Such cells can be useful as lighting panels, although their efficiency has thus far been too low for most applications and their reliability is poor.

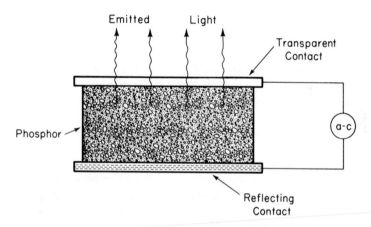

FIGURE 4-9.   An electroluminescent cell excited by an a-c electric field.

There are a number of mechanisms which can play a role in light emission from electroluminescent cells. These include several processes involving ionization of atoms by the high field at an individual phosphor particle or by impact ionization by accelerated carriers. A simple theory for the specific electroluminescent mechanism is difficult to obtain since the material is a powder, not a single crystal. Thus the generation and recombination of excess carriers is not strictly a bulk phenomenon; the effects of the surface of each powder grain and the interaction between grains greatly complicate the theoretical problem.

4.3 Carrier Lifetime and Photoconductivity

When excess electrons and holes are created in a semiconductor, there is a corresponding increase in the conductivity of the sample as indicated by Eq. (3-43). If the excess carriers arise from optical excitation, the resulting increase in conductivity is called *photoconductivity*. This is an important effect, with useful applications in the analysis of semiconductor materials and in the operation of several types of devices. In this section we shall examine the mechanisms by which excess electrons and holes recombine and apply the recombination kinetics to the analysis of photoconductive devices. However, the importance of recombination is not limited to cases in which the excess carriers are created optically. In fact, virtually every semiconductor device depends in some way on the recombination of excess electrons and holes. Therefore, the concepts developed in this section will be used extensively in the analyses of diodes, transistors, lasers, and other devices in later chapters.

4.3.1 Direct Recombination of Electrons and Holes. It was pointed out in Section 3.1.4 that electrons in the conduction band of a semiconductor may make transitions to the valence band (i.e., recombine with holes in the valence band) either directly or indirectly. In direct recombination, an excess population of electrons and holes decays by electrons falling from the conduction band to empty states (holes) in the valence band. Energy lost by an electron in making the transition is given up as a photon. Direct recombination occurs *spontaneously*; that is, the probability that an electron and a hole will recombine is constant in time. As in the case of carrier scattering, this constant probability leads us to expect an exponential solution for the decay of the excess carriers. In this case the rate of decay of electrons at any time t is proportional to the number of electrons remaining at t and the number of holes, with some constant of proportionality for recombination, α_r. The *net* rate at which electrons leave the conduction band by recombination is

(4-4) $$-\frac{dn(t)}{dt} = \alpha_r \, n(t) \, p(t) - \alpha_r n_i^2$$

where $\alpha_r n_i^2$ is the thermal generation rate from Eq. (3-7).

Let us assume the excess electron–hole population is created at $t = 0$, for example by a short flash of light, and the initial excess electron and hole densities Δn and Δp are equal. Then as the electrons and holes recombine in pairs, the instantaneous concentrations of excess carriers $\delta n(t)$ and $\delta p(t)$

are also equal. Thus we can write the total concentrations of Eq. (4-4) in terms of the equilibrium densities n_0 and p_0 and the excess densities $\delta n(t) = \delta p(t)$. Using Eq. (3-24) we have

(4-5)
$$-\frac{d\delta n(t)}{dt} = \alpha_r[n_0 + \delta n(t)][p_0 + \delta p(t)] - \alpha_r n_i^2$$

$$= \alpha_r[(n_0 + p_0)\delta n(t) + \delta n^2(t)]$$

This nonlinear equation would be difficult to solve in its present form. Fortunately, it can be simplified for the case of low-level injection. If the excess carrier densities are small, we can neglect the δn^2 term. Furthermore, if the material is extrinsic, we can usually neglect the term representing the equilibrium density of minority carriers. For example, if the material is p-type ($p_0 \gg n_0$), Eq. (4-5) becomes

(4-6)
$$-\frac{d\delta n(t)}{dt} = \alpha_r p_0\, \delta n(t)$$

The solution to this equation is an exponential decay from the original excess density Δn

(4-7)
$$\delta n(t) = \Delta n\, e^{-\alpha_r p_0 t} = \Delta n\, e^{-t/\tau_n}$$

Excess electrons in a p-type semiconductor recombine with a decay constant $\tau_n = (\alpha_r p_0)^{-1}$, called the *recombination lifetime*. Since the calculation is made in terms of the minority carriers, τ_n is often called the *minority carrier lifetime*. The decay of excess holes in n-type material occurs with $\tau_p = (\alpha_r n_0)^{-1}$. In the case of direct recombination, the excess majority carriers decay at exactly the same rate as the minority carriers.

A numerical example may be helpful in visualizing the approximations made in the analysis of direct recombination. Let us assume a sample of GaAs is doped with 10^{15} acceptors/cm³. The intrinsic carrier concentration of GaAs is approximately 10^7 cm⁻³; thus the minority electron density is $n_0 = n_i^2/p_0 = 0.1$ cm⁻³. Certainly the approximation $p_0 \gg n_0$ is valid in this case. Now if 10^{14} EHP/cm³ are created at $t = 0$, we can calculate the decay of these carriers in time. The approximation $\delta n \ll p_0$ is reasonable, as Fig. 4-10 indicates. This figure shows the decay in time of the excess populations for a carrier recombination lifetime of $\tau_n = \tau_p = 10^{-8}$ sec.

There is a large percentage change in the minority carrier electron density in this example and a small percentage change in the majority hole density. Basically, the approximations of extrinsic material and low-level injection allow us to represent $n(t)$ in Eq. (4-4) by the excess density $\delta n(t)$ and $p(t)$ by the equilibrium value p_0. Figure 4-10 indicates that this is a good approximation for the example. A more general expression for the carrier

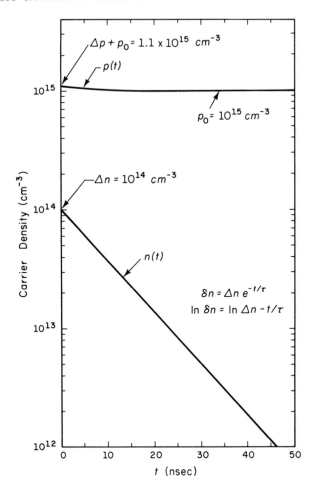

FIGURE 4-10. Decay of excess electrons and holes by recombination, for $\Delta n = \Delta p = 0.1 p_0$, with n_0 negligible, and $\tau = 10$ nsec. The exponential decay of $\delta n(t)$ is linear on this semilogarithmic graph.

lifetime is

(4-8)
$$\tau_n = \frac{1}{\alpha_r(n_0 + p_0)}$$

This expression is valid for n- or p-type material if the injection level is low.

4.3.2 Indirect Recombination; Trapping.

In the column IV semiconductors and in certain compounds, the probability of direct electron–

hole recombination is very small (Appendix III). There is some band gap light given off by materials such as Si and Ge during recombination, but this radiation is very weak and may be detected only by sensitive equipment. The vast majority of the recombination events in indirect materials occur via *recombination levels* within the band gap, and the resulting energy loss by recombining electrons is usually given up to the lattice as heat rather than by the emission of photons. Any impurity or lattice defect can serve as a recombination center if it is capable of receiving a carrier of one type and subsequently capturing the opposite type of carrier, thereby annihilating the pair. For example, Fig. 4-11 illustrates a recombination level E_r which is

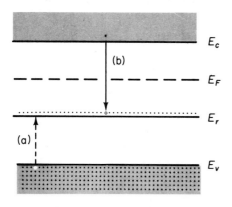

FIGURE 4-11. Capture processes at a recombination level: (a) hole capture at a filled recombination center; (b) electron capture at an empty center.

below E_F at equilibrium and therefore is substantially filled with electrons. When excess electrons and holes are created in this material, each EHP recombines at E_r in two steps: (a) hole capture and (b) electron capture.

Since the recombination centers in Fig. 4-11 are filled at equilibrium, the first event in the recombination process is hole capture. It is important to note that this event is equivalent to an electron at E_r falling to the valence band, leaving behind an empty state in the recombination level. Thus in hole capture, energy is *given up* as heat to the lattice. Similarly, energy is given up when a conduction band electron subsequently falls to the empty state in E_r. When both of these events have occurred, the recombination center is back to its original state (filled with an electron), but an EHP is missing. Thus one EHP recombination has taken place, and the center is ready to participate in another recombination event by capturing a hole.

The carrier lifetime resulting from indirect recombination is somewhat more complicated than is the case for direct recombination, since it is necessary to account for unequal times required for capturing each type of carrier. In particular, recombination is often delayed by the tendency for a captured carrier to be thermally reexcited to its original band before capture of the opposite type of carrier can occur (Section 4.2.1). For example, if electron capture does not follow immediately after hole capture (a) in Fig. 4-11, the hole may be thermally reexcited to the valence band. Energy is required for this process, which is equivalent to a valence band electron being raised to the empty state in the recombination level. This process delays the recombination, since the hole must be captured again before recombination can be completed. At low temperatures there is little thermal energy available for reexcitation of a captured carrier; thus the recombination rate is limited only by the time required for hole capture or electron capture, whichever is longer. As the temperature is raised, however, more thermal energy is available for reexcitation, and the holes spend less time at E_r before being excited back to the valence band. Therefore, the probability of reexcitation is greater at higher temperatures, and we should expect the recombination lifetime τ to become longer (Fig. 4-12).

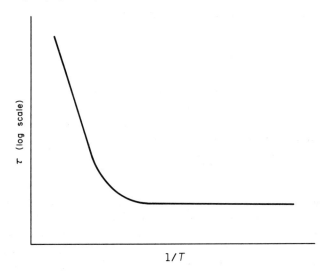

FIGURE 4-12. Inverse temperature dependence of the recombination lifetime for a single recombination level.

When a carrier is trapped temporarily at a center and then is reexcited without recombination taking place, the process is often called *temporary trapping*. Some reexcitation takes place for any recombination center, but it is often negligible compared with the recombination processes (e.g., at low temperatures in Fig. 4-12). Although the nomenclature varies somewhat, it is common to refer to an impurity or defect center as a *trapping center* (or simply *trap*) if, after capture of one type of carrier, the most probable next event is reexcitation. If the most probable next event is capture of the opposite type of carrier, the center is predominately a recombination center. The recombination can be slow or fast, depending on the average time the first carrier is held before the second carrier is captured.

In some cases a slow trapping level will be present along with one or more levels of fast recombination centers. In this case the traps hold carriers for some time before they are reexcited, after which they can recombine via the recombination centers. Such trapping can also occur in direct semiconductors containing certain impurities or defects (e.g., trapping in phosphorescent materials as illustrated in Fig. 4-5). In general, trapping levels located deep in the band gap are slower in releasing trapped carriers than are the levels located near one of the bands. This results from the fact that more energy is required, for example, to reexcite a trapped electron from a center near the middle of the gap to the conduction band than is required to reexcite an electron from a level closer to the conduction band.

As an example of impurity levels in semiconductors, Fig. 4-13 shows the energy level positions of various impurities in Si. In this diagram a

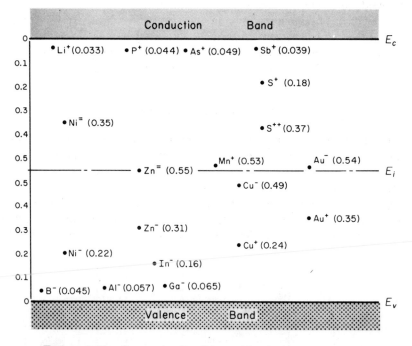

FIGURE 4-13. Energy levels of impurities in Si. The energies are measured from the nearest band edge (E_v or E_c); donor levels are designated by a plus sign and acceptors by a minus sign.†

†References: S. M. Sze and J. C. Irvin, "Resistivity, Mobility, and Impurity Levels in GaAs, Ge, and Si at 300°K," *Solid State Electronics*, vol. 11, pp. 599–602, June 1968. E. Schibli and A. G. Milnes, "Deep Impurities in Silicon," *Materials Science and Engineering*, vol. 2, pp. 173–180, 1967.

superscript indicates whether the impurity is positive (donor) or negative (acceptor) when ionized. Some impurities introduce multiple levels in the band gap; for example, Zn introduces a level (Zn⁻) located 0.31 eV above the valence band and a second level (Zn⁼) near the middle of the gap. Each Zn impurity atom is capable of accepting two electrons from the semiconductor, one in the lower level and then one in the upper level. When the Fermi level is in the upper half of the band gap (n-type material), both Zn levels will be substantially filled at equilibrium, and the impurity serves as a strong trap for holes. When a hole is captured at a Zn⁼ center, the impurity is still singly negative and therefore has a repulsive coulombic barrier for electron capture. Therefore, recombination is unlikely and the hole may be reexcited to the valence band. Since the Zn⁼ level is far from the valence band, a long delay is required, on the average, before reexcitation occurs. Of course, the time a hole spends in the trap increases as the temperature is lowered, since less thermal energy is available.

The effects of recombination and trapping can be measured by a *photoconductive decay* experiment. As Fig. 4-10 shows, a population of excess electrons and holes disappears with a decay constant characteristic of the particular recombination process. The conductivity of the sample during the decay is

$$(4\text{-}9) \qquad \sigma(t) = q[n(t)\mu_n + p(t)\mu_p]$$

Therefore, the time dependence of the carrier densities can be monitored by recording the sample resistance as a function of time. A typical experimental arrangement is shown schematically in Fig. 4-14. A source of short pulses of light is required, along with an oscilloscope for displaying the sample voltage as the resistance varies. The light pulses can be obtained by mechanically "chopping" a steady light (passing the light beam through small openings in a rotating disk); this procedure is adequate if the decay times to

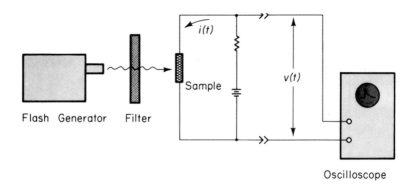

Oscilloscope

FIGURE 4-14. Experimental arrangement for photoconductive decay measurements.

be measured are in the millisecond range. Since the light pulse must turn off faster than the carrier decay time, other methods must be used to generate light flashes for samples with carrier lifetimes of microseconds or less. Microsecond light pulses can be obtained by periodically discharging a capacitor through a flash tube containing a gas such as xenon. For shorter pulses, special techniques such as the use of a pulsed laser (Chapter 7) must be used.

If the light output of the flash generator contains a distribution of wavelengths, it is common to filter out photons with $h\nu \gg E_g$. These high-energy photons would be absorbed at the surface of the sample, and the resulting photoconductive decay would not be characteristic of the bulk sample. An effective filter is a thin sample of the semiconductor material being measured.

Typical photoconductive decay curves are shown in Fig. 4-15.† These oscilloscope traces indicate the effects of trapping superimposed upon fast

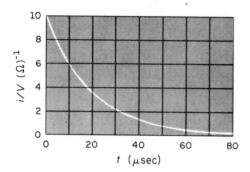

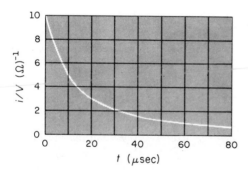

FIGURE 4-15. Conductance traces for two photoconductive decay measurements: (a) exponential decay for simple recombination; (b) recombination with trapping. The dark conductivity is subtracted out in these traces; only the time-varying photoconductance is displayed.

recombination. In (a) the sample is measured at room temperature, in which case thermal reexcitation from the traps is too fast to affect the exponential decay through the recombination centers. The low-temperature trace (b) exhibits a long "tail" on the decay, indicating that a fraction of the generated

---

†*Question:* How would you display conductance (Fig. 4-15) using the circuit of Fig. 4-14 and a typical oscilloscope?

carriers have been trapped and are slowly being released by thermal reexcitation. While the carriers of one type are trapped, the excess conductivity arises from the presence of carriers of the opposite type which are waiting for recombination. (Space charge neutrality requires that the total density of excess electrons and trapped electrons be equal to the density of excess and trapped holes.)

4.3.3 Steady State Carrier Generation; Quasi-Fermi Levels.

In the discussion above we emphasized the transient decay of an excess EHP population. However, the various recombination mechanisms are also important in a sample at thermal equilibrium or with a steady state EHP generation–recombination balance.† For example, a semiconductor at equilibrium experiences thermal generation of EHP at a rate $g(T) = g_i$ described by Eq. (3-7). This generation is balanced by the recombination rate so that the equilibrium densities of carriers n_0 and p_0 are maintained:

$$(4\text{-}10) \qquad g(T) = \alpha_r n_i^2 = \alpha_r n_0 p_0$$

This equilibrium rate balance can include generation from defect centers as well as band-to-band generation.

If a steady light is shone on the sample, an optical generation rate g_{op} will be added to the thermal generation, and the carrier concentrations n and p will increase to new steady state values. We can write the balance between generation and recombination in terms of the equilibrium carrier densities and the departures from equilibrium δn and δp.

$$(4\text{-}11) \qquad g(T) + g_{op} = \alpha_r np = \alpha_r (n_0 + \delta n)(p_0 + \delta p)$$

For steady state recombination and no trapping, $\delta n = \delta p$; thus Eq. (4-11) becomes

$$(4\text{-}12) \qquad g(T) + g_{op} = \alpha_r n_0 p_0 + \alpha_r [(n_0 + p_0)\delta n + \delta n^2]$$

The term $\alpha_r n_0 p_0$ is just equal to the thermal generation rate $g(T)$. Thus, neglecting the δn^2 term for low-level excitation, we can rewrite Eq. (4-12) as

$$(4\text{-}13) \qquad g_{op} = \alpha_r (n_0 + p_0)\delta n = \frac{\delta n}{\tau_n}$$

---

†The term *equilibrium* refers to a condition of no external excitation and no net motion of charge (e.g., a sample at a constant temperature, in the dark, with no fields applied). *Steady state* refers to a nonequilibrium condition in which all processes are constant and are balanced by opposing processes (e.g., a sample with a constant current or a constant optical generation of EHP's just balanced by recombination).

The excess carrier density can be written as

(4-14) $$\delta n = \delta p = g_{op}\tau_n$$

More general expressions are given in Eq. (4-16), which allow for the case $\tau_p \neq \tau_n$, when trapping is present.

As a numerical example, let us assume 10^{13} EHP/cm^3 are created optically every microsecond in a Si sample with $n_0 = 10^{14}$ cm^{-3} and $\tau_n = \tau_p = 2$ μsec. The steady state excess electron (or hole) density is then 2×10^{13} cm^{-3} from Eq. (4-14). While the percentage change in the majority electron density is small, the minority carrier density changes from $p_0 = n_i^2/n_0 = (2.25 \times 10^{20})/10^{14} = 2.25 \times 10^6$ cm^{-3} to 2×10^{13} cm^{-3}.†

It is often desirable to refer to the steady state electron and hole densities in terms of Fermi levels, which can be included in band diagrams for various devices. The Fermi level E_F used in Eqs. (3-25a) and (3-25b) is meaningful only when no excess carriers are present. However, we can write expressions for the steady state densities in the same *form* as the equilibrium expressions by defining separate *quasi-Fermi levels* F_n and F_p for electrons and holes. The resulting carrier density equations

(4-15a) $$n = n_i e^{(F_n - E_i)/kT}$$
(4-15b) $$p = n_i e^{(E_i - F_p)/kT}$$

can be considered as defining relations for the quasi-Fermi levels.‡ In the numerical example above, the steady state electron density is

$$n = n_0 + \delta n = 1.2 \times 10^{14} = (1.5 \times 10^{10})e^{(F_n - E_i)/0.026}$$

where $kT = 0.026$ eV at room temperature. Thus the electron quasi-Fermi level position $F_n - E_i$ is found from

$$F_n - E_i = 0.026 \ln (8 \times 10^3) = 0.234 \text{ eV}$$

and F_n lies 0.234 eV above the intrinsic level. By a similar calculation, the hole quasi-Fermi level lies 0.187 eV below E_i (Fig. 4-16). In this example, the equilibrium Fermi level is $0.026 \ln (6.67 \times 10^3) = 0.229$ eV above the intrinsic level. The quasi-Fermi levels of Fig. 4-16 illustrate dramatically the deviation from equilibrium caused by the optical excitation; the steady state F_n is only slightly above the equilibrium E_F, whereas F_p is greatly displaced below E_F. From the figure it is obvious that the excitation causes a large

---

†Note that the equilibrium equation $n_0 p_0 = n_i^2$ cannot be used with the subscripts removed; that is, $np \neq n_i^2$ when excess carriers are present.

‡In some texts the quasi-Fermi level is called *IMREF*, which is Fermi spelled backward.

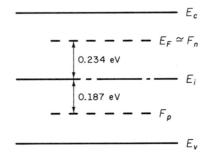

FIGURE 4-16. Quasi-Fermi levels F_n and F_p for a Si sample with $n_0 = 10^{14}$ cm^{-3}, $\tau_p = 2$ μsec, and $g_{op} = 10^{19}$ EHP/cm^3-sec.

percentage change in minority carrier hole concentration and a relatively small change in the electron density.

In summary, the quasi-Fermi levels F_n and F_p are the nonequilibrium analogues of the equilibrium Fermi level E_F. When excess carriers are present, the deviations of F_n and F_p from E_F indicate how far the electron and hole populations are from the equilibrium values n_0 and p_0. A given density of excess EHP's causes a large shift in the minority carrier quasi-Fermi level compared with that for the majority carriers. The separation of the quasi-Fermi levels $F_n - F_p$ is a direct measure of the deviation from equilibrium (at equilibrium $F_n = F_p = E_F$). The concept of quasi-Fermi levels is very useful in visualizing minority and majority carrier densities in devices where these densities vary with position.

4.3.4 Photoconductive Devices.

There are a number of applications for devices which change their resistance when exposed to light. For example, such light detectors can be used in the home to control automatic night lights which turn on at dusk and turn off at dawn. They can also be used to measure illumination levels, as in exposure meters for cameras. Many systems include a light beam aimed at the photoconductor, which signals the presence of an object between the source and detector. Such systems are useful in moving-object counters, burglar alarms, and many other applications. Detectors are used in optical signaling systems in which information is transmitted by a light beam and is received at a photoconductive cell.

Considerations in choosing a photoconductor for a given application include the sensitive wavelength range, time response, and optical sensitivity of the material. In general, semiconductors are most sensitive to photons with energies equal to the band gap or slightly more energetic than band gap. Less energetic photons are not absorbed, and photons with $h\nu \gg E_g$ are absorbed at the surface and contribute little to the bulk conductivity. Therefore, the table of band gaps (Appendix III) indicates the photon energies to which most semiconductor photodetectors respond. For example, CdS ($E_g = 2.42$ eV) is commonly used as a photoconductor in the visible range, and narrow-gap materials such as Ge (0.67 eV) and InSb (0.18 eV) are useful in the infrared portion of the spectrum. Some photoconductors respond to excitations of carriers from impurity levels within the band gap and therefore are sensitive to photons of less than band-gap energy.

The optical sensitivity of a photoconductor can be evaluated by examining the steady state excess carrier densities generated by an optical generation rate g_{op}. If the mean time each carrier spends in its respective band before capture is τ_n and τ_p, we have

(4-16) $$\delta n = \tau_n g_{op}, \quad \text{and} \quad \delta p = \tau_p g_{op}$$

and the photoconductivity change is

(4-17) $$\Delta\sigma = q g_{op}(\tau_n \mu_n + \tau_p \mu_p)$$

For simple recombination, τ_n and τ_p will be equal. If trapping is present, however, one of the carriers may spend little time in its band before being trapped. From Eq. (4-17) it is obvious that for maximum photoconductive response, we want high mobilities and long lifetimes. Some semiconductors are especially good candidates for photoconductive devices because of their high mobility; for example, InSb has an electron mobility of about 10^5 cm^2/V-sec and therefore is used as a sensitive infrared detector in many applications.

The time response of a photoconductive cell is limited by the recombination times, the degree of carrier trapping, and the time required for carriers to drift through the device in an electric field. Often these properties can be adjusted by proper choice of material and device geometry, but in some cases improvements in response time are made at the expense of sensitivity. For example, the drift time can be reduced by making the device short, but this substantially reduces the responsive area of the device. In addition, it is often desirable that the device have a large dark resistance, and for this reason, shortening the length may not be practical. There is usually a compromise between sensitivity, response time, dark resistance, and other requirements in choosing a device for a particular application.

Some semiconductor materials are useful as detectors in polycrystalline or powdered form rather than in single-crystal form. Most CdS detectors are made by depositing a thin layer of material on a ceramic substrate. The thickness of the semiconductor layer is matched to the penetration depth of photons in the wavelength range of interest. To increase the dark resistance, CdS cells are often deposited in a folded line pattern (Fig. 4-17) to increase the active length of the device. Such devices are most useful for applications involving steady or slowly varying light signals.

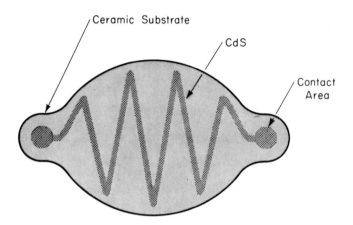

FIGURE 4-17. CdS photoconductive cell.

4.4 Diffusion of Carriers

When excess carriers are created nonuniformly in a semiconductor, the electron and hole densities vary with position in the sample. Any such spatial variation (*gradient*) in n and p calls for a net motion of the carriers from regions of high carrier density to regions of low carrier density. This type of motion is called *diffusion* and represents an important charge transport process in semiconductors. The two basic processes of current conduction are diffusion due to a carrier gradient and drift in an electric field.

4.4.1 Diffusion Processes. When a bottle of perfume is opened in one corner of a closed room, the scent is soon detected throughout the room. If there is no convection or other net motion of the air, the scent spreads by diffusion. It is important to note that no conventional driving force exists in this example, which impels the scented air molecules to move throughout the volume. The diffusion is simply the natural result of the *random motion* of the individual molecules. Consider, for example, a volume of arbitrary shape with scented air molecules inside and unscented molecules outside the volume (Fig. 4-18). All the molecules are undergoing random thermal motion and collisions with other molecules. Thus each molecule moves in an arbitrary direction until it collides with another air molecule, after which it moves in a new direction. This is analogous to the random scattering of carriers discussed in Section 3.4.1. By similar arguments, we can assign some mean free time $\bar{t}$ as the average time between collisions for

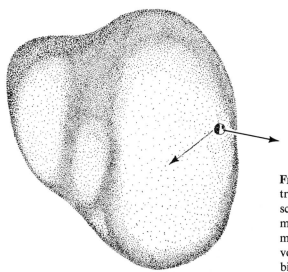

FIGURE 4-18. An arbitrary volume containing scented air molecules. A molecule on the edge can move into or out of the volume with equal probability.

the molecules. If the motion is truly random, a molecule at the edge of the volume has equal probabilities of moving into or out of the volume on its next step (assuming the curvature of the surface is negligible on the molecular scale). Therefore, after a mean free time $\bar{t}$, half the molecules at the edge will have moved into the volume and half will have moved out of the volume. The net effect is that the volume containing scented molecules has increased. This process will continue until the molecules are uniformly distributed in the room. Only then will a given volume gain as many molecules as it loses in a given time. In other words, net diffusion will continue as long as gradients exist in the distribution of scented molecules.

Carriers in a semiconductor diffuse in a carrier gradient by random thermal motion and scattering from the lattice and impurities. For example, a pulse of excess electrons injected at $x = 0$ at time $t = 0$ will spread out in time as shown in Fig. 4-19. Initially, the excess electrons are concentrated at $x = 0$; as time passes, however, electrons diffuse to regions of low electron density until finally $n(x)$ is constant.

We can calculate the rate at which the electrons diffuse in a one-dimensional problem by considering an arbitrary distribution $n(x)$ such as Fig. 4-20a. Since the mean free path $\bar{l}$ between collisions is a small incremental distance, we can divide x into segments $\bar{l}$ wide, with $n(x)$ evaluated at the center of each segment (Fig. 4-20b).

The electrons in segment (1) to the left of x_0 in Fig. 4-20b have equal chances of moving left or right, and in a mean free time $\bar{t}$ one-half of them will move into segment (2). The same is true of electrons within one mean free path of x_0 to the right; one-half of these electrons will move through x_0

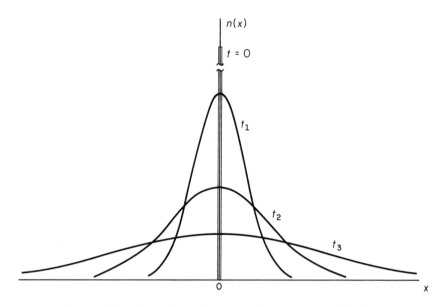

FIGURE 4-19. Spreading of a pulse of electrons by diffusion.

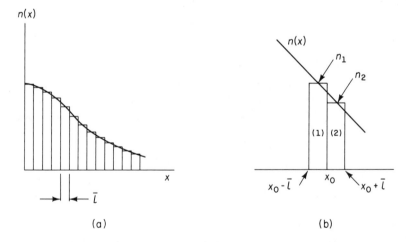

FIGURE 4-20. An arbitrary electron concentration gradient in one dimension: (a) division of $n(x)$ into segments of length equal to a mean free path for the electrons; (b) expanded view of two of the segments centered at x_0.

from right to left in a mean free time. Therefore, the *net* number of electrons passing x_0 from left to right in one mean free time is $\frac{1}{2}(n_1\bar{l}A) - \frac{1}{2}(n_2\bar{l}A)$, where the area perpendicular to x is A. The rate of electron flow in the $+x$-direction per unit area [the electron flux density $\phi_n(x)$] is given by

$$(4\text{-}18) \qquad \phi_n(x) = \frac{\bar{l}}{2\bar{t}}(n_1 - n_2)$$

Since the mean free path $\bar{l}$ is a small differential length, the difference in electron density $(n_1 - n_2)$ can be written as

$$(4\text{-}19) \qquad n_1 - n_2 = \frac{n(x) - n(x + \Delta x)}{\Delta x}\bar{l}$$

where x is taken at the center of segment (1) and $\Delta x = \bar{l}$. In the limit of small Δx (i.e., small mean free path $\bar{l}$ between scattering collisions), Eq. (4-18) can be written in terms of the carrier gradient $dn(x)/dx$

$$(4\text{-}20) \qquad \phi_n(x) = \frac{\bar{l}^2}{2\bar{t}}\lim_{\Delta x \to 0}\frac{n(x) - n(x + \Delta x)}{\Delta x} = -\frac{\bar{l}^2}{2\bar{t}}\frac{dn(x)}{dx}$$

The quantity $\bar{l}^2/2\bar{t}$† is called the *electron diffusion coefficient* D_n, with units cm²/sec. The minus sign in Eq. (4-20) arises from the definition of the derivative; it simply indicates that the net motion of electrons due to diffusion is in the direction of *decreasing* electron density. This is the result we expect, since net diffusion occurs from regions of high particle density to regions of low particle density. By identical arguments, we can show that holes in a hole density gradient move with a diffusion constant D_p. Thus

$$(4\text{-}21\text{a}) \qquad \phi_n(x) = -D_n\frac{dn(x)}{dx}$$

$$(4\text{-}21\text{b}) \qquad \phi_p(x) = -D_p\frac{dp(x)}{dx}$$

The diffusion current crossing a unit area (the current density) is the particle flux density multiplied by the charge of the carrier

$$(4\text{-}22\text{a}) \qquad J_n(\text{diff.}) = -(-q)D_n\frac{dn(x)}{dx} = +qD_n\frac{dn(x)}{dx}$$

$$(4\text{-}22\text{b}) \qquad J_p(\text{diff.}) = -(+q)D_p\frac{dp(x)}{dx} = -qD_p\frac{dp(x)}{dx}$$

---

†If motion in three dimensions is allowed (as in the case of a real semiconductor), the diffusion would be smaller in the x-direction. Actually, the diffusion constant should be calculated from the true energy distributions and scattering mechanisms. Diffusion constants are usually determined experimentally for a particular material, as described in Section 4.4.5.

It is important to note that electrons and holes move together in a carrier gradient [Eqs. (4-21)], but the resulting currents are in opposite directions [Eqs. (4-22)] because of the opposite charge of electrons and holes.

4.4.2 Diffusion and Drift of Carriers. If an electric field is present in addition to the carrier gradient, the current densities will each have a drift component and a diffusion component

(4-23a)
$$J_n(x) = q\mu_n n(x)\, \mathscr{E}(x) + qD_n \frac{dn(x)}{dx}$$
$$\text{drift} \qquad\qquad \text{diffusion}$$

(4-23b)
$$J_p(x) = q\mu_p p(x)\, \mathscr{E}(x) - qD_p \frac{dp(x)}{dx}$$

and the total current density is the sum of the contributions due to electrons and holes

(4-24)
$$J(x) = J_n(x) + J_p(x)$$

We can best visualize the relation between the particle flow and the currents of Eqs. (4-23) by considering a diagram such as shown in Fig. 4-21. In this figure an electric field is assumed to be in the x-direction, along with carrier densities $n(x)$ and $p(x)$ which decrease with increasing x. Thus the derivatives in Eqs. (4-21) are negative, and diffusion takes place in the $+ x$-direction. The resulting electron and hole diffusion currents [J_p (diff.) and J_n (diff.)] are in opposite directions, according to Eqs. (4-22). Holes drift in the direction of the electric field [ϕ_p (drift)], whereas electrons drift in the opposite direction because of their negative charge. The resulting drift current is in the

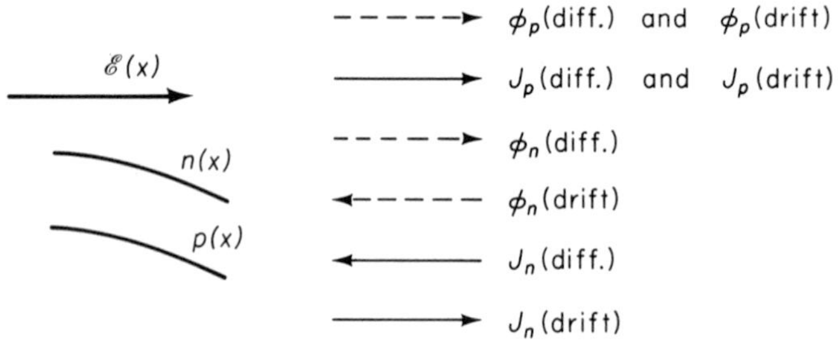

FIGURE 4-21. Drift and diffusion directions for electrons and holes in a carrier gradient and an electric field. Particle flow directions are indicated by dashed arrows, and the resulting currents are indicated by solid arrows.

+x-direction in each case. Note that the drift and diffusion components of the current are additive for holes when the field is in the direction of decreasing hole density, whereas the two components are subtractive for electrons under similar conditions. The total current may be due primarily to the flow of electrons or holes, depending on the relative concentrations and the relative magnitudes and directions of electric field and carrier gradients.

At equilibrium, no net current flows in a semiconductor. Thus any fluctuation which would begin a diffusion current sets up an electric field which redistributes carriers by drift. An examination of the requirements for equilibrium indicates that the diffusion coefficient and mobility must be related by

$$(4\text{-}25) \qquad\qquad \frac{D}{\mu} = \frac{kT}{q}$$

for either carrier type. This important equation is called the *Einstein relation*. It allows us to calculate either D or μ from a measurement of the other. Table 4-1 lists typical values of D and μ for several semiconductors at room temperature.

TABLE 4-1. Diffusion coefficient and mobility of electrons and holes for several semiconductors at 300°K.

	D_n (cm²/sec)	D_p (cm²/sec)	μ_n (cm²/V-sec)	μ_p (cm²/V-sec)
Ge	100	50	3900	1900
Si	35	12.5	1350	480
GaAs	220	10	8500	400

As Eq. (4-25) implies, a decrease in mobility caused by the introduction of impurities or lattice defects results in a smaller diffusion constant. We can see this is a reasonable result by reexamining the definition of the diffusion coefficient. Since D is proportional to $\bar{l}^2/\bar{t} \propto \bar{l}v_{th}$, where v_{th} is the mean thermal velocity $\bar{l}/\bar{t}$, we expect D to vary directly with the mean free path $\bar{l}$ at a given temperature. Thus a decrease in $\bar{l}$ caused by increased scattering results in a smaller diffusion constant as well as a decreased mobility. Writing D in this form also allows us to demonstrate the Einstein relation from the definitions of D and μ. If we take the average kinetic energy of the carriers to be approximately kT, Eqs. (3-40) and (4-20) give

$$(4\text{-}26) \qquad\qquad \frac{D}{\mu} = \frac{\frac{1}{2}\bar{l}v_{th}}{q\bar{t}/m^*} = \frac{1}{q}\left(\frac{1}{2}m^*v_{th}^2\right) \cong \frac{kT}{q}$$

We have neglected several averages over appropriate energy and velocity distributions in Eq. (4-26), but this approximate approach does indicate the basic relationship between D and μ. A more rigorous proof of the Einstein relation is discussed in Prob. 4.9.

In discussing the motion of carriers in an electric field, we should indicate the influence of the field on the energies of electrons in the band diagrams. Assuming an electric field $\mathscr{E}(x)$ in the x-direction, we can draw the energy bands as in Fig. 4-22, to include the change in potential energy of electrons

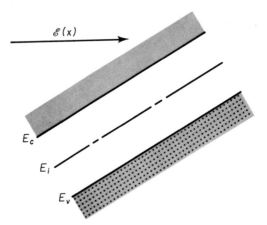

FIGURE 4-22. Energy band diagram of a semiconductor in an electric field $\mathscr{E}(x)$.

in the field. Since electrons drift in a direction opposite to the field, we expect the potential energy for electrons to increase in the direction of the field. The electrostatic potential $\mathscr{V}(x)$ varies in the opposite direction, since it is defined in terms of positive charges and is related to the electron potential energy $E(x)$ by $\mathscr{V}(x) = E(x)/(-q)$.

From the definition of electric field

$$(4\text{-}27) \qquad \mathscr{E}(x) = -\frac{d\mathscr{V}(x)}{dx}$$

we can relate $\mathscr{E}(x)$ to the electron potential energy in the band diagram by choosing some reference in the band for the electrostatic potential. We are interested only in the spatial variation in $\mathscr{V}(x)$ for Eq. (4-27), so any point in the bands will serve the purpose. Choosing E_i as a convenient reference, we can relate the electric field to this reference by

$$(4\text{-}28) \qquad \mathscr{E}(x) = -\frac{d\mathscr{V}(x)}{dx} = -\frac{d}{dx}\left[\frac{E_i}{(-q)}\right] = \frac{1}{q}\frac{dE_i}{dx}$$

Therefore, the variation of band energies with $\mathscr{E}(x)$ as drawn in Fig. 4-22 is correct. The direction of the slope in the bands relative to $\mathscr{E}$ is simple to remember: Since the diagram indicates electron energies, we know the slope in the bands must be such that electrons move "downhill" in the field.

We can use the results of Eqs. (4-23), (4-25), and (4-28) to demonstrate the power of the concept of quasi-Fermi levels in semiconductors [see Eq. (4-15)]. If we take the general case of nonequilibrium electron density with drift and diffusion, we must write the total electron current as

$$(4\text{-}29) \qquad J_n(x) = q\mu_n n(x)\,\mathcal{E}(x) + qD_n \frac{dn(x)}{dx}$$

where the gradient in electron density is

$$(4\text{-}30) \qquad \frac{dn(x)}{dx} = \frac{d}{dx}[n_i e^{(F_n - E_i)/kT}] = \frac{n(x)}{kT}\left(\frac{dF_n}{dx} - \frac{dE_i}{dx}\right)$$

Using Eq. (4-30) and the Einstein relation, the total electron current becomes

$$(4\text{-}31\text{a}) \qquad J_n(x) = q\mu_n n(x)\,\mathcal{E}(x) + \mu_n n(x)\left[\frac{dF_n}{dx} - \frac{dE_i}{dx}\right]$$

But Eq. (4-28) indicates that the subtractive term in the brackets is just $q\mathcal{E}(x)$, giving a direct cancellation of $q\mu_n n(x)\,\mathcal{E}(x)$

$$(4\text{-}31\text{b}) \qquad J_n(x) = \mu_n n(x)\frac{dF_n}{dx}$$

Thus, the processes of electron drift and diffusion are summed up by the spatial variation of the quasi-Fermi level. The same derivation can be made for holes, and we can write the current due to drift and diffusion in the form of a *modified Ohm's law*

$$(4\text{-}32\text{a}) \qquad J_n(x) = q\mu_n n(x)\frac{d(F_n/q)}{dx} = \sigma_n(x)\frac{d(F_n/q)}{dx}$$

$$(4\text{-}32\text{b}) \qquad J_p(x) = q\mu_p p(x)\frac{d(F_p/q)}{dx} = \sigma_p(x)\frac{d(F_p/q)}{dx}$$

Therefore, any drift, diffusion, or combination of the two in a semiconductor results in a current proportional to the gradients of the two quasi-Fermi levels. Conversely, a lack of current implies constant quasi-Fermi levels throughout the sample.

4.4.3 Diffusion and Recombination; The Continuity Equation. In the discussion of diffusion of excess carriers, we have thus far neglected the important effects of recombination. These effects must be in-

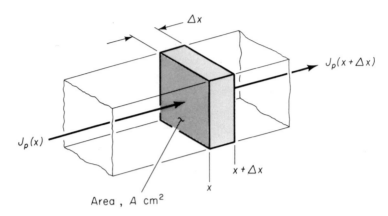

FIGURE 4-23. Current entering and leaving a volume $\Delta x A$.

cluded in a description of conduction processes, however, since recombination can cause a variation in the carrier density. For example, consider a differential length Δx of a semiconductor sample with area A in the yz-plane (Fig. 4-23). The hole current density leaving the volume, $J_p(x + \Delta x)$, can be larger or smaller than the current entering, $J_p(x)$, depending on the generation and recombination of carriers taking place within the volume. The net increase in hole density per unit time, $\partial p / \partial t$, is the difference between the hole flux per unit volume entering and leaving, minus the recombination rate. We can convert hole current density to hole particle flux density by dividing J_p by q. The current densities are already expressed per unit area; thus dividing $J_p(x)/q$ by Δx gives the number of carriers entering the volume $\Delta x A$ per unit time, and $(1/q)J_p(x + \Delta x)/\Delta x$ is the number leaving

$$(4\text{-}33) \qquad \frac{\partial p}{\partial t}\bigg|_{x \to x + \Delta x} = \frac{1}{q} \frac{J_p(x) - J_p(x + \Delta x)}{\Delta x} - \frac{\delta p}{\tau_p}$$

$$\begin{array}{ccc} \text{Rate of} & & \text{increase of hole density} & \text{recombination} \\ \text{hole buildup} & = & \text{in } \Delta x A \text{ per unit time} & - & \text{rate} \end{array}$$

As Δx approaches zero, we can write the current change in derivative form

$$(4\text{-}34\text{a}) \qquad \frac{\partial p(x, t)}{\partial t} = \frac{\partial \delta p}{\partial t} = -\frac{1}{q} \frac{\partial J_p}{\partial x} - \frac{\delta p}{\tau_p}$$

The expression (4-34a) is called the *continuity equation* for holes. For electrons we can write

$$(4\text{-}34\text{b}) \qquad \frac{\partial \delta n}{\partial t} = \frac{1}{q} \frac{\partial J_n}{\partial x} - \frac{\delta n}{\tau_n}$$

since the electronic charge is negative.

When the current is carried strictly by diffusion (negligible drift), we can replace the currents in Eqs. (4-34) by the expressions for diffusion current; for example, for electron diffusion we have

(4-35)
$$J_n(\text{diff.}) = qD_n\frac{\partial \delta n}{\partial x}$$

Substituting this into Eq. (4-34b) we obtain the *diffusion equation* for electrons

(4-36a)
$$\frac{\partial \delta n}{\partial t} = D_n\frac{\partial^2 \delta n}{\partial x^2} - \frac{\delta n}{\tau_n}$$

and similarly for holes

(4-36b)
$$\frac{\partial \delta p}{\partial t} = D_p\frac{\partial^2 \delta p}{\partial x^2} - \frac{\delta p}{\tau_p}$$

These equations are useful in solving transient problems of diffusion with recombination. For example, a pulse of electrons in a semiconductor (Fig. 4-19) spreads out by diffusion and disappears by recombination. To solve for the electron distribution in time, $n(x, t)$, we would begin with the diffusion equation, Eq. (4-36a).

4.4.4 Steady State Carrier Injection; Diffusion Length.

In many problems a steady state distribution of excess carriers is maintained, such that the time derivatives in Eqs. (4-36) are zero. In the steady state case the diffusion equations become

(4-37a)
$$\frac{d^2\delta n}{dx^2} = \frac{\delta n}{D_n\tau_n} \equiv \frac{\delta n}{L_n^2}$$

(4-37b)
$$\frac{d^2\delta p}{dx^2} = \frac{\delta p}{D_p\tau_p} \equiv \frac{\delta p}{L_p^2}, \qquad (\textit{steady state})$$

where $L_n \equiv \sqrt{D_n\tau_n}$ is called the electron *diffusion length*, and L_p is the diffusion length for holes. We no longer need partial derivatives, since the time variation is zero for steady state.

The physical significance of the diffusion length can be understood best by an example. Let us assume excess holes are somehow injected into a semi-infinite semiconductor bar at $x = 0$, and the steady state hole injection maintains a constant excess hole density at the injection point $\delta p(x = 0) = \Delta p$. The injected holes diffuse along the bar, recombining with a characteristic lifetime τ_p. In steady state we expect the distribution of excess holes to decay to zero for large values of x, because of the recombination (Fig. 4-24). For

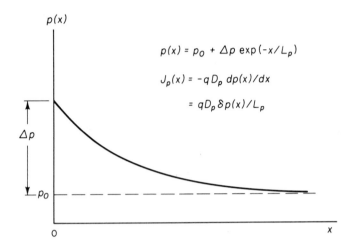

FIGURE 4-24. Injection of holes at $x = 0$, giving a steady state hole distribution $p(x)$ and a resulting diffusion current density $J_p(x)$.

this problem we use the steady state diffusion equation for holes, Eq. (4-37b). The solution to this equation has the form

$$(4\text{-}38) \qquad \delta p(x) = C_1 e^{x/L_p} + C_2 e^{-x/L_p}$$

We can evaluate C_1 and C_2 from the boundary conditions. Since recombination must reduce $\delta p(x)$ to zero for large values of x, $\delta p = 0$ at $x = \infty$, and therefore $C_1 = 0$. Similarly, the condition $\delta p = \Delta p$ at $x = 0$ gives $C_2 = \Delta p$, and the solution is

$$(4\text{-}39) \qquad \delta p(x) = \Delta p e^{-x/L_p}$$

The injected excess hole density dies out exponentially in x due to recombination, and the diffusion length L_p represents the distance at which the excess hole density is reduced to $1/e$ of its value at the point of injection. We can show that L_p *is the average distance a hole diffuses before recombining.* To calculate an average diffusion length, we must obtain an expression for the probability that an injected hole recombines in a particular interval dx. The probability that a hole injected at $x = 0$ *survives* to x without recombination is $\delta p(x)/\Delta p = \exp(-x/L_p)$, the ratio of the steady state densities at x and 0. On the other hand, the probability that a hole at x will *recombine* in the subsequent interval dx is

$$(4\text{-}40) \qquad \frac{\delta p(x) - \delta p(x + dx)}{\delta p(x)} = \frac{-(d\delta p(x)/dx)\, dx}{\delta p(x)} = \frac{1}{L_p} dx$$

Thus the total probability that a hole injected at $x = 0$ will recombine in a given dx is the product of the two probabilities

$$(4\text{-}41) \qquad (e^{-x/L_p})\left(\frac{1}{L_p}\,dx\right) = \frac{1}{L_p}\,e^{-x/L_p}\,dx$$

Then, using the usual averaging techniques described by Eq. (2-21), the average distance a hole diffuses before recombining is

$$(4\text{-}42) \qquad \langle x \rangle = \int_0^{\infty} x\,\frac{e^{-x/L_p}}{L_p}\,dx = L_p$$

The steady state distribution of excess holes causes diffusion, and therefore a hole current, in the direction of decreasing density. From Eqs. (4-22b) and (4-39) we have

$$(4\text{-}43) \quad J_p(x) = -qD_p\frac{dp}{dx} = -qD_p\frac{d\delta p}{dx} = q\frac{D_p}{L_p}\Delta pe^{-x/L_p} = q\frac{D_p}{L_p}\delta p(x)$$

Since $p(x) = p_0 + \delta p(x)$, the space derivative involves only the excess density. We notice that since $\delta p(x)$ is proportional to its derivative for an exponential distribution, the diffusion current at any x is just proportional to the excess density δp at that position.

Although this example seems rather restricted, its usefulness will become apparent in Chapter 5 in the discussion of p-n junctions. The injection of minority carriers across a junction often leads to exponential distributions as in Eq. (4-39), with the resulting diffusion current of Eq. (4-43).

4.4.5 The Haynes–Shockley Experiment.

One of the classic semiconductor experiments is the demonstration of drift and diffusion of minority carriers. This experiment, first performed by J. R. Haynes and W. Shockley in 1951 at the Bell Telephone Laboratories, illustrates clearly the important role of minority carriers in semiconductors. The experiment allows independent measurement of the minority carrier mobility μ and diffusion constant D, and it can also be used to calculate the recombination lifetime τ. The basic principles of the Haynes–Shockley experiment are as follows: A pulse of holes is created in an n-type bar (for example) which contains an electric field (Fig. 4-25); as the pulse drifts in the field and spreads out by diffusion, the hole density is monitored at some point down the bar; the time required for the holes to drift a given distance in the field gives a measure of the mobility; and the spreading of the pulse during a given time is used to calculate the diffusion constant.

In Fig. 4-25 a pulse of excess carriers is created by a light flash at some

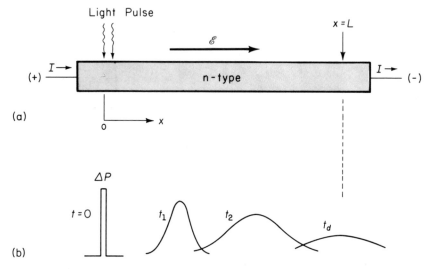

FIGURE 4-25. Drift and diffusion of a hole pulse in an n-type bar: (a) sample geometry; (b) position and shape of the pulse for several times during its drift down the bar.

point $x = 0$ in an n-type semiconductor ($n_0 \gg p_0$). We assume the excess carriers have a negligible effect on the electron density but change the hole density significantly. The holes drift in the direction of the electric field and eventually reach the point $x = L$, where they are monitored. By measuring the drift time t_d, we can calculate the drift velocity v_d, and therefore the hole mobility

$$(4\text{-}44) \qquad \qquad v_d = \frac{L}{t_d}$$

$$(4\text{-}45) \qquad \qquad \mu_p = \frac{v_d}{\mathscr{E}}$$

Thus the hole mobility can be calculated directly from a measurement of the drift time for the pulse as it moves down the bar.

As the pulse drifts in the $\mathscr{E}$ field it also spreads out by diffusion. By measuring the spread in the pulse, we can calculate D_p. To predict the distribution of holes in the pulse as a function of time, let us first reexamine the case of diffusion of a pulse *without drift, neglecting recombination* (Fig. 4-19). The equation which the hole distribution must satisfy is the time-dependent diffusion equation, Eq. (4-36b). For the case of negligible recombination (τ_p long compared with the times involved in the diffusion), we can write the diffusion equation as

(4-46)
$$\frac{\partial \delta p(x, t)}{\partial t} = D_p \frac{\partial^2 \delta p(x, t)}{\partial x^2}$$

The function which satisfies this equation is called a *gaussian distribution* (Prob. 4.12)

(4-47)
$$\delta p(x, t) = \left[\frac{\Delta P}{2\sqrt{\pi D_p t}} \right] e^{-x^2/4 D_p t}$$

where ΔP is the number of holes per unit area created over a negligibly small distance δ at $t = 0$ ($\Delta P = \Delta p \delta$, where δ is very small). The factor in brackets indicates that the peak value of the pulse (at $x = 0$) decreases with time, and the exponential factor predicts the spread of the pulse in the positive and negative x-directions (Fig. 4-26). If we designate the peak value of the

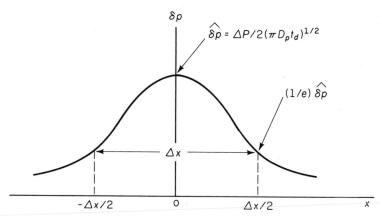

FIGURE 4-26.   Calculation of D_p from the shape of the δp distribution after time t_d. No drift or recombination is included.

pulse as $\hat{\delta p}$ at any time (say t_d), we can use Eq. (4-47) to calculate D_p from the value of δp at some point x. The most convenient choice is the point $\Delta x/2$, at which δp is down by $1/e$ of its peak value $\hat{\delta p}$. At this point we can write

(4-48)
$$e^{-1} \hat{\delta p} = \hat{\delta p} e^{-(\Delta x/2)^2/4 D_p t_d}$$

(4-49)
$$D_p = \frac{(\Delta x)^2}{16 t_d}$$

Therefore, if we can somehow measure the width of the pulse Δx after it has spread by diffusion for t_d sec, we can obtain the diffusion constant D_p from Eq. (4-49). In the Haynes–Shockley experiment Δx is measured with the pulse centered at $x = L$ (Fig. 4-25), t_d sec after injection. We can recalculate $\delta p(x, t)$ from the continuity equation by including drift as well as diffusion, or more simply, we can consider the drift of the pulse separately from its diffusion and use superposition in drawing the pulse shapes of Fig. 4-25b. Considering diffusion separately implies that in drawing a gaussian pulse at any time t, we take the center of the pulse as the origin of a dimensional variable $x' = x - v_d t$ (i.e., the origin for x' drifts with the pulse, remaining at the instantaneous position of the peak). This allows calculation of a pulse width Δx as in Fig. 4-26 and Eq. (4-49).

A more detailed illustration of the Haynes–Shockley experiment is given in Fig. 4-27. Often a hole injection contact is used instead of a light pulse at point (1), and a contact which collects holes is used to monitor the pulse at point (2). As we shall see in Chapter 5, a properly constructed forward-biased p-n junction can serve as a hole injector, while a reverse-biased junction can be used for the collector (2). In the original version of the Haynes–Shockley experiment, point contacts (sharp metal wires in contact with the semiconductor), having many of the same injection and collection

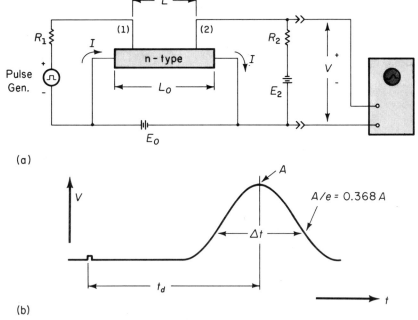

(a)

(b)

FIGURE 4-27. The Haynes–Shockley experiment: (a) circuit schematic; (b) typical trace on the oscilloscope screen.

properties as junctions, were used at (1) and (2) on a Ge sample. The current collected as the pulse moves past point (2) is displayed on the oscilloscope screen as a voltage across R_2 (Fig. 4-27b). We assume that the pulses obtained from the pulse generator are very short compared with the drift time t_d and that the time between pulses is longer than t_d.

When a pulse is injected at (1), there is a small conductive feed-through "pip" on the oscilloscope screen. This is due to the fact that R_1, R_2, and the sample are part of a voltage divider, and R_2 must receive some small part of the pulse generator voltage. This small pip can be used to determine the $t = 0$ reference. Usually the pulse generator has an output which can be used to trigger the oscilloscope at the time the pulse is injected. After time t_d the center of the pulse is under point (2), and the shape of the pulse is displayed on a time scale as in Fig. 4-27b.

The drift time t_d can be measured directly from the trace, and the drift velocity and mobility are calculated from

(4-50) $$v_d = \frac{L}{t_d}, \qquad \mu_p = \frac{v_d}{\mathscr{E}} = \frac{L/t_d}{E_0/L_0}$$

The oscilloscope trace displays voltage vertically and time horizontally. However, we can convert positions on the time scale to distance by using the measured value of v_d, and assuming the collector at (2) behaves linearly, we can take the vertical scale as being proportional to the value of δp under the collector at any time t. Thus when the trace is down to $1/e$ of its peak value A, the excess hole density is $1/e$ of its peak value at that time. Measuring the width of the pulse on the oscilloscope screen, we have

(4-51) $$\Delta x = \Delta t v_d, \qquad D_p = \frac{(\Delta t \, v_d)^2}{16 t_d} = \frac{(\Delta t L)^2}{16 t_d^3}$$

One problem with displaying the pulse as a function of time as shown in Fig. 4-27b is that the peak and the $1/e$ point pass under the collector at different times. Unless the drift velocity v_d is fast compared with diffusion, the pulse shape displayed on the oscilloscope will be distorted. Since some time is required for the entire pulse to drift past the collector at point (2), the trailing edge of the pulse will exhibit more spread due to diffusion by the time it is displayed than is the case for the leading edge. (*Question*: Which side of the oscilloscope trace of Fig. 4-27b corresponds to the leading edge of the pulse in the sample?) This distortion can be reduced by increasing the battery voltage E_0, thereby increasing the electric field and the drift velocity v_d.

In addition to drift and diffusion, the excess holes experience recombination during the time t_d. Therefore, the hole pulse is reduced by recombination as it drifts. This loss of signal in time can be used to calculate τ_p (Prob. 4.14).

Although the experiment described here involves injection of holes into n-type material, analogous comments could be made for the drift and diffusion of an electron pulse in a p-type sample. The battery, pulse generator, and collector voltages in Fig. 4-27a would be reversed for p-type material. In either case the experiment reveals the properties of *minority* carriers as they drift down the bar and diffuse.

Before leaving this subject, we should indicate the relative roles of majority and minority carriers in the drift of a pulse. As the minority holes in the pulse of Fig. 4-25 drift with the electric field, space charge neutrality requires that at each point in time the excess majority carrier (electron) concentration in the pulse must match that of the holes [see Eq. (3-27)]. Space charge neutrality cannot be maintained precisely in the drifting pulse; however, the requirement is very nearly met, and $\delta p \simeq \delta n$ at each point. This means that the excess electron and hole concentrations in the pulse *move together* down the bar. This does not mean *individual* electrons move in the $+\mathcal{E}$ direction, since that would violate the basic equations of electrodynamics for negative charges. What does occur is that electrons enter and leave the pulse at the proper rate to maintain $\delta n \simeq \delta p$ at each point in the distribution. A careful treatment of the continuity and space charge neutrality equations indicates that pulses made up of electrons and holes propagate in the direction which the *minority* carriers take in the field. If the material is not strongly extrinsic (i.e., if n_0 and p_0 are comparable), the pulse will move with modified values of mobility and diffusion constant, which are combinations of the values for electrons and holes. This is called *ambipolar drift and diffusion*. If the material were exactly intrinsic ($n_0 = p_0 = n_i$), neither electrons nor holes would be the minority carriers. In that case a pulse of excess carriers would not propagate in an $\mathcal{E}$ field but would simply spread out by diffusion.

READING LIST

V. F. WEISSKOPF, "How Light Interacts with Matter," *Scientific American*, vol. 219, no. 3, pp. 60–71, September 1968.

W. SHOCKLEY, *Electrons and Holes in Semiconductors*. New York: Van Nostrand Reinhold Co., 1950.

T. L. MARTIN, Jr., and W. F. LEONARD, *Electrons and Crystals*. Belmont, Calif.: Brooks/Cole Publishing Company, 1970.

J. P. McKELVEY, *Solid State and Semiconductor Physics*. New York: Harper & Row, Publishers, 1966.

A. VAN DER ZIEL, *Solid State Physical Electronics* (2nd ed.). Englewood Cliffs, N.J.: Prentice-Hall, Inc., 1968.

J. I. PANKOVE, *Optical Processes in Semiconductors*. Englewood Cliffs, N. J.: Prentice-Hall, Inc., 1971.

PROBLEMS

4.1 A 0.46 μm-thick sample of GaAs is illuminated with monochromatic light of $hv = 2$ eV. The absorption constant α is 5×10^4 cm^{-1}. The power incident on the sample is 10 mW.

  (a) Find the total energy absorbed by the sample per second (J/sec).

  (b) Find the rate of excess thermal energy given up by the electrons to the lattice before recombination (J/sec).

  (c) Find the number of photons per second given off from recombination events.

4.2 Show by a simple two-dimensional sketch of a ZnS lattice how the charge of Cu activator ions can be compensated by the following: Al, Cl, S vacancy, and Zn interstitial.

4.3 We wish to calculate the carrier lifetime from the trace in Fig. 4-15. The *half-life* of an exponential decay is the time required for the time-dependent variable (i/v in this case) to drop from any given value to half that value. Measure the half-life for several intervals along the trace of Fig. 4-15a. Is the decay exponential? If so, determine the lifetime τ from the measured half-life. Repeat for the minority carrier decay in Fig. 4-10. Is the decay exponential when trapping is present (Fig. 4-15b)?

4.4 A direct semiconductor with $n_i = 10^{10}$ cm^{-3} is doped with 10^{14} cm^{-3} donors. Its low-level carrier lifetime τ is $\tau_n = \tau_p = 10^{-7}$ sec.

  (a) If a sample of this material is uniformly exposed to a steady optical generation rate of $g_{op} = 2 \times 10^{21}$ EHP/cm^3-sec, calculate the excess carrier density $\Delta n = \Delta p$. *Note*: this excitation rate is not low level.

  (b) If the carrier lifetime is defined as the excess carrier density divided by the recombination rate, what is τ at this excitation level?

4.5 A strongly n-type semiconductor is illuminated by a steady light which creates g_{op} EHP/cm^3-sec. The equilibrium carrier densities of the sample are n_0 and p_0, and the recombination coefficient is α_r. At time $t = 0$, an additional flash creates $\Delta p = \Delta n$ EHP/cm^3. Show that the hole density at any time t after the flash is

$$p(t) = p_0 + \Delta p e^{-\alpha_r n_0 t} + \frac{g_{op}}{\alpha_r n_0}$$

(Continued)

(Continued 4.5)
Assume uniform, low-level excitation and negligible heating. Begin with the following rate balance equation: The excess hole degradation rate is the instantaneous recombination rate minus the total generation rate.

4.6 (a) An n-type semiconductor sample is doped with N_d donors/cm³. A certain impurity has a single acceptor level located exactly at the center of the band gap. What concentration of this acceptor is required to compensate the semiconductor such that $n = n_i$ after compensation? Neglect the small difference between E_i and $E_g/2$ calculated in Prob. 3.6.

(b) Zinc is a double acceptor in Si; each Zn impurity can accept one electron at the lower level shown in Fig. 4-13 and a second electron at the upper level. What density of Zn atoms is required to compensate a Si sample which is originally n-type with $N_d = 10^{16}$ cm⁻³? In Chapter 5 we shall see that samples can be doped by diffusion of impurities at elevated temperatures. Assuming Zn can be diffused into Si such that the Zn concentration N_{Zn} reaches its solid solubility at the diffusion temperature (Appendix V), determine the appropriate diffusion temperature for compensation of this sample.

4.7 The bar of n-type Ge ($N_d = 10^{14}$ cm⁻³) discussed in Prob. 3.9b has a cross-sectional area of 10^{-2} cm² and a length of 1 cm. The carrier lifetimes are $\tau_n = \tau_p = 100$ μsec. Calculate the dark resistance and the resistance with an illumination which generates 10^{17} EHP/cm³-sec uniformly throughout the sample.

4.8 Assume a photoconductor in the shape of a bar of length L and area A has a constant voltage V applied, and it is illuminated such that g_{op} EHP/cm³-sec are generated uniformly throughout. If $\mu_n \gg \mu_p$, we can assume the optically induced change in current ΔI is dominated by the mobility μ_n and lifetime τ_n for electrons. Show that $\Delta I = qALg_{op}\tau_n/\tau_t$ for this photoconductor, where τ_t is the transit time of electrons drifting down the length of the bar.

4.9 A semiconductor sample is doped such that the conduction band electron density varies linearly in the x-direction, $n(x) = n_i(1 + Gx)$.

(a) Remembering that the Fermi level is invariant at equilibrium, find the expression for the electric field $\mathscr{E}(x)$.

(b) If $n(x)$ varies in the sample, how is equilibrium maintained?

(c) Use your answer to (b) to derive the Einstein relation.

4.10 (a) Calculate the separation in the quasi-Fermi levels $F_n - F_p$ for the illuminated sample described in Prob. 4.4.

(b) Assume the sample is illuminated over a narrow range of its length. In a band diagram, sketch the quasi-Fermi levels in the illuminated zone and for several diffusion lengths on either side of the illumination. Use a band gap of 1 eV in the sketch, and assume space charge neutrality.

4.11 The current required to feed the hole injection at $x = 0$ in Fig. 4-24 is obtained by evaluating Eq. (4-43) at $x = 0$. The result is $I_p(x = 0) = qAD_p \Delta p/L_p$. Show that this current can be calculated by integrating the charge stored in the steady state hole distribution $\delta p(x)$ and then dividing by the average hole lifetime τ_p. Explain why this approach gives $I_p(x = 0)$.

4.12 Show by direct substitution that the gaussian distribution, Eq. (4-47), is a solution to the diffusion equation, Eq. (4-46). Since this equation neglects recombination, show that the total number of holes remains constant at $A\Delta P$ for any time t, where A is the cross-sectional area of the sample. How would you modify Eq. (4-47) to include recombination?

4.13 An n-type Ge sample is used in the Haynes–Shockley experiment shown in Fig. 4-27. The length of the sample is 1 cm, and the probes (1) and (2) are separated by 0.95 cm. The battery voltage E_0 is 2 V. A pulse arrives at point (2) 0.25 msec after injection at (1); the width of the pulse Δt is 117 μsec. Calculate the hole mobility and diffusion constant, and check the results against the Einstein relation.

4.14 We wish to use the Haynes–Shockley experiment to calculate the hole lifetime τ_p in an n-type sample. Assume the peak voltage of the pulse displayed on the oscilloscope screen is proportional to the hole concentration under the collector terminal at time t_d, and that the displayed pulse can be approximated as gaussian. The electric field is varied and the following data taken: For $t_d = 200$ μsec, the peak is 20 mV; for $t_d = 50$ μsec, the peak is 80 mV. What is τ_p?

p-n JUNCTIONS **5**

Most semiconductor devices contain at least one junction between p-type and n-type material. These p-n junctions are fundamental to the performance of functions such as rectification, amplification, switching, and other operations in electronic circuits. In this chapter we shall discuss the equilibrium state of the junction and the flow of electrons and holes across a junction under steady state and transient conditions. With the background provided in this chapter on junction properties, we can then proceed to discuss specific devices in later chapters.

5.1 Fabrication of p-n Junctions

Before investigating their electrical properties, we shall discuss how p-n junctions are made. The technology of junction fabrication is, of course, a broad subject which includes the accumulated knowledge and experience of many research and manufacturing groups. Without attempting a thorough description of these manufacturing techniques, we can investigate some of the more basic methods of forming junctions and making contacts to them in mountings suitable for devices. The type of junction we seek is a change from n-type to p-type material within a *single crystal*. Therefore, we may change the dopant from donors to acceptors during the growth of the crystal. Alternatively, we may introduce impurities of one type into regions of a crystal which was grown with doping of the opposite type. In either case, only the doping is changed at the junction; there are no gross changes in the lattice structure itself.

5.1.1 Grown Junctions. A method of junction fabrication which was popular in the early days of semiconductor device development is the *grown junction* technique. In this method the dopant is abruptly changed in the melt during the process of crystal growth (Fig. 5-1). As an example, let us assume an n-type Si crystal containing 10^{14} phosphorus atoms/cm³ is grown from a melt. The growth process is stopped temporarily while sufficient boron atoms are added to the melt to result in 6×10^{14} boron atoms/cm³ in

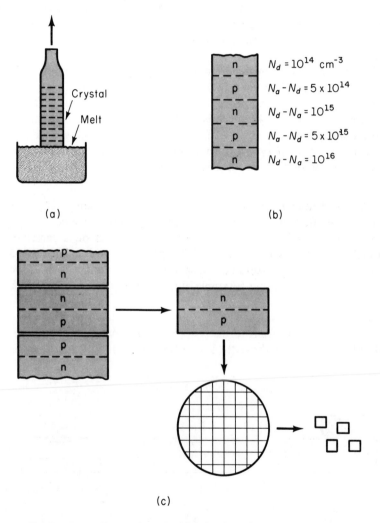

(a)   (b)

(c)

Figure 5-1.   Grown junction fabrication: (a) junction formation during the growth of an ingot from the melt; (b) donor and acceptor densities for a series of grown junctions in an ingot; (c) slicing of a wafer from the ingot and dicing of the wafer into small squares for junction devices.

the crystal. This is a sufficient concentration of acceptors to compensate the original donors and leave a net acceptor density of $N_a - N_d = 5 \times 10^{14}$ cm^{-3}. This process (called *counterdoping*) can be continued as more donors and acceptors are alternately added, until the melt becomes contaminated. When the growth is completed, individual sections are sliced from the ingot (Fig. 5-1c) to form wafers of Si containing single p-n junctions. These wafers are then separated into small squares (dice) for individual devices. This separation can be accomplished by cutting the wafer into squares or, more simply, by scribing the wafer with a diamond stylus and then breaking the dice apart.

The grown junction technique has been largely supplanted by the more flexible methods of introducing junctions into crystals after growth. An important exception, however, is the *epitaxial* growth of p-n junctions (Sections 1.3.3 and 1.3.4), which is used widely in integrated circuits and for other applications. An epitaxial layer of one type can be grown on a substrate of the opposite type to form a sharp, single-crystal junction. We shall discuss the use of epitaxial junctions in more detail in Chapter 10.

5.1.2 Alloyed Junctions. A convenient technique for making p-n junctions in the research laboratory and for some production processes is the alloying of a metal containing doping atoms on a semiconductor with the opposite type of dopant. This process was used in the 1950's to produce millions of diodes and transistors and has not yet been abandoned. As an example of alloying, a sample of n-type Ge can be heated with a pellet of In on it to form a small local melt from which a p-n junction grows on the parent crystal (Fig. 5-2). As the In and Ge are heated to about 160°C, an alloy of the two materials forms (Fig. 5-2b). The proportion of In to Ge in this alloy melt is well above the solid solubility of In in Ge (see Appendix V

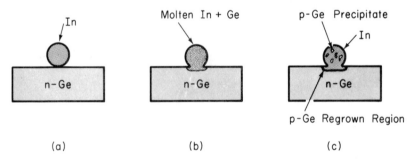

(a)   (b)   (c)

FIGURE 5-2. The alloying process: (a) pellet of In in contact with n-type Ge; (b) molten phase of In and Ge mixture during heating; (c) cross sectional view of an alloyed junction, showing p-type regrown region and precipitation of Ge in the In.

for solid solubilities of impurities in Ge and Si). As the temperature is lowered, Ge grows out of the alloy mixture.† At the interface between the cooling mixture and the Ge crystal, a *regrown region* of Ge is formed. This regrown region has the same lattice structure as the Ge parent crystal, but it now has a high concentration of In atoms and is therefore p-type. The regrowing of the p-type Ge crystal onto the underlying Ge substrate is similar to the process of liquid epitaxy discussed in Section 1.3.3. In the present case, however, the Ge substrate serves not only as the seed crystal for the growing layer but also provides the Ge in the melt. The resulting p-n junction consists of the interface between the regrown p-type Ge region and the n-type Ge sample.

   Alloyed junctions are usually made in a reducing atmosphere to prevent oxidation of the crystal surface and the alloy during heating. Even a very thin layer of oxide at the surface of the semiconductor can inhibit the alloying process. A typical hydrogen atmosphere alloy station for small-scale laboratory work is illustrated in Fig. 5-3. In this arrangement, current is passed through a thin strip of metal (such as nichrome or tungsten) suitable for use

(a)   (b)

FIGURE 5-3. Laboratory alloy station: (a) an alloying operation using a heater strip in a hydrogen atomosphere; (b) a closeup view of the heater strip and a header. Hydrogen gas flows through the glass envelope covering the heater strip in (a).

---

   †This process is analogous to the familiar problem of dissolving sugar in tea. A solubility limit is reached with iced tea, such that no more sugar goes into solution, even with vigorous stirring. More sugar will dissolve in hot tea, however, and it is possible to cool the hot solution to obtain tea which is *supersaturated* with sugar (more than normal for the lower temperature). In cooling, however, some sugar may precipitate out of the mixture.

as a heater because of its high melting point. The sample to be alloyed is placed on a carbon or ceramic strip which rests on the heater. Hydrogen gas flows through the glass envelope during alloying and is released through the open bottom of the system. This opening accommodates rapid gas expansion in case of an accidental hydrogen–oxygen explosion within the glass envelope. Alloying can also be done in inert gas atmospheres or in a vacuum.

Alloy stations can be used to attach semiconductor samples onto *headers* of suitable size and with proper electrical leads for particular device applications. A header is a device mounting which has a flat surface (usually a gold-plated metal) onto which the sample is alloyed and possesses electrical leads for making contact to the device. Before the p-n junction is made, the n-type Ge sample can be contacted to the header by alloying with a solder preform containing a column V impurity (e.g., Pb–Sn solder containing about 1% Sb). The alloyed region at the header interface is heavily doped n-type (often referred to as n^+) to match the n-type sample. Then the junction is made with the In pellet, which has a lower melting temperature than the Pb–Sn solder. Finally, a wire must be connected from the In to one of the header leads. This entire operation is accomplished with the aid of a microscope mounted above the glass envelope to permit observation of the process. Use of a laboratory-scale alloy station is discussed in more detail in Section 6.2.5 as an example of tunnel diode fabrication.

5.1.3 Diffused Junctions.

In terms of volume of production, the most common technique for forming p-n junctions is the impurity diffusion process. This method, along with the use of selective masking to control junction geometry, makes possible the wide variety of devices available in the form of IC's. Selective diffusion is an impressive technique in its controllability, accuracy, and versatility, as we shall see in Chapter 10. As a preliminary illustration of this point, however, we can remember that selective diffusion and other IC techniques are sufficiently accurate to allow the fabrication of dozens of p-n junction devices, properly interconnected, on a Si chip the size of this o.

The diffusion of impurities into a solid is basically the same type of process as discussed in Section 4.4 in terms of excess carriers. In each case, diffusion is a result of random motion, and particles diffuse in the direction of decreasing concentration gradient. The random motion of impurity atoms in a solid is, of course, rather limited unless the temperature is high. Thus diffusion of doping impurities into a semiconductor such as Si is accomplished at temperatures in the neighborhood of 1000°C. At these temperatures, many atoms in the semiconductor move out of their lattice sites, leaving vacancies into which impurity atoms can move. Most of the common doping impurities diffuse by this type of vacancy motion and occupy lattice positions in the crystal after it is cooled.

If a wafer of n-type Si is placed in a furnace at 1000°C, in a gaseous atmosphere containing a high concentration of boron atoms, a sharp impurity concentration gradient occurs at the surface of the sample. Since the temperature is high enough to allow slow random motion of atoms, boron establishes a certain solid solubility concentration at the surface and begins to diffuse into the sample (Fig. 5-4). The diffusion constant, describing the

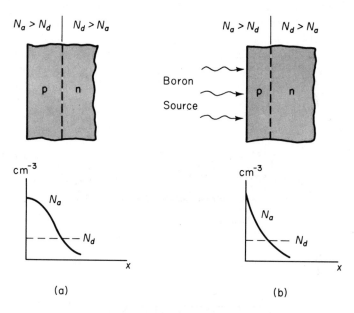

FIGURE 5-4.   Impurity density profiles for diffusion: (a) diffusion with a limited source at the surface; (b) constant source.

ability of an impurity to wander into the sample, is known for most of the common impurities in many semiconductors as a function of temperature; thus the distribution of impurities in the sample at any time during the diffusion can be calculated from a solution of the diffusion equation with the appropriate boundary conditions. If the source of boron atoms at the surface of the sample mentioned above is limited (e.g., a given number of atoms deposited on the Si surface before diffusion), a gaussian distribution as described by Eq. (4-47) is obtained (Fig. 5-4a). On the other hand, if the boron atoms are supplied continuously, such that the concentration at the surface is maintained at a constant value, the distribution shown in Fig. 5-4b results. This distribution appears almost exponential with distance into the sample; actually, the impurity profile in Fig. 5-4b follows what is called a *complementary error function*. In some cases the diffusant impurity is deposited in a shallow *predeposition* diffusion (Fig. 5-4b), and the source is then removed for a *drive-in* diffusion. If no impurities evaporate or otherwise escape from the surface, the result is a gaussian distribution (Fig. 5-4a). In either case, there is some point in the sample at which the introduced acceptor density just equals the background donor density in the originally n-type sample. This point is the location of the p-n junction; to the left of this point in each sample of Fig. 5-4, acceptor atoms predominate and the material is p-type, whereas to the right of the junction, the background donor atoms predominate and the material is n-type. The depth of the junction beneath the surface of the sample can be controlled accurately by the time and temperature of the diffusion (Prob. 5.22).

A typical diffusion furnace for Si device production is shown schematically in Fig. 5-5a. The furnace core surrounds a silica tube through which

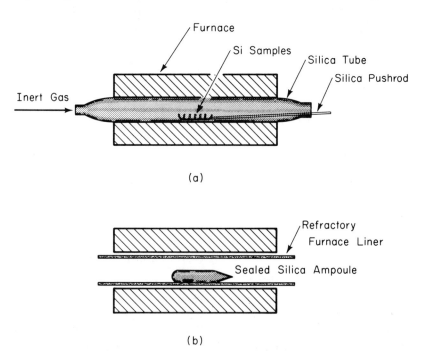

(a)

(b)

FIGURE 5-5.   Diffusion furnaces: (a) open-tube diffusion system; (b) closed-tube diffusion system.

an inert gas flows. The Si samples are placed in the tube during diffusion, and the impurity atoms are introduced into the gas flow over the samples. Common impurity source materials for diffusions in Si are B_2O_3, BBr_3, and BCl_3 for boron; phosphorus sources include P_2O_5 and $POCl_3$. Solid sources are placed in the silica tube upstream from the sample or in a separate heating zone of the furnace; gaseous sources can be metered directly into the gas flow system; and with liquid sources the inert carrier gas is bubbled through the liquid before being introduced into the furnace tube. The Si wafers are held in a silica "boat" which can be pushed into position in the furnace and removed by a silica rod. This arrangement is called an *open-tube* diffusion system.

It is important to remember the degree of cleanliness required in these processing steps. Since typical doping densities represent one part per million or less, cleanliness and purity of materials is critically important. Thus the impurity source and carrier gas in Fig. 5-5a must be extremely pure; the silica tube, sample holder, and pushrod must be cleaned and etched in hydrofluoric acid (HF) before use (once in use, the tube cleanliness can be maintained if no unwanted impurities are introduced); finally, the Si wafers themselves must undergo an elaborate cleaning procedure before diffusion, including a final etch containing HF to remove any SiO_2 from the surface (this may be done selectively, as discussed below).

Although open-tube diffusion is the most common type for Si device fabrication, the *closed-tube* method (Fig. 5-5b) is useful for many applications. In this method the sample and the impurity source are placed inside a silica tube (ampoule) which is then evacuated or filled with a low pressure of inert gas. The tube is sealed off with a H_2–O_2 torch and is placed in the furnace for diffusion. The impurity atoms provide a continuous source for the diffusion. This method is cumbersome in that a tube must be evacuated and sealed off for each diffusion. On the other hand, it has the advantage that the impurity atoms are restricted to the sealed tube. Thus a single furnace can be used for a series of diffusions, including many types of impurities. This is not the case for an open-tube furnace, which must be used only for a single type of impurity.

Selective diffusions can be made in Si by opening "windows" in a layer of SiO_2 grown on the surface of the wafer. Hydrofluoric acid removes any unmasked SiO_2, but does not disturb the Si itself. It is convenient that Si oxidizes in the form of SiO_2, since this makes possible the basic oxide masking process needed for Si device fabrication. Wafers of Si can be placed in a furnace containing O_2 gas to grow an SiO_2 layer (less than one micron thick) on the surface. The oxidized wafer is in effect encapsulated by silica glass, which blocks penetration by most impurity atoms. By removing the oxide in desired regions of the wafer, diffusion can be carried out at these controlled locations (Fig. 5-6). The process used for selectively removing the oxide involves covering the surface of the oxide with an acid-resistant coating except where the diffusion windows are needed; then an HF etch removes the SiO_2 at the uncoated locations. The coating is a photosensitive organic material called *photoresist* (*PR*), which can be polymerized by ultraviolet light. If the light is passed through a *contact mask* containing the desired pattern, the

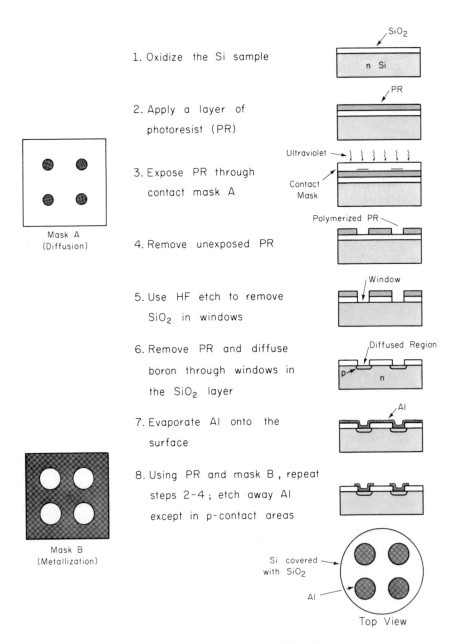

1. Oxidize the Si sample

2. Apply a layer of photoresist (PR)

3. Expose PR through contact mask A

4. Remove unexposed PR

5. Use HF etch to remove SiO_2 in windows

6. Remove PR and diffuse boron through windows in the SiO_2 layer

7. Evaporate Al onto the surface

8. Using PR and mask B, repeat steps 2–4; etch away Al except in p-contact areas

Mask A
(Diffusion)

Mask B
(Metallization)

FIGURE 5-6. Steps in the fabrication of diffused p-n junctions. For clarity, only four diodes per wafer are shown, and the relative thicknesses of the oxide, PR, and Al layers are exaggerated.

coating can be polymerized everywhere except where the pattern is to appear. Then the unpolymerized areas can be removed prior to the HF etching, which removes the SiO_2 in the uncoated areas. This technique is similar to methods of *photolithography*. The masks for this process are made in the form of large-scale drawings which are then reduced photographically to the proper size. The close tolerances available with this technique make possible the intricate diffusion patterns necessary for Si IC's. We shall discuss mask fabrication in more detail in Section 10.2.1.

The example of Fig. 5-6 illustrates the selective diffusion of many isolated p-n junctions on a single wafer. In most cases the oxide layer is left on the wafer, and a metal such as Al is evaporated onto the windows to contact the diffused layers.† Then the individual devices can be separated by scribing and breaking the wafer, and the final steps of the process are mounting of individual devices on headers and connecting leads to the Al contact regions. Very accurate lead bonders are available for bonding Au or Al wire (about one mil in diameter) to the device and then to a post on the header.

The procedure of Fig. 5-6 is basic to the production of transistors and IC's, as we shall see in Chapters 8 and 10. However, individually packaged diffused diodes are usually fabricated by more straightforward methods. For example, a method of making glass-encapsulated diodes is illustrated in Fig. 5-7. After diffusion of a p-n junction in an entire Si wafer, each side of the wafer is plated with Ni to facilitate contact to the p and n regions. The wafer is then scribed and broken into many individual diode chips. Each chip is mounted on a small metal rod which is coated with an appropriate solder to contact the Ni layer on the chip. A second metal rod (often terminated in a solder-coated spring to allow for thermal expansion and contraction) is brought into contact with the top of the chip to form the second lead of the diode. The entire device is heated, and the glass envelope is sealed at the lower glass bead. The result is a glass-encapsulated p-n junction with leads attached to each side of the junction. The various steps in this process are automated easily, and such devices can be produced rapidly.

5.1.4 Ion Implantation.
A useful alternative to high-temperature diffusion is the direct implantation of energetic ions into the semiconductor. In this process a beam of impurity ions is accelerated to kinetic energies

---

†In many cases another oxidation and masking step is used between steps 6 and 7 of Fig. 5-6 to reduce the size of the contact windows; if the Al region is smaller than the diffused area being contacted, the chances of bypassing the junction with a short circuit are reduced.

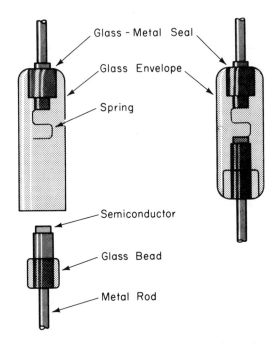

FIGURE 5-7. Typical production-line fabrication of a glass-encapsulated diode.

ranging from several keV to several MeV and is directed onto the surface of the semiconductor. As the impurity atoms enter the crystal, they give up their energy to the lattice in collisions and finally come to rest at some average penetration depth, called the *range*. Depending on the impurity and its implantation energy, the range in a given semiconductor may vary from a few hundred angstroms to about one micron.

An obvious advantage of implantation is that it can be done at relatively low temperatures; this means that doping layers can be implanted without disturbing previously diffused regions. Ions can be implanted through oxide layers but are generally blocked by metal layers. Therefore, the metallization techniques described above can be used to define a pattern through which implantation can take place. Very shallow (tenths of a micron) and well-defined doping layers can be achieved by this method. As we shall see in later chapters, many devices require thin doping regions and may be improved by ion implantation techniques. Furthermore, it is possible to implant impurities which do not diffuse conveniently into semiconductors, and the implanted doping density can be extended beyond the normal solid solubility in the crystal. Another possibility is the implantation of a *buried layer* below the surface. At implantation energies in the MeV range, many impurity atoms stop well below the surface of the semiconductor; this effect can be used, for example, to bury a p-type layer a fraction of a micron below the surface of an n-type crystal.

There are difficulties associated with this method of junction formation. One problem is lattice damage resulting from collisions between the ions and the lattice atoms. However, most of this damage can be removed in Si by heating the crystal to a few hundred degrees centigrade. This process is called *annealing*. Another problem is the fact that crystal lattices have certain "open" directions through which the penetrating ions can move. For example, the $\langle 100 \rangle$ and $\langle 110 \rangle$ directions in the diamond lattice are very open. The relatively easy motion of energetic particles down an open crystal direction is called *channeling*. Since the range of implanted ions varies with the orientation of the crystal, channeling provides an extra difficulty (but also added flexibility) in device processing. Channeling can be reduced by tilting the crystal so that the beam enters at a slight angle to the open directions; alternatively, the angle of incidence of the ions can be randomized somewhat by passing the beam through an amorphous layer (e.g., SiO_2) on the surface of the crystal.

At present, ion implantation is used primarily for special-purpose doping; examples are fabrication of very thin layers, doping requiring low processing temperatures, doping with impurities which cannot be diffused, and doping involving very sharp boundaries. In these special cases, ion implantation provides a doping method which cannot be matched by other techniques.

5.2 Equilibrium Conditions

In this chapter we wish to develop both a useful mathematical description of the p-n junction and a strong qualitative understanding of its properties. There must be some compromise in these two goals, since a complete mathematical treatment would obscure the essentially simple physical features of junction operation; on the other hand, a completely qualitative description would not be useful in making calculations. The approach, therefore, will be to describe the junction mathematically while neglecting small effects which add little to the basic solution. In Section 5.6 we shall include several deviations from the simple theory.

The mathematics of p-n junctions is greatly simplified for the case of the *step junction*, which has uniform p doping on one side of a sharp junction and uniform n doping on the other side. This model represents the alloyed junction quite well; the diffused junction, however, is actually *graded* ($N_d - N_a$ varies over a significant distance on either side of the junction). After the basic ideas of junction theory are explored for the step junction, we can make the appropriate corrections to extend the theory to the graded junction. In these discussions we shall assume one-dimensional current flow in samples of uniform cross-sectional area.

In this section we investigate the properties of the step junction at equilibrium (i.e., with no external excitation and no net currents flowing in the device). We shall find that the difference in doping on each side of the junction causes a potential difference between the two types of material. This is a reasonable result, since we would expect some charge transfer because of diffusion between the p material (many holes) and the n material (many electrons). In addition, we shall find that there are four components of current which flow across the junction due to the drift and diffusion of electrons and holes. These four components combine to give zero net current for the equilibrium case. However, the application of bias to the junction increases some of these current components with respect to others, giving net current flow. If we understand the nature of these four current components, a sound view of p-n junction operation, with or without bias, will follow.

5.2.1 The Contact Potential.

Let us consider separate regions of p- and n-type semiconductor material, brought together to form a junction (Fig. 5-8). This is not a practical way of forming a device, but it does allow us to discover the requirements of equilibrium at a junction. Before they are joined, the n material has a large density of electrons and few holes, whereas the converse is true for the p material. Upon joining the two regions (Fig. 5-8b), we expect diffusion of carriers to take place because of the large carrier density gradients at the junction. Thus holes diffuse from the p side into the n side, and electrons diffuse from n to p. The resulting diffusion current cannot build up indefinitely, however, because an opposing electric field is created at the junction (Fig. 5-8b). If the two regions were boxes of red air molecules and green molecules (perhaps due to appropriate types of pollution), eventually there would be a homogeneous mixture of the two after the boxes were joined. This cannot occur in the case of the charged particles in a p-n junction because of the development of space charge and the electric field $\mathscr{E}$. If we consider that electrons diffusing from n to p leave behind uncompensated† donor ions (N_d^+) in the n material, and holes leaving the p region leave behind uncompensated acceptors (N_a^-), it is easy to visualize

---

†We recall that neutrality is maintained in the bulk materials of Fig. 5-8a by the presence of one electron for each ionized donor $(n = N_d^+)$ in the n material and one hole for each ionized acceptor $(p = N_a^-)$ in the p material (neglecting minority carrier densities). Thus, if electrons leave n, some of the positive donor ions near the junction are left uncompensated, as in Fig. 5-8b. The donors and acceptors are fixed in the lattice, in contrast to the mobile electrons and holes.

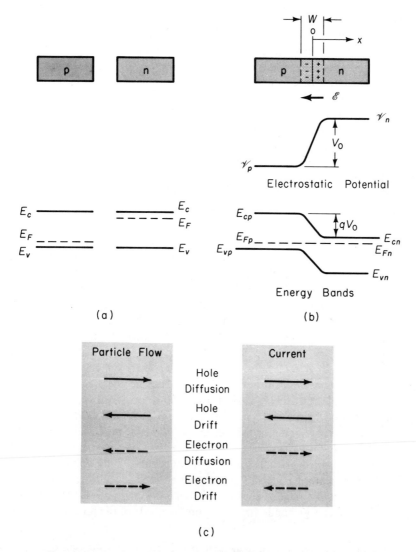

FIGURE 5-8. Properties of an equilibrium p-n junction: (a) isolated, neutral regions of p-type and n-type material and energy bands for the isolated regions; (b) junction, showing space charge in the transition region W, the resulting electric field $\mathscr{E}$ and contact potential V_0, and the separation of the energy bands; (c) directions of the four components of particle flow within the transition region, and the resulting current directions.

the development of a region of positive space charge near the n side of the junction and negative charge near the p side. The resulting electric field is directed from the positive charge toward the negative charge. Thus $\mathscr{E}$ is in the direction opposite to that of diffusion current for each type of carrier (recall electron current is opposite to the direction of electron flow). Therefore, the field creates a drift component of current from n to p, opposing the diffusion current (Fig. 5-8c).

Since we know that no *net* current can flow across the junction at equilibrium, the current due to the drift of carriers in the $\mathscr{E}$ field must exactly cancel the diffusion current. Furthermore, since there can be no net buildup of electrons or holes on either side as a function of time, the drift and diffusion currents must cancel for *each* type of carrier

(5-1a) $$J_p \,(\text{drift}) + J_p \,(\text{diff.}) = 0$$

(5-1b) $$J_n \,(\text{drift}) + J_n \,(\text{diff.}) = 0$$

Therefore, the electric field $\mathscr{E}$ builds up to the point where the net current is zero at equilibrium. The electric field appears in some region W about the junction, and there is an equilibrium potential difference V_0 across W. In the electrostatic potential diagram of Fig. 5-8b, there is a gradient in potential in the direction opposite to $\mathscr{E}$, in accordance with the fundamental relation $\mathscr{E}(x) = -d\mathscr{V}(x)/dx.$† We assume the electric field is zero in the neutral regions outside W. Thus there is a constant potential $\mathscr{V}_n$ in the neutral n material, a constant $\mathscr{V}_p$ in the neutral p material, and a potential difference $V_0 = \mathscr{V}_n - \mathscr{V}_p$ between the two. The region W is called the *transition region*,‡ and the potential difference V_0 is called the *contact potential*. The contact potential appearing across W is a *built-in* potential barrier, in that it is necessary to the maintenance of equilibrium at the junction; it does not imply any external potential. Indeed, the contact potential cannot be measured by placing a voltmeter across the device, because new contact potentials are formed at each probe, just canceling V_0. By definition V_0 is an equilibrium quantity, and no net current can result from it.

The contact potential separates the bands as in Fig. 5-8b; the valence and conduction energy bands are higher on the p side of the junction than on

---

†When we write $\mathscr{E}(x)$, we refer to the value of $\mathscr{E}$ as computed in the x-direction. This value will of course be negative, since it is directed opposite to the true direction of $\mathscr{E}$ as shown in Fig. 5-8b.

‡Other names for this region are the *space charge region*, since space charge exists within W while neutrality is maintained outside this region, and the *depletion region*, since W is almost depleted of carriers compared with the rest of the crystal. The contact potential V_0 is also called the *diffusion potential*, since it represents a potential barrier which diffusing carriers must surmount in going from one side of the junction to the other.

the n side by the amount qV_0.† The separation of the bands at equilibrium is just that required to make the Fermi level constant throughout the device. We shall show that this is the case in the following section. However, we could predict the lack of spatial variation of the Fermi level from thermodynamic arguments. In addition, we note from Eqs. (4-32) that any gradient in the quasi-Fermi level implies a net current. Since $E_F = F_n = F_p$ at equilibrium, and since the net current must be zero, we conclude that E_F must be constant across the junction. Thus if we know the band diagram, including E_F, for each separate material (Fig. 5-8a), we can find the band separation for the junction simply by drawing a diagram such as Fig. 5-8b with the Fermi levels aligned (Prob. 5.3).

To obtain a quantitative relationship between V_0 and the doping densities on each side of the junction, we must use the requirements for equilibrium in the drift and diffusion current equations. The drift and diffusion components of the hole current just cancel at equilibrium

$$(5-2) \qquad J_p(x) = q \left[\underset{\text{drift}}{\mu_p p(x) \mathscr{E}(x)} - \underset{\text{diffusion}}{D_p \frac{dp(x)}{dx}} \right] = 0$$

This equation can be rearranged to obtain

$$(5-3) \qquad \frac{\mu_p}{D_p} \mathscr{E}(x) = \frac{1}{p(x)} \frac{dp(x)}{dx}$$

where the x-direction is arbitrarily taken from p to n. The electric field can be written in terms of the gradient in the potential, $\mathscr{E}(x) = -d\mathscr{V}(x)/dx$, so that Eq. (5-3) becomes

$$(5-4) \qquad -\frac{q}{kT} \frac{d\mathscr{V}(x)}{dx} = \frac{1}{p(x)} \frac{dp(x)}{dx}$$

with the use of the Einstein relation for μ_p/D_p. This equation can be solved by integration over the appropriate limits. In this case we are interested in the potential on either side of the junction, $\mathscr{V}_p$ and $\mathscr{V}_n$, and the hole density just at the edge of the transition region on either side, p_p and p_n. For a step junction it is reasonable to take the electron and hole densities in the neutral regions outside the transition region as their equilibrium values. Since we have assumed a one-dimensional geometry, p and $\mathscr{V}$ can be taken reasonably

---

†The electron energy diagram of Fig. 5-8b is related to the electrostatic potential diagram by $-q$, the negative charge on the electron. Since $\mathscr{V}_n$ is a higher potential than $\mathscr{V}_p$ by the amount V_0, the electron energies on the n side are *lower* than those on the p side by qV_0.

as functions of x only. Integration of Eq. (5-4) gives

(5-5)
$$-\frac{q}{kT}\int_{\mathscr{V}_p}^{\mathscr{V}_n} d\mathscr{V} = \int_{p_p}^{p_n} \frac{1}{p}\, dp$$

(5-6)
$$-\frac{q}{kT}(\mathscr{V}_n - \mathscr{V}_p) = \ln p_n - \ln p_p = \ln \frac{p_n}{p_p}$$

The potential difference $\mathscr{V}_n - \mathscr{V}_p$ is the contact potential V_0 (Fig. 5-8b). Thus we can write V_0 in terms of the equilibrium hole densities on either side of the junction

(5-7)
$$V_0 = \frac{kT}{q} \ln \frac{p_p}{p_n}$$

If we consider the step junction to be made up of material with N_a acceptors/cm^3 on the p side and a concentration of N_d donors on the n side, we can write Eq. (5-7) as

(5-8)
$$V_0 = \frac{kT}{q} \ln \frac{N_a}{n_i^2/N_d} = \frac{kT}{q} \ln \frac{N_a N_d}{n_i^2}$$

by considering the majority carrier density to be the doping density on each side.

Another useful form of Eq. (5-7) is

(5-9)
$$\frac{p_p}{p_n} = e^{qV_0/kT}$$

By using the equilibrium condition $p_p n_p = n_i^2 = p_n n_n$, we can extend Eq. (5-9) to include the electron densities on either side of the junction

(5-10)
$$\frac{p_p}{p_n} = \frac{n_n}{n_p} = e^{qV_0/kT}$$

This relation will be very valuable in calculation of the I–V characteristics of the junction.

5.2.2 Equilibrium Fermi Levels.

We have observed that the Fermi level must be constant throughout the device at equilibrium. This observation can be demonstrated easily using the results of the previous section. Since we have assumed that p_n and p_p are given by their equilibrium values outside the transition region, we can write Eq. (5-9) in terms of the basic definitions of these quantities using Eq. (3-19)

(5-11)
$$\frac{p_p}{p_n} = e^{qV_0/kT} = \frac{N_v e^{-(E_{Fp}-E_{vp})/kT}}{N_v e^{-(E_{Fn}-E_{vn})kT}}$$

(5-12)
$$e^{qV_0/kT} = e^{(E_{Fn}-E_{Fp})/kT} e^{(E_{vp}-E_{vn})/kT}$$

The Fermi level and valence band energies are written with subscripts to indicate the p side and the n side of the junction.

From Fig. 5-8b the energy bands on either side of the junction are separated by the contact potential V_0 times the electronic charge q; thus the energy difference $E_{vp} - E_{vn}$ is just qV_0. Equation (5-12) indicates that the Fermi levels on either side of the junction are indeed equal at equilibrium $(E_{Fn} - E_{Fp} = 0)$. When bias is applied to the junction, the potential barrier is raised or lowered from the value of the contact potential, and the Fermi levels on either side of the junction are shifted with respect to each other.

5.2.3 Space Charge at a Junction.
Within the transition region, electrons and holes are in transit from one side of the junction to the other. Some electrons diffuse from n to p, and some are swept by the electric field from p to n (and conversely for holes); there are, however, very few carriers within the transition region at any given time, since the electric field serves to sweep out carriers which have wandered into W. To a good approximation, we can consider the space charge within the transition region as due only to the uncompensated donor and acceptor ions. The charge density within W is plotted in Fig. 5-9b. Neglecting carriers within the space charge region, the charge density on the n side is just q times the density of donor ions N_d, and the negative charge density on the p side is $-q$ times the density of acceptors N_a.

Since the dipole about the junction must have an equal number of charges on either side $(Q_+ = |Q_-|)$,† the transition region may extend into the p and n regions unequally, depending on the relative doping of the two sides. For example, if the p side is more lightly doped than the n side $(N_a < N_d)$, the space charge region must extend further into the p material than into the n, to "uncover" an equivalent amount of charge. For a sample of cross-sectional area A, the total uncompensated charge on either side of the junction is

(5-13)
$$qAx_{p0}N_a = qAx_{n0}N_d$$

---

†A simple way of remembering this equal charge requirement is to note that electric flux lines must begin and end on charges of opposite sign. Therefore, if Q_+ and Q_- were not of equal magnitude, the electric field would not be contained within W but would extend farther into the p or n regions until the enclosed charges became equal.

(a)

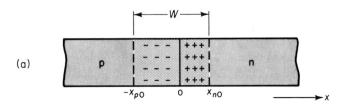

(b)

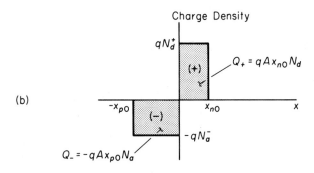

(c)

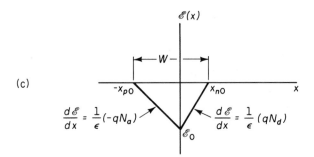

FIGURE 5-9. Space charge and electric field distribution within the transition region of a p-n junction with $N_d > N_a$: (a) the transition region, with $x = 0$ defined at the metallurgical junction; (b) charge density within the transition region, neglecting the free carriers; (c) the electric field distribution, where the reference direction for $\mathscr{E}$ is arbitrarily taken as the $+x$ direction.

where x_{p0} is the penetration of the space charge region into the p material, and x_{n0} is the penetration into n. The total width of the transition region (W) is the sum of x_{p0} and x_{n0}.

To calculate the electric field distribution within the transition region, we begin with the point form of *Gauss's law*, which relates the gradient of the electric field to the local space charge at any point x

$$(5\text{-}14) \qquad \frac{d\mathscr{E}(x)}{dx} = \frac{q}{\epsilon}(p - n + N_d^+ - N_a^-)$$

This equation is greatly simplified within the transition region if we neglect the contribution of the carriers $(p-n)$ to the space charge. With this approximation we have two regions of constant space charge

$$(5\text{-}15a) \qquad \frac{d\mathscr{E}}{dx} = \frac{q}{\epsilon}N_d, \qquad 0 < x < x_{n0}$$

$$(5\text{-}15b) \qquad \frac{d\mathscr{E}}{dx} = -\frac{q}{\epsilon}N_a, \qquad -x_{p0} < x < 0$$

assuming complete ionization of the impurities ($N_d^+ = N_d$, and $N_a^- = N_a$). We can see from these two equations that a plot of $\mathscr{E}(x)$ vs. x within the transition region has two slopes, positive ($\mathscr{E}$ increasing with x) on the n side and negative ($\mathscr{E}$ becoming more negative as x increases) on the p side. There is some maximum value of the field $\mathscr{E}_0$ at $x = 0$ (the metallurgical junction between the p and n materials), and $\mathscr{E}(x)$ is everywhere negative within the transition region (Fig. 5-9c). These conclusions come from Gauss's law, but we could predict the qualitative features of Fig. 5-9 without equations. We expect the electric field $\mathscr{E}(x)$ to be negative throughout W, since we know that the $\mathscr{E}$ field actually points in the $-x$-direction, from n to p (i.e., from the positive charges of the transition region dipole toward the negative charges). The electric field is assumed to go to zero at the edges of the transition region, since we are neglecting any small $\mathscr{E}$ field in the neutral n or p regions. Finally, there must be a maximum $\mathscr{E}_0$ at the junction, since this point is between the charges Q_+ and Q_- on either side of the transition region. All the electric flux lines pass through the $x = 0$ plane, so this is the obvious point of maximum electric field.

The value of $\mathscr{E}_0$ can be found by integrating either part of Eq. (5-15) with appropriate limits (see Fig. 5-9c in choosing the limits of integration).

$$(5\text{-}16a) \qquad \int_{\mathscr{E}_0}^{0} d\mathscr{E} = \frac{q}{\epsilon}N_d \int_{0}^{x_{n0}} dx, \qquad 0 < x < x_{n0}$$

$$(5\text{-}16b) \qquad \int_{0}^{\mathscr{E}_0} d\mathscr{E} = -\frac{q}{\epsilon}N_a \int_{-x_{p0}}^{0} dx, \qquad -x_{p0} < x < 0$$

Therefore, the maximum value of the electric field is

$$(5\text{-}17) \qquad \mathscr{E}_0 = -\frac{q}{\epsilon} N_d x_{n0} = -\frac{q}{\epsilon} N_a x_{p0}$$

It is simple to relate the electric field to the contact potential V_0, since the $\mathscr{E}$ field at any x is the negative of the potential gradient at that point

$$(5\text{-}18) \qquad \mathscr{E}(x) = -\frac{d\mathscr{V}(x)}{dx}, \qquad \text{or } -V_0 = \int_{-x_{p0}}^{x_{n0}} \mathscr{E}(x)\, dx$$

Thus the negative of the contact potential is simply the area under the $\mathscr{E}(x)$ vs. x triangle. This relates the contact potential to the width of the depletion region

$$(5\text{-}19) \qquad V_0 = -\frac{1}{2}\mathscr{E}_0 W = \frac{1}{2}\frac{q}{\epsilon} N_d x_{n0} W$$

Since the balance of charge requirement is $x_{n0}N_d = x_{p0}N_a$, and W is simply $x_{p0} + x_{n0}$, we can write $x_{n0} = WN_a/(N_a + N_d)$ in Eq. (5-19)

$$(5\text{-}20) \qquad V_0 = \frac{1}{2}\frac{q}{\epsilon}\frac{N_a N_d}{N_a + N_d} W^2$$

By solving for W, we have an expression for the width of the transition region in terms of the contact potential, the doping densities, and known constants q and ϵ.

$$(5\text{-}21) \qquad W = \left[\frac{2\epsilon V_0}{q}\left(\frac{N_a + N_d}{N_a N_d}\right)\right]^{1/2} = \left[\frac{2\epsilon V_0}{q}\left(\frac{1}{N_a} + \frac{1}{N_d}\right)\right]^{1/2}$$

There are several useful variations of Eq. (5-21); for example, V_0 can be written in terms of the doping densities with the aid of Eq. (5-8)

$$(5\text{-}22) \qquad W = \left[\frac{2\epsilon kT}{q^2}\left(\ln\frac{N_a N_d}{n_i^2}\right)\left(\frac{1}{N_a} + \frac{1}{N_d}\right)\right]^{1/2}$$

We can also calculate the penetration of the transition region into the n and p materials

$$(5\text{-}23a) \qquad x_{p0} = \frac{WN_d}{N_a + N_d} = \frac{W}{1 + N_a/N_d} = \left\{\frac{2\epsilon V_0}{q}\left[\frac{N_d}{N_a(N_a + N_d)}\right]\right\}^{1/2}$$

$$(5\text{-}23b) \qquad x_{n0} = \frac{WN_a}{N_a + N_d} = \frac{W}{1 + N_d/N_a} = \left\{\frac{2\epsilon V_0}{q}\left[\frac{N_a}{N_d(N_a + N_d)}\right]\right\}^{1/2}$$

As expected, Eqs. (5-23) predict that the transition region extends farther into the side with the lighter doping. For example, if $N_a \ll N_d$, x_{p0} is large compared with x_{n0}. This agrees with our qualitative argument that a deep penetration is necessary in lightly doped material to "uncover" the same amount of space charge as for a short penetration into heavily doped material.

Another important result of Eq. (5-21) is that the transition width W varies as the square root of the potential across the region. In the derivation to this point, we have considered only the equilibrium contact potential V_0. In the next section we shall see that an applied voltage can increase or decrease the potential across the transition region by aiding or opposing the equilibrium electric field. Therefore, Eq. (5-21) predicts that an applied voltage will increase or decrease the width of the transition region as well.

5.3 Forward- and Reverse-Biased Junctions; Steady State Conditions

One useful feature of a p-n junction is that current flows quite freely in the p to n direction when the p region has a positive external voltage bias relative to n (forward bias and forward current), whereas virtually no current flows when p is made negative relative to n (reverse bias and reverse current). This asymmetry of the current flow makes the p-n junction diode very useful as a *rectifier*. As an example of rectification, suppose a sinusoidal a-c generator is placed in series with a resistor and a diode which passes current only in one direction. The resulting current through the resistor will reflect only half of the sinusoidal signal—for example, only the positive half-cycles. The rectified sine wave has an average value and can be used, for example, to charge a storage battery; on the other hand, the input sine wave has no average value. Diode rectifiers are useful in many applications in electronic circuits, particularly in "waveshaping" (making use of the nonlinear nature of the diode to change the shape of time-varying signals). We shall discuss p-n junction rectifiers further in Chapter 6.

While rectification is an important application, it is only the beginning of a host of uses for the biased junction. As we shall see in Chapter 6, biased p-n junctions can be used as voltage-variable capacitors, photocells, light emitters, and many more devices which are basic to modern electronics. Two or more junctions can be used to form transistors and controlled switches.

In this section we begin with a qualitative description of current flow in a biased junction. With the background of the previous section, the basic features of current flow are relatively simple to understand, and these qualitative concepts form the basis for the analytical description of forward and reverse currents in a junction.

5.3.1 Qualitative Description of Current Flow at a Junction.

We assume an applied voltage bias V appears across the transition region of the junction rather than in the neutral n and p regions. Of course, there will be some voltage drop in the neutral material, if a current flows through it. But in most p-n junction devices, the length of each region is small compared with its area, and the doping is usually moderate to heavy; thus the resistance is small in each neutral region, and only a small voltage drop can be maintained outside the space charge (transition) region. For almost all calculations it is valid to assume that an applied voltage appears entirely across the transition region. We shall take V to be positive when the external bias is positive on the p side relative to the n side.

Since an applied voltage changes the electrostatic potential barrier and thus the electric field within the transition region, we would expect changes in the various components of current at the junction (Fig. 5-10). In addition, the separation of the energy bands is affected by the applied bias, along with the width of the depletion region. Let us begin by examining qualitatively the effects of bias on the important features of the junction.

The *electrostatic potential barrier* at the junction is lowered by a forward bias V_f from the equilibrium contact potential V_0 to the smaller value $V_0 - V_f$. This lowering of the potential barrier occurs because a forward bias (p positive with respect to n) raises the electrostatic potential on the p side relative to the n side. For a reverse bias ($V = -V_r$) the opposite occurs; the electrostatic potential of the p side is depressed relative to the n side, and the potential barrier at the junction becomes larger ($V_0 + V_r$).

The *electric field* within the transition region can be deduced from the potential barrier. Alternatively, we notice that the field decreases with forward bias, since the applied electric field opposes the built-in field. With reverse bias the field at the junction is increased by the applied field, which is in the same direction as the equilibrium field.

The change in electric field at the junction calls for a change in the *transition region width* W, since it is still necessary that a proper number of positive and negative charges (in the form of uncompensated donor and acceptor ions) be exposed for a given value of the $\mathscr{E}$ field. Thus we would expect the width W to decrease under forward bias (smaller $\mathscr{E}$, fewer uncompensated charges) and to increase under reverse bias. Equations (5-21) and (5-23) can be used to calculate W, x_{p0}, and x_{n0} if V_0 is replaced by the new barrier height $V_0 - V$.

The *separation of the energy bands* is a direct function of the electrostatic potential barrier at the junction. The height of the electron energy barrier is simply the electronic charge q times the height of the electrostatic potential barrier. Thus the bands are separated less $[q(V_0 - V_f)]$ under forward bias than at equilibrium, and more $[q(V_0 + V_r)]$ under reverse bias. We assume the Fermi level deep inside each neutral region is essentially the equilibrium

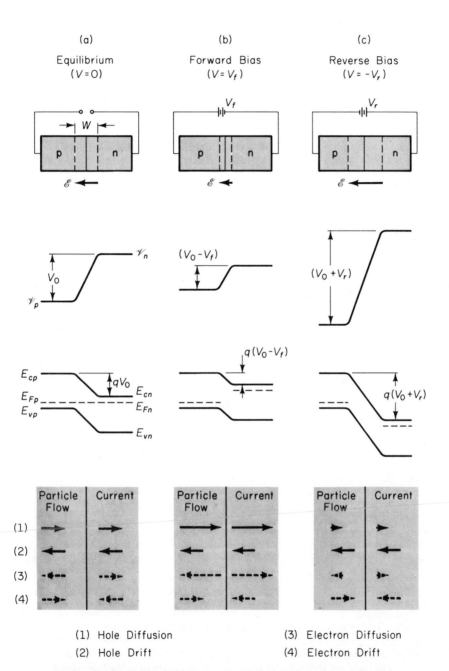

(a)
Equilibrium
$(V = 0)$

(b)
Forward Bias
$(V = V_f)$

(c)
Reverse Bias
$(V = -V_r)$

(1) Hole Diffusion   (3) Electron Diffusion
(2) Hole Drift   (4) Electron Drift

FIGURE 5-10. Effects of bias at a p-n junction; transition region width and electric field, electrostatic potential, energy band diagram, and particle flow and current directions within W for (a) equilibrium, (b) forward bias, and (c) reverse bias.

value (we shall return to this assumption later); therefore, the shifting of the energy bands under bias implies a separation of the Fermi levels on either side of the junction. Under forward bias, the Fermi level on the n side E_{F_n} is above E_{F_p} by the energy qV_f; for reverse bias, E_{F_p} is qV_r joules higher than E_{F_n}. *In energy units of electron volts, the Fermi levels in the two neutral regions are separated by the applied voltage.*

The *diffusion current* is composed of majority carrier electrons on the n side surmounting the potential energy barrier to diffuse to the p side, and holes surmounting their barrier from p to n.† There is a distribution of energies for electrons in the n-side conduction band (Fig. 3-16), and some electrons in the high-energy "tail" of the distribution have enough energy to diffuse from n to p at equilibrium in spite of the barrier. With forward bias, however, the barrier is lowered (to $V_0 - V_f$), and many more electrons in the n-side conduction band have sufficient energy to diffuse from n to p over the smaller barrier. Therefore, the electron diffusion current can be quite large with forward bias. Similarly, more holes can diffuse from p to n under forward bias because of the lowered barrier. For reverse bias the barrier becomes so large ($V_0 + V_r$) that virtually no electrons in the n-side conduction band or holes in the p-side valence band have enough energy to surmount it. Therefore, the diffusion current is usually negligible for reverse bias.

The *drift current* is relatively insensitive to the height of the potential barrier. This sounds strange at first, since we normally think in terms of material with ample carriers, and therefore we expect drift current to be simply proportional to the applied field. The reason for this apparent anomaly is the fact that the drift current is limited *not* by *how fast* carriers are swept down the barrier, *but* rather *how often*. For example, minority carrier electrons on the p side which wander into the transition region will be swept down the barrier by the $\mathscr{E}$ field, giving rise to the electron component of drift current. However, this current is small not because of the size of the barrier, but because there are very few minority electrons in the p side to participate. Every electron on the p side which diffuses to the transition region will be swept down the potential energy hill, whether the hill is large or small. The electron drift current does not depend on how fast an individual electron is swept from p to n, but rather on how many electrons are swept down the

---

†Remember that the potential energy barriers for electrons and holes are directed oppositely. The barrier for electrons is apparent from the energy band diagram, which is always drawn for electron energies. For holes, the potential energy barrier at the junction has the same shape as the electrostatic potential barrier (the conversion factor between electrostatic potential and hole energy is $+q$). A simple check of these two barrier directions can be made by asking the directions in which carriers are swept by the $\mathscr{E}$ field within the transition region—a hole is swept in the direction of $\mathscr{E}$, from n to p (swept down the potential "hill" for holes); an electron is swept opposite to $\mathscr{E}$, from p to n (swept down the potential energy "hill" for electrons).

barrier per second. Similar comments apply regarding the drift of minority holes from the n side to the p side of the junction. To a good approximation, therefore, the electron and hole drift currents at the junction are independent of the applied voltage.

The supply of minority carriers on each side of the junction required to participate in the drift component of current is generated by thermal excitation of electron-hole pairs. For example, an EHP created near the junction on the p side provides a minority electron in the p material. If the EHP is generated within a diffusion length L_n of the transition region, this electron can diffuse to the junction and be swept down the barrier to the n side. The resulting current due to drift of generated carriers across the junction is commonly called the *generation current* since its magnitude depends entirely on the rate of generation of EHP's. As we shall discuss later, this generation current can be increased greatly by optical excitation of EHP's near the junction (the p-n junction *photodiode*).

The *total current* crossing the junction is composed of the sum of the diffusion and drift components. As Fig. 5-10 indicates, the electron and hole diffusion currents are both directed from p to n (although the particle flow directions are opposite to each other), and the drift currents are from n to p. The *net* current crossing the junction is zero at equilibrium, since the drift and diffusion components cancel for each type of carrier (the equilibrium electron and hole components need not be equal, as in Fig. 5-10, as long as the net hole current and the net electron current are each zero). Under reverse bias, both diffusion components are negligible because of the large barrier at the junction, and the only current is the relatively small (and essentially voltage-independent) generation current from n to p. This generation current is shown in Fig. 5-11, in a sketch of a typical *I–V* plot for a p-n junction. In this figure the positive direction for the current I is taken from p to n, and the applied voltage V is positive when the positive battery terminal is connected to p and the negative terminal to n. The only current flowing in this p-n junction diode for negative V is the small current I (gen.) due to carriers generated in the transition region or minority carriers which diffuse to the junction and are collected. The current at $V = 0$ (equilibrium) is zero since the generation and diffusion currents cancel[†]

$$(5\text{-}24) \qquad I = I\,(\text{diff.}) - |\,I\,(\text{gen.})\,| = 0, \qquad \text{for } V = 0$$

[†]The total current I is the sum of the generation and diffusion components. However, these components are oppositely directed, I (diff.) being positive and I (gen.) being negative for the chosen reference direction. To avoid confusion of signs, we use here the magnitude of the drift current $|\,I\,(\text{gen.})\,|$ and include its negative sign in Eq. (5-24). Thus when we write the term $-|\,I\,(\text{gen.})\,|$, there is no doubt that the generation current is in the negative current direction. This approach emphasizes the fact that the two components of current add with opposite signs to give the total current.

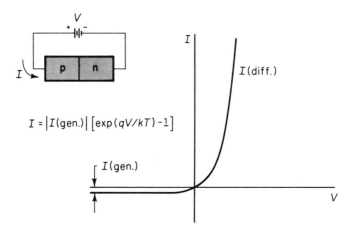

FIGURE 5-11. *I-V* characteristic of a p-n junction.

As we shall see in the next section, an applied forward bias $V = V_f$ increases the probability that a carrier can diffuse across the junction, by the factor $\exp(qV_f/kT)$. Thus the diffusion current under forward bias is given by its equilibrium value multiplied by $\exp(qV/kT)$; similarly, for reverse bias the diffusion current is the equilibrium value reduced by the same factor, with $V = -V_r$. Since the equilibrium diffusion current is equal in magnitude to $|I(\text{gen.})|$, the diffusion current with applied bias is simply $|I(\text{gen.})| \exp(qV/kT)$. The total current I is then the diffusion current minus the absolute value of the generation current

$$(5\text{-}25) \qquad I = |I(\text{gen.})|(e^{qV/kT} - 1)$$

In Eq. (5-25) the applied voltage V can be positive or negative, $V = V_f$ or $V = -V_r$. When V is positive and greater than a few kT/q ($kT/q = 0.026$ V at room temperature), the exponential term is much greater than unity. The current thus increases exponentially with forward bias. When V is negative (reverse bias), the exponential term approaches zero and the current is $-|I(\text{gen.})|$, which is in the n to p (negative) direction. This negative generation current is also called the *reverse saturation current*. The striking feature of Fig. 5-11 is the nonlinearity of the *I–V* characteristic. Current flows relatively freely in the forward direction of the diode, but almost no current flows in the reverse direction.

5.3.2 Carrier Injection. From the discussion in the previous section we expect the minority carrier density on each side of a p-n junction to vary with the applied bias because of variations in the diffusion of carriers across the junction. The equilibrium ratio of hole densities on each side

(5-26)
$$\frac{p_p}{p_n} = e^{qV_0/kT}$$

becomes with bias (Prob. 5.5)

(5-27)
$$\frac{p(-x_{p0})}{p(x_{n0})} = e^{q(V_0-V)/kT}$$

This equation is obtained by the same mathematics as Eq. (5-9), but with the altered barrier $V_0 - V$; it relates the steady state hole densities on the two sides of the transition region with either forward or reverse bias (V positive or negative). For low-level injection we can neglect changes in the majority carrier densities, which vary only slightly with bias compared with their equilibrium values. With this simplification we can write the ratio of Eq. (5-26) to (5-27) as

(5-28)
$$\frac{p(x_{n0})}{p_n} = e^{qV/kT}, \qquad taking \; p(-x_{p0}) = p_p$$

With forward bias, Eq. (5-28) suggests a greatly increased minority carrier hole density at the edge of the transition region on the n side $p(x_{n0})$ than was the case at equilibrium. Conversely, the hole density $p(x_{n0})$ under reverse bias (V negative) is reduced below the equilibrium value p_n. The exponential increase of the hole density at x_{n0} with forward bias is an example of *minority carrier injection*. As Fig. 5-12 suggests, a forward bias V results in a steady state injection of excess holes into the n region and electrons into the p region. We can easily calculate the excess hole density Δp_n at the edge of the transition region x_{n0} by subtracting the equilibrium hole density from Eq. (5-28)

(5-29)
$$\Delta p_n = p(x_{n0}) - p_n = p_n(e^{qV/kT} - 1)$$

and similarly for excess electrons on the p side

(5-30)
$$\Delta n_p = n(-x_{p0}) - n_p = n_p(e^{qV/kT} - 1)$$

From our study of diffusion of excess carriers in Section 4.4.4, we expect that injection leading to a steady density of Δp_n excess holes at x_{n0} will produce a *distribution* of excess holes in the n material. As the holes diffuse deeper into the n region, they recombine with electrons in the n material, and the resulting excess hole distribution is obtained as a solution of the diffusion equation, Eq. (4-37). If the n region is long compared with the hole diffusion length L_p, the solution is exponential, as in Eq. (4-39). Similarly, the injected electrons in the p material diffuse and recombine, giving an exponential distribution of excess electrons. For convenience, let us define two new coordinates (Fig. 5-12): Distances measured in the x-direction in the n material from x_{n0} will be designated x_n; distances in the p material measured

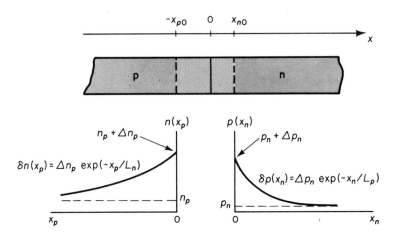

FIGURE 5-12. Minority carrier distributions on the two sides of the transition region for a forward-biased p-n junction; definitions of distances x_n and x_p measured from the transition region edges.

in the $-x$-direction with $-x_{p0}$ as the origin will be called x_p. This convention will simplify the mathematics considerably. We can write the diffusion equation as in Eq. (4-37) for each side of the junction and solve for the distributions of excess carriers (δn and δp) assuming long p and n regions

$$(5\text{-}31a) \qquad \delta n(x_p) = \Delta n_p e^{-x_p/L_n} = n_p(e^{qV/kT} - 1)e^{-x_p/L_n}$$

$$(5\text{-}31b) \qquad \delta p(x_n) = \Delta p_n e^{-x_n/L_p} = p_n(e^{qV/kT} - 1)e^{-x_n/L_p}$$

The hole diffusion current at any point x_n in the n material can be calculated from Eq. (4-43)

$$(5\text{-}32) \qquad I_p(x_n) = -qAD_p\frac{d\delta p(x_n)}{dx_n} = qA\frac{D_p}{L_p}\Delta p_n e^{-x_n/L_p} = qA\frac{D_p}{L_p}\delta p(x_n)$$

where A is the cross-sectional area of the junction. Thus the hole diffusion current at each position x_n is proportional to the excess hole density at that point.[†] The total hole current injected into the n material at the junction can be obtained simply by evaluating Eq. (5-32) at x_{n0}

$$(5\text{-}33) \qquad I_p(x_n = 0) = \frac{qAD_p}{L_p}\Delta p_n = \frac{qAD_p}{L_p}p_n(e^{qV/kT} - 1)$$

[†]With carrier injection due to bias, it is clear that the equilibrium Fermi levels cannot be used to describe carrier densities in the device. It is necessary to use the concept of quasi-Fermi levels, taking into account the spatial variations of the carrier densities (Prob. 5.6).

By a similar analysis, the injection of electrons into the p material leads to an electron current at the junction of

(5-34) $\qquad I_n(x_p = 0) = -\dfrac{qAD_n}{L_n}\Delta n_p = -\dfrac{qAD_n}{L_n}n_p(e^{qV/kT} - 1)$

The minus sign in Eq. (5-34) means that the electron current is opposite to the x_p-direction; that is, the true direction of I_n is in the $+x$-direction, adding to I_p in the total current (Fig. 5-13). If we neglect recombination in the transition region, we can consider that each injected electron reaching $-x_{p0}$ must pass through x_{n0}. Thus the total diode current I at x_{n0} can be calculated as the sum of $I_p(x_n = 0)$ and $-I_n(x_p = 0)$. If we take the $+x$-direction as the reference direction for the total current I, we must use a minus sign with $I_n(x_p)$ to account for the fact that x_p is defined in the $-x$-direction

(5-35) $\qquad I = I_p(x_n = 0) - I_n(x_p = 0) = \dfrac{qAD_p}{L_p}\Delta p_n + \dfrac{qAD_n}{L_n}\Delta n_p$

(5-36) $\qquad I = qA\left(\dfrac{D_p}{L_p}p_n + \dfrac{D_n}{L_n}n_p\right)(e^{qV/kT} - 1)$

Equation (5-36) is the *diode equation*, having the same form as the qualitative relation Eq. (5-25). Nothing in the derivation excludes the possibility that the bias voltage V can be negative; thus the diode equation describes the total current through the diode for either forward or reverse bias. We can calculate the reverse saturation current by letting $V = -V_r$

(5-37) $\qquad I = qA\left(\dfrac{D_p}{L_p}p_n + \dfrac{D_n}{L_n}n_p\right)(e^{-qV_r/kT} - 1)$

If V_r is larger than a few kT/q, the total current is just the reverse current $-|I\,(\text{gen.})|$

(5-38) $\qquad I = -qA\left(\dfrac{D_p}{L_p}p_n + \dfrac{D_n}{L_n}n_p\right) = -|I\,(\text{gen.})|$

Another simple and instructive way of calculating the total current is to consider the injected current as supplying the carriers for the excess distributions (Fig. 5-13b). For example, $I_p(x_n = 0)$ must supply enough holes per second to maintain the steady state exponential distribution $\delta p(x_n)$ as the holes recombine. The total positive charge stored in the excess carrier distribution at any instant of time is

(5-39) $\qquad Q_p = qA\displaystyle\int_0^\infty \delta p(x_n)\,dx_n = qA\,\Delta p_n \int_0^\infty e^{-x_n/L_p}\,dx_n = qAL_p\,\Delta p_n$

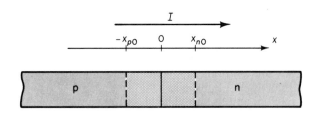

(a)

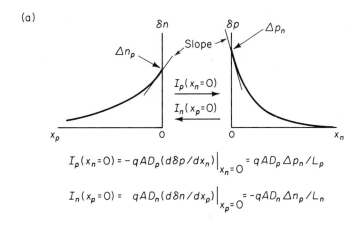

$$I_p(x_n=0) = -qAD_p(d\delta p/dx_n)\Big|_{x_n=0} = qAD_p \Delta p_n/L_p$$

$$I_n(x_p=0) = qAD_n(d\delta n/dx_p)\Big|_{x_p=0} = -qAD_n \Delta n_p/L_n$$

(b)

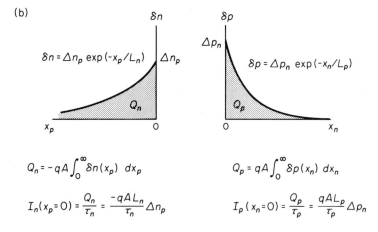

$$Q_n = -qA\int_0^\infty \delta n(x_p)\,dx_p \qquad\qquad Q_p = qA\int_0^\infty \delta p(x_n)\,dx_n$$

$$I_n(x_p=0) = \frac{Q_n}{\tau_n} = \frac{-qAL_n}{\tau_n}\Delta n_p \qquad\qquad I_p(x_n=0) = \frac{Q_p}{\tau_p} = \frac{qAL_p}{\tau_p}\Delta p_n$$

$$I = I_p(x_n=0) - I_n(x_p=0) = qA\,(D_p\Delta p_n/L_p + D_n\Delta n_p/L_n)$$

$$= qA\left(\frac{D_p p_n}{L_p} + \frac{D_n n_p}{L_n}\right)\left[\exp(qV/kT) - 1\right]$$

FIGURE 5-13. Two methods for calculating junction current from the excess minority carrier distributions: (a) diffusion currents at the edges of the transition region; (b) charge in the distributions divided by the minority carrier lifetimes.

The average lifetime of a hole in the n-type material is τ_p. Thus, on the average, this entire charge distribution recombines and must be replenished every τ_p sec. The injected hole current at $x_n = 0$ needed to maintain the distribution is simply the total charge divided by the average time of replacement

$$(5\text{-}40) \qquad I_p(x_n = 0) = \frac{Q_p}{\tau_p} = qA\frac{L_p}{\tau_p}\Delta p_n = qA\frac{D_p}{L_p}\Delta p_n$$

using $D_p/L_p = L_p/\tau_p$.

This is the same result as Eq. (5-33), which was calculated from the diffusion currents. Similarly, we can calculate the negative charge stored in the distribution $\delta n(x_p)$ and divide by τ_n to obtain the injected electron current in the p material. This method, called the *charge control approximation*, illustrates the important fact that the minority carriers injected into either side of a p-n junction diffuse into the neutral material and recombine with the majority carriers. The minority carrier current [for example, $I_p(x_n)$] decreases exponentially with distance into the neutral region. Thus several diffusion lengths away from the junction, most of the total current is carried by the majority carriers. We shall discuss this point in more detail in the following section.

In summary, we can calculate the current at a p-n junction in two ways (Fig. 5-13): (a) from the slopes of the excess minority carrier distributions at the two edges of the transition region and (b) from the steady state charge stored in each distribution. We add the hole current injected into the n material $I_p(x_n = 0)$ to the electron current injected into the p material $I_n(x_p = 0)$, after including a minus sign with $I_n(x_p)$ to conform with the conventional definition of positive current in the $+x$-direction. We are able to add these two currents because of the assumption that no recombination takes place within the transition region.† Thus we effectively have the total electron and hole current at one point in the device (x_{n0}). Since the total current must be constant throughout the device (despite variations in the current components), I as described by Eq. (5-36) is the total current at every position x in the diode.

One implication of Eq. (5-36) is that the total current at the junction is dominated by injection of carriers from the more heavily doped side into the side with lesser doping. For example, if the p material is very heavily doped and the n region is lightly doped, the minority carrier density on the p side (n_p) is negligible compared with the minority carrier density on the n side (p_n). Thus the diode equation can be approximated by injection of holes only, as in Eq. (5-33). This means that the charge stored in the minority carrier distributions is due mostly to holes on the n side (Fig. 5-14). This structure is called a p^+-n junction, where the $+$ superscript simply means

---

†This assumption will be discussed further in Section 5.6.2.

heavy doping. Another characteristic of the p^+-n or n^+-p structure is that the transition region extends primarily into the lightly doped region, as we found in the discussion of Eq. (5-23). Having one side heavily doped is a useful arrangement for many practical devices, as we shall see in our discussions of switching diodes and transistors. This type of junction is common in devices which are fabricated by counterdoping. For example, an n-type Si sample with $N_d = 10^{14}$ cm^{-3} can be used as the substrate for an alloyed or diffused junction. If the doping of the p region is greater than 10^{19} cm^{-3} (typical of alloyed junctions), the structure is definitely p^+-n, with n_p more than five orders of magnitude smaller than p_n. Since this configuration is common in device technology, we shall return to it in much of the following discussion.

In this discussion of carrier injection and minority carrier distributions, we have primarily assumed forward bias. The distributions for reverse bias can be obtained from the same equations (Fig. 5-14b), if a negative value of V is introduced. For example, if $V = -V_r$ (p negatively biased with respect to n), we can approximate Eq. (5-29) as

$$(5-41) \qquad \Delta p_n = p_n(e^{q(-V_r)/kT} - 1) \simeq -p_n, \qquad for \ V_r \gg kT/q$$

and similarly $\Delta n_p \simeq -n_p$.

Thus for a reverse bias of more than a few tenths of a volt, the minority carrier density at each edge of the transition region becomes essentially zero as the excess density approaches the negative of the equilibrium density. The excess minority carrier densities in the neutral regions are still given by Eq. (5-31), so that depletion of carriers below the equilibrium values extends approximately a diffusion length beyond each side of the transition region. This reverse-bias depletion of minority carriers can be thought of as *minority carrier extraction*, analogous to the injection of forward bias. Physically, extraction occurs because minority carriers at the edges of the depletion region are swept down the barrier at the junction to the other side and are not replaced by an opposing diffusion of carriers. For example, when holes at x_{n0} are swept across the junction to the p side by the $\mathscr{E}$ field, a gradient in the hole distribution in the n material exists, and holes in the n region diffuse toward the junction. The steady state hole distribution in the n region has the inverted exponential shape of Fig. 5-14b. It is important to remember that although the reverse saturation current occurs at the junction by drift of carriers down the barrier, this current is fed from each side by diffusion toward the junction of minority carriers in the neutral regions. The rate of carrier drift across the junction (reverse saturation current) depends on the rate at which holes arrive at x_{n0} (and electrons at x_{p0}) by diffusion from the neutral material. These minority carriers are supplied by thermal generation, and we can show (Prob. 5.7) that the expression for the

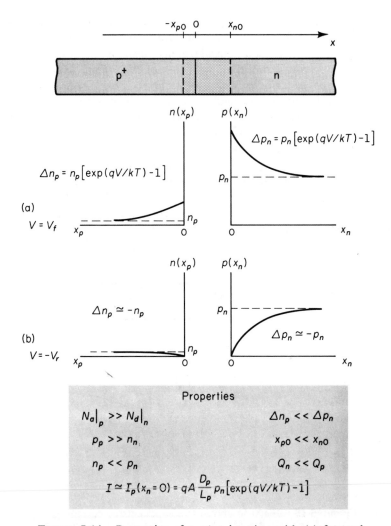

FIGURE 5-14. Properties of a p^+-n junction with (a) forward bias and (b) reverse bias.

reverse current, Eq. (5-38), represents the rate at which carriers are generated thermally within a diffusion length of each side of the transition region.

5.3.3 Minority and Majority Carrier Currents.
We have cal-culated the total diode current by considering only the minority carrier diffusion currents, evaluated at the edges of the transition region. This was

possible because minority carriers are injected into each side under forward bias and extracted from each side for reverse bias. *The drift of minority carriers can be neglected* in the neutral regions *outside W*, because the minority carrier density is small compared with that of the majority carriers. If the minority carriers contribute to the total current at all, their contribution must be through diffusion (dependent on the gradient of the carrier density) rather than through drift (dependent on the magnitude of the carrier density). Even a very small density of minority carriers can have an appreciable effect on the current if the spatial variation in the density is large.

Calculation of the majority carrier currents in the two neutral regions is simple, once we have found the minority carrier current. Since the total current I must be constant throughout the device, the majority carrier component of current is just the difference between I and the minority component (Fig. 5-15). In the p^+-n junction, the total current is equal to the hole diffusion

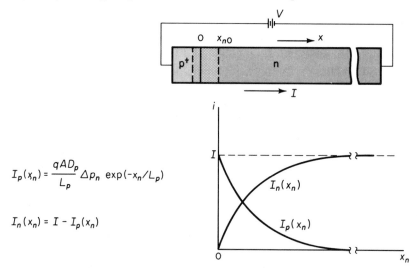

$$I_p(x_n) = \frac{qAD_p}{L_p} \Delta p_n \, \exp(-x_n/L_p)$$

$$I_n(x_n) = I - I_p(x_n)$$

FIGURE 5-15. Electron and hole components of current in the n region of a forward-biased p^+-n junction.

current at the edge of the transition region x_{n0}. Since $I_p(x_n)$ is proportional to the excess hole density at each position in the n material [Eq. (5-32)], it decreases exponentially in x_n with the decreasing $\delta p(x_n)$. Thus the electron component of current must increase appropriately with x_n to maintain the total current I. Far from the junction, the current in the n material is carried almost entirely by electrons. The physical explanation of this is that electrons must flow in from the n material (and ultimately from the negative terminal of the battery), to resupply electrons lost by recombination in the excess hole distribution near the junction. If the junction were p-n instead of p^+-n, so that electrons as well as holes were injected across the transition region, the electron current $I_n(x_n)$ would include sufficient electron flow to supply not only recombination near x_{n0}, but also injection of electrons into the p region. Of course, the flow of electrons in the n material toward the junction constitutes a current in the $+x$-direction, contributing to the total current I.

Once we have calculated the electron current $I_n(x_n)$, we can find its drift and diffusion components. The simplest approach is first to calculate the diffusion component and then subtract this from $I_n(x_n)$ to obtain the drift component. If the material outside the transition region is essentially neutral as we have assumed, we can apply the requirements of space charge neutrality (one positive charge for each negative charge). This means that the excess electron density at each point x_n must equal the excess hole density

$$(5\text{-}42) \qquad \delta n(x_n) = \delta p(x_n) = \Delta p_n e^{-x_n/L_p}$$

When holes are injected into n, resulting in an exponential hole distribution, the n-type material supplies electrons to match the excess holes, and the excess electron density has the same exponential distribution. Of course, the percentage change in the electron density is small for low-level injection, but the absolute change is important to the requirements of neutrality.

Since the excess electron distribution in x_n matches the hole distribution, the two types of carriers will diffuse together in the $+x_n$-direction (Fig. 5-16a). The resulting currents will be oppositely directed, however, because

(a)

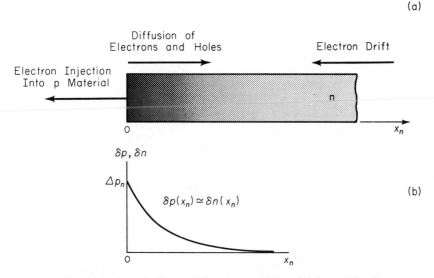

(b)

FIGURE 5-16. Drift and diffusion of carriers in the n side of a forward-biased p-n junction: (a) particle flow directions of drift and diffusion components; (b) distribution of excess electrons and holes.

of the difference in sign of the diffusing charges

(5-43)
$$I_n(x_n)_{\text{diff.}} = qAD_n\frac{d\delta n(x_n)}{dx_n} = qAD_n\frac{d\delta p(x_n)}{dx_n}$$

$$= -\frac{qAD_n}{L_p}\Delta p_n e^{-x_n/L_p} = -\frac{D_n}{D_p}I_p(x_n)$$

Thus the electrons in the exponential distribution diffuse with diffusion constant D_n and create a current component in the $-x$-direction. Like the hole diffusion current, the current of Eq. (5-43) is significant only within a few diffusion lengths of the junction. Far from the junction the drift current dominates

(5-44)
$$I_n(x_n)_{\text{drift}} = I_n(x_n) - I_n(x_n)_{\text{diff.}}$$

From Fig. 5-16 we see that the electrons drifting toward the junction from the external contact must supply electrons for (1) *recombination* with holes in the excess carrier distribution, (2) *diffusion* of electrons in the $+x$-direction within the distribution, and (3) *injection* of electrons into the p material.

For the special case of a p^+-n junction (no electron injection) and $D_n = D_p$, the electron and hole diffusion currents just cancel, and the electron drift current equals the total current everywhere in x_n (Fig. 5-17a). In

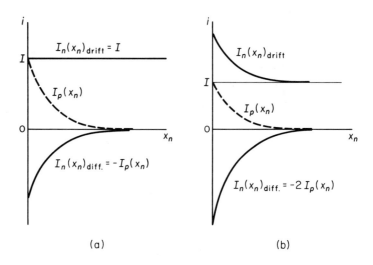

(a)   (b)

FIGURE 5-17. Drift and diffusion components of the electron current in the n region of a forward-biased p^+-n junction: (a) $D_n = D_p$; (b) $D_n = 2D_p$.

general, however, $D_n \neq D_p$ and the electron drift current near the junction
may be larger or smaller than I as required to make the total electron current
zero at the p^+-n junction (Fig. 5-17b).

Before leaving the discussion of majority carrier currents, we should note
that the electric field in the neutral regions cannot be zero as we previously
assumed; otherwise, there would be no drift currents. Thus our assumption
that all of the applied voltage appears across the transition region is not
completely accurate. On the other hand, the majority carrier densities are
usually large in the neutral regions, so that only a small $\mathscr{E}$ field is needed to
drive the drift currents. Thus the assumption that junction voltage equals
applied voltage is acceptable for most calculations.

5.4 Reverse-Bias Breakdown

We have found that a p-n junction biased in the reverse direction exhibits
a small, essentially voltage-independent saturation current. This is true until
a critical reverse bias is reached, for which *reverse breakdown* occurs (Fig.
5-18). At this critical voltage (V_{br}) the reverse current through the diode

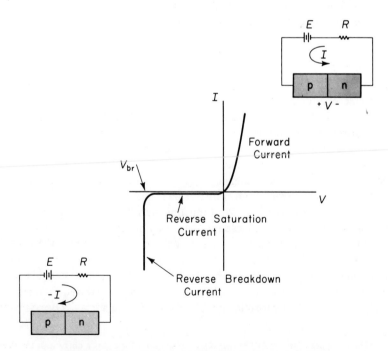

FIGURE 5-18.   Reverse breakdown in a p-n junction.

increases sharply, and relatively large currents can flow with little further increase in voltage. The existence of a critical breakdown voltage introduces almost a right-angle appearance to the reverse characteristic of most diodes.

There is nothing inherently destructive about reverse breakdown. If the current is limited to a reasonable value by the external circuit, the p-n junction can be operated in reverse breakdown as safely as in the forward-bias condition. For example, the maximum reverse current which can flow in the device of Fig. 5-18 is $(E - V_{br})/R$; the series resistance R can be chosen to limit the current to a safe level for the particular diode used. If the current is not limited externally, the junction can be damaged by excessive reverse current, which overheats the device as the maximum power rating is exceeded. It is important to remember, however, that such destruction of the device is not necessarily due to mechanisms unique to reverse breakdown; similar results occur if the device passes excessive current in the forward direction.† As we shall see in the next chapter, useful devices called *breakdown diodes* are designed to operate in the reverse breakdown regime of their characteristics.

Reverse breakdown can occur by two mechanisms, each of which requires a critical electric field in the junction transition region. The first mechanism, called the *Zener effect*, is operative at low voltages (up to a few volts reverse bias). If the breakdown occurs at higher voltages (from a few volts to thousands of volts), the mechanism is *avalanche breakdown*. We shall discuss these two mechanisms in this section.

5.4.1 Zener Breakdown. When a heavily doped junction is reverse biased, the energy bands become crossed at relatively low voltages (i.e., the n-side conduction band appears opposite the p-side valence band). As Fig. 5-19 indicates, the crossing of the bands aligns the large number of empty states in the n-side conduction band opposite the many filled states of the p-side valence band. If the barrier separating these two bands is narrow, tunneling of electrons can occur, as discussed in Section 2.4.4. Tunneling of electrons from the p-side valence band to the n-side conduction band constitutes a reverse current from n to p; this is the *Zener effect*.

The basic requirements for tunneling current are a large number of electrons separated from a large number of empty states by a narrow barrier of finite height. Since the tunneling probability depends upon the width of the barrier (d in Fig. 5-19), it is important that the metallurgical junction be sharp and the doping high, so that the transition region W extends only a very short distance from each side of the junction. These requirements can be met, for example, by forming an alloyed p region in a heavily doped n-type

---

†The dissipated power (IV) in the junction is greater for a given current in the breakdown regime than would be the case for forward bias, simply because V is greater.

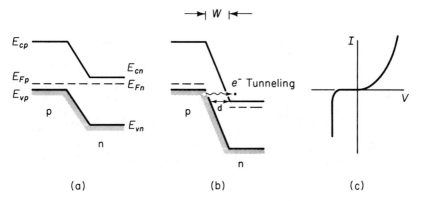

FIGURE 5-19. The Zener effect: (a) heavily doped junction at equilibrium; (b) reverse bias with electron tunneling from p to n; (c) *I-V* characteristic.

sample. If the junction is not abrupt, or if either side of the junction is lightly doped, the transition region W will be too wide for tunneling.

As the bands are crossed (at a few tenths of a volt for a heavily doped junction), the tunneling distance d may be too large for appreciable tunneling. However, d becomes smaller as the reverse bias is increased, as we can demonstrate geometrically (Prob. 5.12). This assumes that the transition region width W does not increase appreciably with reverse bias. For low voltages and heavy doping on each side of the junction, this is a good assumption. However, if Zener breakdown does not occur with reverse bias of a few volts, avalanche breakdown will become dominant.

In the simple covalent bonding model, the Zener effect can be thought of as *field ionization* of the host atoms at the junction. That is, the reverse bias of a heavily doped junction causes a large electric field within W; at a critical field strength, electrons participating in covalent bonds may be torn from the bonds by the field and accelerated to the n side of the junction. The electric field required for this type of ionization is on the order of 10^6 V/cm.

5.4.2 Avalanche Breakdown.
For lightly doped junctions electron tunneling is negligible, and instead, the breakdown mechanism involves the *impact ionization* of host atoms by energetic carriers. Normal lattice-scattering events can result in the creation of EHP's if the carrier being scattered has sufficient energy. For example, if the electric field $\mathscr{E}$ in the transition region is large, an electron entering from the p side may be accelerated to high enough kinetic energy to cause an ionizing collision with the lattice (Fig. 5-20a). A single such interaction results in *carrier multiplication*; the original electron and the generated electron are both swept to the n side of the junc-

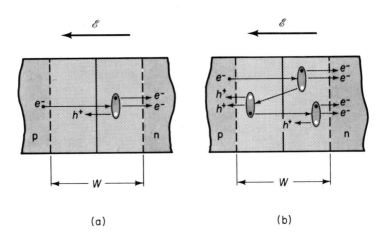

(a)   (b)

FIGURE 5-20. Electron-hole pairs created by impact ionization: (a) a single ionizing collision by an incoming electron; (b) primary, secondary, and tertiary collisions.

tion, and the generated hole is swept to the p side. The degree of multiplication can become very high if carriers generated within the transition region also have ionizing collisions with the lattice. For example, an incoming electron may have a collision with the lattice and create an EHP; each of these carriers has a chance of creating a new EHP, and each of those can also create an EHP, and so forth (Fig. 5-20b). This is an *avalanche* process, since each incoming carrier can initiate the creation of a large number of new carriers.

We can make an approximate analysis of avalanche multiplication by assuming a carrier of either type has a probability P of having an ionizing collision with the lattice while being accelerated a distance W through the transition region. Thus for n_{in} electrons entering from the p side, there will be Pn_{in} ionizing collisions and an EHP (secondary carriers) for each collision. After the Pn_{in} collisions by the primary electrons, we have the primary plus the secondary electrons, $n_{in}(1 + P)$. After a collision, each EHP moves effectively a distance W within the transition region. For example, if an EHP is created at the center of the region, the electron drifts a distance $W/2$ to n and the hole $W/2$ to p. Thus the probability that an ionizing collision will occur due to the motion of the secondary carriers is still P in this simplified model. For $n_{in}P$ secondary pairs there will be $(n_{in}P)P$ ionizing collisions and $n_{in}P^2$ tertiary pairs. Summing up the total number of electrons out of the region at n after many collisions we have

(5-45) $$n_{out} = n_{in}(1 + P + P^2 + P^3 + \ldots)$$

assuming no recombination. In a more comprehensive theory we would include recombination as well as different probabilities for ionizing collisions by electrons and holes. In our simple theory, the electron multiplication M_n is

$$(5\text{-}46) \qquad M_n = \frac{n_{\text{out}}}{n_{\text{in}}} = 1 + P + P^2 + P^3 + \cdots = \frac{1}{1 - P}$$

as can be verified by direct division. As the probability of ionization approaches unity, the carrier multiplication (and therefore the reverse current through the junction) increases without limit. Actually, the limit on the current will be dictated by the external circuit. The relation between multiplication and P was easy to write in Eq. (5-46); however, the relation of P to parameters of the junction is much more complicated. Physically, we expect the ionization probability to increase with increasing electric field, and therefore to depend on the reverse bias and the doping densities (through W). We can find more physical insight into the avalanche mechanism by examining the critical breakdown voltage as a function of the material parameters.

The critical voltage at which breakdown occurs (V_{br}) can be found from the value of $\mathscr{E}$ in the transition region necessary to impart the appropriate energy to carriers drifting in the field. A detailed analysis of this problem is complicated, but we can approximate V_{br} by finding the field necessary to impart an energy in excess of the band gap energy E_g to a carrier drifting across the transition region. Since a carrier loses energy in each collision with the lattice, we can assume that for avalanche the field must impart E_g joules to the carrier within approximately one mean free path $\bar{l}$.† Thus the magnitude of the electric field for breakdown $\mathscr{E}_{br}$ must be

$$q\mathscr{E}_{br}\bar{l} = E_g$$

(5-47)

$$\mathscr{E}_{br} = \frac{E_g}{q\bar{l}}$$

One source of heat in a junction during avalanche breakdown involves lattice collisions by carriers having energies greater than E_g. In addition to EHP generation, each collision of this type contributes thermal energy to the lattice.

The electric field is not uniform in the transition region (Fig. 5-9c); therefore, we must evaluate the critical breakdown field at the maximum

---

†The actual requirement is less stringent than this, since carriers can acquire the required E_g joules cumulatively over more than one mean free path. There are also other complications in an exact calculation of the critical field. Thus Eq. 5-47 should be considered as a qualitative argument, indicating an approximately linear relationship between the critical field and the band gap of the material.

value $\mathscr{E}_0$, given by Eq. (5-19). Substituting $V_0 - V$ for V_0 in this expression to include bias and letting $V = -V_{br}$, with $V_{br} \gg V_0$, we obtain

$$\mathscr{E}_{br} = \frac{2V_{br}}{W} = \frac{E_g}{q\bar{l}}$$

(5-48)
$$V_{br} = \frac{E_g W}{2q\bar{l}} = \frac{E_g}{2q\bar{l}}\left[\frac{2\epsilon V_{br}}{q}\left(\frac{1}{N_a} + \frac{1}{N_d}\right)\right]^{1/2}$$

$$= \frac{E_g^2 \epsilon}{2q^3\bar{l}^2}\left(\frac{1}{N_a} + \frac{1}{N_d}\right)$$

Although the analysis leading to Eq. (5-48) is oversimplified, two important physical insights can be gained from it. First, the avalanche breakdown voltage is greater for materials with wide band gaps. Thus, for example, Si devices have generally higher values of V_{br} than Ge devices. Second, the breakdown voltage is dictated by the doping density on the *most lightly doped* side of the junction. If either N_a or N_d is small, the breakdown voltage will be high. One finds experimentally that a plot of V_{br} vs. inverse doping density has a smaller slope than the linear dependence of Eq. (5-48) implies. This discrepancy is due to the simplifying assumptions of the derivation.

Measurements of carrier multiplication M in junctions near breakdown lead to an empirical relation

(5-49)
$$M = \frac{1}{1 - (V/V_{br})^n}$$

where the exponent **n** varies from about 3 to 6, depending on the type of material used for the junction.

5.5 Transient and a-c Conditions

We have considered the properties of p-n junctions under equilibrium conditions and with steady state current flow. Most of the basic concepts of junction devices can be obtained from these properties, except for the important behavior of junctions under transient or a-c conditions. Since most solid state devices are used for switching or for processing a-c signals, we cannot claim to understand p-n junctions without knowing at least the basics of time-dependent processes. Unfortunately, a complete analysis of these effects involves more mathematical manipulation than is appropriate for an introductory discussion. Basically, the problem involves solving the various current flow equations in two simultaneous variables, space and time.

We can, however, obtain the basic results for several special cases which represent typical time-dependent applications of junction devices.

In this section we investigate the important influence of excess carriers in transient and a-c problems. The switching of a diode from its forward state to its reverse state is analyzed to illustrate a typical transient problem. Finally, these concepts are applied to the case of small a-c signals to determine the equivalent capacitance of a p-n junction.

5.5.1 Time Variation of Stored Charge.

Another look at the excess carrier distributions of a p-n junction under bias (e.g., Fig. 5-12) tells us that any change in current must lead to a change of charge stored in the carrier distributions. Since time is required in building up or depleting a charge distribution, however, the stored charge must inevitably lag behind the current in a time-dependent problem. This is inherently a capacitive effect, as we shall see in Section 5.5.3.

For a proper solution of a transient problem, we must use the time-dependent continuity equations, Eqs. (4-34). We can obtain each component of the current at position x and time t from these equations; for example, from Eq. (4-34a) we can write

$$(5\text{-}50) \qquad -\frac{\partial J_p(x, t)}{\partial x} = q\frac{\delta p(x, t)}{\tau_p} + q\frac{\partial p(x, t)}{\partial t}$$

To obtain the instantaneous current density, we can integrate both sides at time t to obtain

$$(5\text{-}51) \qquad J_p(0) - J_p(x) = q\int_0^x \left[\frac{\delta p(x, t)}{\tau_p} + \frac{\partial p(x, t)}{\partial t}\right] dx$$

For injection into a long n region from a p^+ region, we can take the current at $x_n = 0$ to be all hole current, and J_p at $x_n = \infty$ to be zero (Fig. 5-15). Then the total injected current, including time variations, is

$$i(t) = i_p(x_n = 0, t) = \frac{qA}{\tau_p}\int_0^\infty \delta p(x_n, t)\, dx_n + qA\frac{\partial}{\partial t}\int_0^\infty \delta p(x_n, t)\, dx_n$$

$$(5\text{-}52) \qquad i(t) = \frac{Q_p(t)}{\tau_p} + \frac{dQ_p(t)}{dt}$$

This result indicates that the hole current injected across the p^+-n junction (and therefore approximately the total diode current) is determined by two charge storage effects: (1) the usual recombination term Q_p/τ_p in which the excess carrier distribution is replaced every τ_p sec, and (2) a charge buildup (or depletion) term dQ_p/dt which allows for the fact that the distribution of excess carriers can be increasing or decreasing in a time-dependent problem. For steady state the dQ_p/dt term is zero, and Eq. (5-52) reduces to Eq. (5-40) as expected. In fact, we could have written Eq. (5-52) intuitively rather than having obtained it from the continuity equation, since it is reasonable that the hole current injected at any given time must supply minority carriers for recombination and for whatever variations occur in the total stored charge.

We can solve for the stored charge as a function of time for a given current transient. For example, the step turn-off transient (Fig. 5-21a), in

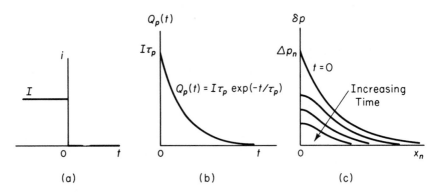

(a)　　　　　　　　　(b)　　　　　　　　　(c)

FIGURE 5-21.　Effects of a step turn-off transient in a p⁺-n diode: (a) current through the diode; (b) decay of stored charge in the n-region; (c) excess hole distribution in the n-region as a function of time during the transient.

which a current I is suddenly removed at $t = 0$, leaves the diode with stored charge. Since the excess holes in the n region must die out by recombination with the matching excess electron population, some time is required for $Q_p(t)$ to reach zero. Solving Eq. (5-52) with Laplace transforms, with $i(t > 0) = 0$ and $Q_p(0) = I\tau_p$, we obtain

$$0 = \frac{1}{\tau_p}Q_p(s) + sQ_p(s) - I\tau_p$$

$$Q_p(s) = \frac{I\tau_p}{s + 1/\tau_p}$$

(5-53)　　$$Q_p(t) = I\tau_p e^{-t/\tau_p}$$

As expected, the stored charge dies out exponentially from its initial value $I\tau_p$ with a time constant equal to the hole lifetime in the n material.

An important implication of Fig. 5-21 is that even though the current is suddenly terminated, the voltage across the junction persists until Q_p

disappears. Since the excess hole density can be related to junction voltage by formulas derived in Section 5.3.2, we can presumably solve for $v(t)$. We already know that at any time during the transient, the excess hole density at $x_n = 0$ is

(5-54) $$\Delta p_n(t) = p_n(e^{qv(t)/kT} - 1)$$

so that finding $\Delta p_n(t)$ will easily give us the transient voltage. Unfortunately, it is not simple to obtain $\Delta p_n(t)$ exactly from our expression for $Q_p(t)$. The problem is that the hole distribution does not remain in the convenient exponential form it has in steady state. As Fig. 5-21c suggests, the quantity $\delta p(x_n, t)$ becomes markedly nonexponential as the transient proceeds. For example, since the injected hole current is proportional to the gradient of the hole distribution at $x_n = 0$ (Fig. 5-13a), zero current implies zero gradient. Thus the slope of the distribution must be exactly zero at $x_n = 0$ throughout the transient.† This zero slope at the point of injection distorts the exponential distribution, particularly in the region near the junction. As time progresses in Fig. 5-21c, δp (and therefore δn) decreases as the excess electrons and holes recombine. To find the exact expression for $\delta p(x_n, t)$ during the transient would require a rather difficult solution of the time-dependent continuity equation.

An approximate solution for $v(t)$ can be obtained by assuming an exponential distribution for δp at every instant during the decay. This type of *quasi-steady state* approximation neglects distortion due to the slope requirement at $x_n = 0$ and the effects of diffusion during the transient. Thus we would expect the calculation to give rather crude results. On the other hand, such a solution can give us a feeling for the variation of junction voltage during the transient. If we take

(5-55) $$\delta p(x_n, t) = \Delta p_n(t) \, e^{-x_n/L_p}$$

we have for the stored charge at any instant

(5-56) $$Q_p(t) = qA \int_0^{\infty} \Delta p_n(t) \, e^{-x_n/L_p} \, dx_n = qAL_p \, \Delta p_n(t)$$

Relating $\Delta p_n(t)$ to $v(t)$ by Eq. (5-54) we have

(5-57) $$\Delta p_n(t) = p_n(e^{qv(t)/kT} - 1) = \frac{Q_p(t)}{qAL_p}$$

---

†We notice that, while the *magnitude* of δp cannot change instantaneously, the *slope* must go to zero immediately. This can occur in a small region near the junction with negligible redistribution of charge at $t = 0$.

Thus in the quasi-steady state approximation, the junction voltage varies according to

$$(5\text{-}58) \qquad v(t) = \frac{kT}{q} \ln\left(\frac{I\tau_p}{qAL_p p_n} e^{-t/\tau_p} + 1\right)$$

during the turn-off transient of Fig. 5-21. This analysis, while not accurate in its details, does indicate clearly that the voltage across a p-n junction cannot be changed instantaneously, and that stored charge can present a problem in a diode intended for switching application.

Many of the problems of stored charge can be reduced by designing a p^+-n diode (for example) with a very narrow n region. If the n region is shorter than a hole diffusion length, very little charge is stored. Thus, little time is required to switch the diode on and off. This type of structure, called the *narrow base diode*, is considered in Section 6.1.2 and in Prob. 6.5. The switching process can be made still faster by purposely adding recombination centers, such as Au atoms in Si, to increase the recombination rate.

5.5.2 Reverse Recovery Transient. In most switching applications a diode is switched from forward conduction to a reverse-biased state, and vice versa. The resulting stored charge transient is somewhat more complicated than for a simple turn-off transient, and therefore it requires slightly more analysis. An important result of this example is that a reverse current much larger than the normal reverse saturation current can flow in a junction during the time required for readjustment of the stored charge.

Let us assume a p^+-n junction is driven by a square wave generator which periodically switches from $+E$ to $-E$ volts (Fig. 5-22a). While E is positive the diode is forward biased, and in steady state the current I_f flows through the junction. If E is much larger than the small forward voltage of the junction, the source voltage appears almost entirely across the resistor, and the current is approximately $i = I_f \simeq E/R$. After the generator voltage is reversed ($t > 0$), the current must initially reverse to $i = -I_r \simeq -E/R$. The reason for this unusually large reverse current through the diode is that the stored charge (and hence the junction voltage) cannot be changed instantaneously. Therefore, just as the current is reversed, the junction voltage remains at the small forward-bias value it had before $t = 0$. A voltage loop equation then tells us that the large reverse current $-E/R$ must flow temporarily. While the current is negative through the junction, the slope of the $\delta p(x_n)$ distribution must be positive at $x_n = 0$.

As the stored charge is depleted from the neighborhood of the junction (Fig. 5-22b), we can find the junction voltage again from Eq. (5-54). As long as Δp_n is positive, the junction voltage $v(t)$ is positive and small; thus

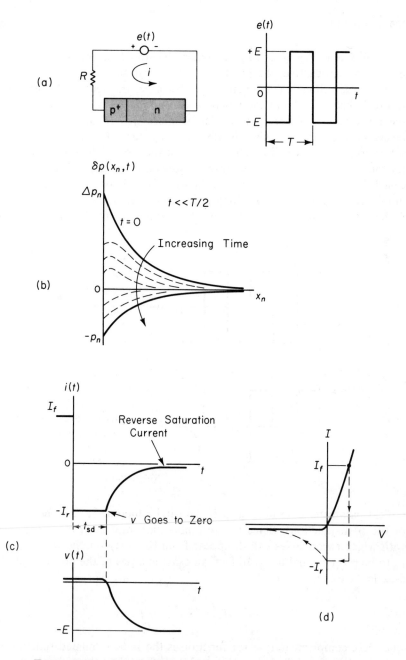

FIGURE 5-22. Storage delay time in a p⁺-n diode: (a) circuit and input square wave; (b) hole distribution in the n-region as a function of time during the transient; (c) variation of current and voltage with time; (d) sketch of transient current and voltage on the device *I-V* characteristic.

$i \simeq -E/R$ until Δp_n goes to zero. When the stored charge is depleted and Δp_n becomes negative, the junction exhibits a negative voltage. Since the reverse-bias voltage of a junction can be large, the source voltage begins to divide between R and the junction. As time proceeds, the magnitude of the reverse current becomes smaller as more of $-E$ appears across the reverse-biased junction, until finally the only current is the small reverse saturation current which is characteristic of the diode. The time t_{sd} required for the stored charge (and therefore the junction voltage) to become zero is called the *storage delay time*. This delay time is an important figure of merit in evaluating diodes for switching applications. It is usually desirable that t_{sd} be small compared with the switching times required (Fig. 5-23). The

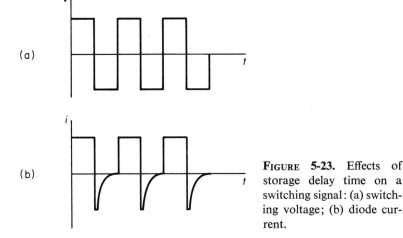

(a)

(b)

FIGURE 5-23. Effects of storage delay time on a switching signal: (a) switching voltage; (b) diode current.

critical parameter determining t_{sd} is the carrier lifetime (τ_p for the example of the p+-n junction). Since the recombination rate determines the speed with which excess holes can disappear from the n region, we would expect t_{sd} to be proportional to τ_p. In fact, an exact analysis of the problem of Fig. 5-22 leads to the result

(5-59)
$$t_{sd} = \tau_p \left(\text{erfc} \frac{I_f}{I_f + I_r} \right)^2$$

where the complementary error function is the same tabulated function we mentioned in the diffusion problem of Fig. 5-4b. For the example of Fig. 5-22, $I_f = I_r$ (in magnitude) $= E/R$. The value of erfc $\frac{1}{2}$ is approximately $\frac{1}{2}$; thus $t_{sd} \simeq \tau_p/4$ for this case. Although the exact solution leading to Eq. (5-59) is too lengthy for us to consider here, an approximate result can be obtained from the quasi-steady state assumption (Prob. 5.16).

An important result of Eq. (5-59) is that τ_p can be calculated in a straightforward way from a measurement of storage delay time. In fact, measurement of t_{sd} from an experimental arrangement such as Fig. 5-22a is a common method of measuring lifetimes. In some cases this is a more convenient technique than the photoconductive decay measurement discussed in Section 4.3.2.

As in the case of the turn-off transient of the previous section, the storage delay time can be reduced by introducing recombination centers into the diode material, thus reducing the carrier lifetimes, or by utilizing the narrow base diode configuration.

5.5.3 Capacitance of p-n Junctions.

There are basically two types of capacitance associated with a junction: (1) the *junction capacitance* due to the dipole in the transition region and (2) the *charge storage capacitance* arising from the lagging behind of voltage as current changes, due to charge storage effects.† Both of these capacitances are important, and they must be considered in designing p-n junction devices for use with time-varying signals. The junction capacitance (1) is dominant under reverse-bias conditions, and the charge storage capacitance (2) is largest when the junction is forward biased. In many applications of junction devices, the capacitance is a limiting factor in the usefulness of the device; on the other hand, there are important applications in which the capacitance discussed here can be useful in circuit applications and in providing important information about the structure of the p-n junction.

The junction capacitance of a diode is easy to visualize from the charge distribution in the transition region (Fig. 5-9). The uncompensated acceptor ions on the p side provide a negative charge, and an equal positive charge results from the ionized donors on the n side of the transition region. The capacitance of the resulting dipole is slightly more difficult to calculate than is the usual parallel plate capacitance, but we can obtain it in a few steps.

Instead of the common expression $C = |Q/v|$, which applies to capacitors in which charge is a linear function of voltage, we must use the more general definition

$$(5\text{-}60) \qquad\qquad C = \left| \frac{dQ}{dv} \right|$$

since the charge Q on each side of the transition region varies nonlinearly with the applied voltage. We can demonstrate this nonlinear dependence by reviewing the equations for the width of the transition region (W) and the

---

†The capacitance (1) above is also referred to as *transition region capacitance* or *depletion layer capacitance;* (2) is often called the *diffusion capacitance.*

resulting charge. The equilibrium value of W was found in Eq. (5-21) to be

(5-61)
$$W = \left[\frac{2\epsilon V_0}{q}\left(\frac{N_a + N_d}{N_a N_d}\right)\right]^{1/2} \quad (equilibrium)$$

Since we are dealing with the nonequilibrium case with voltage v applied, we must use the altered value of the electrostatic potential barrier $(V_0 - v)$, as discussed in relation to Fig. 5-10. The proper expression for the width of the transition region is then

(5-62)
$$W = \left[\frac{2\epsilon(V_0 - v)}{q}\left(\frac{N_a + N_d}{N_a N_d}\right)\right]^{1/2} \quad (with\ bias)$$

In this expression the applied voltage v can be either positive or negative to account for forward or reverse bias. As expected, the width of the transition region is increased for reverse bias and is decreased under forward bias. Since the uncompensated charge Q on each side of the junction varies with the transition region width, variations in the applied voltage result in corresponding variations in the charge, as required for a capacitor. The value of Q can be written in terms of the doping density and transition region width on each side of the junction (Fig. 5-9)

(5-63)
$$|Q| = qAx_{n0}N_d = qAx_{p0}N_a$$

Relating the total width of the transition region W to the individual widths x_{n0} and x_{p0} from Eq. (5-23) we have

(5-64)
$$x_{n0} = \frac{N_a}{N_a + N_d}W, \qquad x_{p0} = \frac{N_d}{N_a + N_d}W$$

and therefore the charge on each side of the dipole is

(5-65)
$$|Q| = qA\frac{N_d N_a}{N_d + N_a}W = A\left[2q\epsilon(V_0 - v)\frac{N_d N_a}{N_d + N_a}\right]^{1/2}$$

Thus the charge is indeed a nonlinear function of applied voltage. From this expression and the definition of capacitance in Eq. (5-60), we can calculate the junction capacitance C_j. Since the voltage which varies the charge in the transition region is the barrier height $(V_0 - v)$, we must take the derivative with respect to this potential difference

(5-66)
$$C_j = \left|\frac{dQ}{d(V_0 - v)}\right| = \frac{A}{2}\left[\frac{2q\epsilon}{(V_0 - v)}\frac{N_d N_a}{N_d + N_a}\right]^{1/2}$$

The quantity C_j is a *voltage-variable capacitance*, since C_j is proportional to $(V_0 - v)^{-1/2}$. There are several important applications for variable capacitors, including use in tuned circuits. The p-n junction device which makes use of the voltage-variable properties of C_j is called a *varactor*. We shall discuss this device further in Section 6.1.4.

Although the dipole charge is distributed in the transition region of the junction, the form of the parallel plate capacitor formula is obtained from the expressions for C_j and W

$$(5-67) \qquad C_j = \epsilon A\left[\frac{q}{2\epsilon(V_0 - v)}\frac{N_d N_a}{N_d + N_a}\right]^{1/2} = \frac{\epsilon A}{W}$$

In analogy with the parallel plate capacitor, the transition region width W corresponds with the plate separation of the conventional capacitor.

In the case of an asymmetrically doped junction, the transition region extends primarily into the least heavily doped side, and the capacitance is determined by only one of the doping densities. For a p$^+$-n junction, $N_a \gg N_d$ and $x_{n0} \simeq W$, while x_{p0} is negligible. The capacitance is then

$$(5-68) \qquad C_j = \frac{A}{2}\left[\frac{2q\epsilon}{V_0 - v}N_d\right]^{1/2} \qquad \text{for p}^+\text{-n}$$

It is therefore possible to obtain the doping density of the lightly doped n region from a measurement of junction capacitance. For example, in a reverse-biased junction the applied voltage $v = -v_r$ can be made much larger than the contact potential V_0, so that the latter becomes negligible. If the area of the junction can be measured, a reliable value of N_d results from a measurement of C_j. However, these equations were obtained by assuming a sharp step junction. Certain modifications must be made in the case of a graded junction (Section 5.6.4 and Prob. 5.21).

The junction capacitance dominates the reactance of a p-n junction under reverse bias; for forward bias, however, the charge storage capacitance C_s becomes dominant. To calculate the capacitance due to charge storage effects, we must perform a small-signal a-c analysis of the injected carrier distribution. To simplify the mathematics, we shall assume a p$^+$-n junction and treat the excess hole distribution in the n material in the quasi-steady state approximation. As usual, the instantaneous current through the junction is related to the instantaneous value of stored charge by Eq. (5-52). If we assume an exponential distribution of excess holes for the quasi-steady state approximation, we have

$$(5-69) \qquad Q_p(t) = qA\int_0^\infty \Delta p_n e^{-x_n/L_p}\,dx_n = qAL_p\,\Delta p_n(t)$$

Substituting this expression into Eq. (5-52) gives us the approximate relationship between the current and the excess hole density at the edge of the transition region for any instant of time

$$(5\text{-}70) \qquad i(t) = qAL_p \left[\frac{\Delta p_n(t)}{\tau_p} + \frac{d\,\Delta p_n(t)}{dt} \right]$$

As in the transient problems of the preceding sections, we can relate $\Delta p_n(t)$ to the instantaneous junction voltage by Eq. (5-54). In most cases it is a reasonable assumption that the forward-bias voltage is large compared with kT/q (recall that kT/q is only about 0.026 eV at room temperature). Thus if we approximate Eq. (5-54) by

$$(5\text{-}71) \qquad \Delta p_n(t) = p_n(e^{qv(t)/kT} - 1) \simeq p_n e^{qv(t)/kT}, \qquad v(t) \gg \frac{kT}{q}$$

and write the instantaneous voltage as a sum of d-c and a-c components

$$(5\text{-}72) \qquad v(t) = V(\text{d-c}) + v(\text{a-c})$$

we obtain for $\Delta p_n(t)$

$$(5\text{-}73) \qquad \Delta p_n(t) = p_n e^{q[V(\text{d-c}) + v(\text{a-c})]/kT} = p_n e^{[qV(\text{d-c})/kT]} e^{[qv(\text{a-c})/kT]}$$

The factor $p_n \exp[qV(\text{d-c})/kT]$ is just the value of Δp_n in steady state [i.e., with $v(\text{a-c}) = 0$]. Thus

$$(5\text{-}74) \qquad \Delta p_n(t) = \Delta p_n(\text{d-c}) e^{qv(\text{a-c})/kT}$$

If the a-c component of the junction voltage is small compared with kT/q, we can approximate the exponential term to first order†

$$(5\text{-}75) \qquad \Delta p_n(t) = \Delta p_n(\text{d-c}) \left[1 + \frac{qv(\text{a-c})}{kT} \right]$$

Substituting this expression into Eq. (5-70) gives us the relation between current and voltage

$$i(t) = I(\text{d-c}) + i(\text{a-c})$$

$$(5\text{-}76) \qquad = \frac{qAL_p}{\tau_p} \Delta p_n(\text{d-c}) \left[1 + \frac{q}{kT} v(\text{a-c}) + \frac{q\tau_p}{kT} \frac{dv(\text{a-c})}{dt} \right]$$

---

†The expansion of $\exp(x)$ is the series $1 + x + x^2/2! + x^3/3! + \cdots$. If the exponent x is much less than 1, the higher-order terms become negligible, and we can approximate the series by the zero-order and first-order terms: $e^x \simeq 1 + x$.

As expected, the steady state current is given by

(5-77)
$$I(\text{d-c}) = qAL_p \frac{\Delta p_n(\text{d-c})}{\tau_p}$$

On the other hand, the expression for the a-c component of current [the time-varying terms in Eq. (5-76)] involves a term proportional to the a-c voltage and a term proportional to the time derivative of the a-c voltage

(5-78)
$$i(\text{a-c}) = \frac{q}{kT} I(\text{d-c}) v(\text{a-c}) + \frac{q}{kT} I(\text{d-c}) \tau_p \frac{dv(\text{a-c})}{dt}$$

(5-79)
$$i(\text{a-c}) = G_s v(\text{a-c}) + C_s \frac{dv(\text{a-c})}{dt}$$

where

$$G_s = \frac{q}{kT} I(\text{d-c}) \quad \text{and} \quad C_s = G_s \tau_p$$

The factor G_s is an a-c conductance, and the multiplier of the derivative of voltage is the charge storage capacitance C_s. We have assumed forward bias in this derivation [Eq. (5-71)]; C_s is not important for reverse bias, where the capacitance of the transition region C_j dominates (Prob. 5.17).

The charge storage capacitance can be a serious limitation for p-n junctions in high-frequency circuits. As in the case of the switching performance discussed in the two preceding sections, the high-frequency a-c response of a junction can be improved by reducing the carrier lifetime. Since C_s is proportional to τ_p, a short hole lifetime can make the forward-biased capacitance of a p$^+$-n junction acceptably small for many applications.

5.6 Deviations from the Simple Theory

The approach we have taken in studying p-n junctions has focused on the basic principles of operation, neglecting secondary effects. This allows for a relatively uncluttered view of carrier injection and other junction properties, and illuminates the essential features of diode operation. To complete the description, however, we must now fill in a few details which can affect the operation of junction devices under special circumstances.

Most of the deviations from the simple theory can be treated by fairly straightforward modifications of the basic equations. In this section we shall investigate the most important deviations and alter the theory wherever possible. In a few cases, we shall simply indicate the approach to be taken and the result. The most important alterations to the simple diode theory are the effects of contact potential and changes in majority carrier density on carrier injection, recombination and generation within the transition region, ohmic effects, and the effects of graded junctions.

5.6.1 Effects of Contact Potential on Carrier Injection.

If the forward-bias *I–V* characteristics of various semiconductor diodes are compared, it becomes clear that something is missing in the simple diode equation, Eq. (5-36). For example, Fig. 5-24 compares the low-temperature

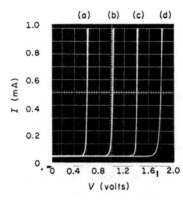

FIGURE **5-24.** *I-V* characteristics of heavily doped p-n junction diodes at 77°K, illustrating the effects of contact potential on the forward current: (a) Ge, $E_g \simeq 0.7$ eV; (b) Si, $E_g \simeq 1.1$ eV; (c) GaAs, $E_g \simeq 1.5$ eV; (d) GaAsP, $E_g \simeq 2.0$ eV.

characteristics of heavily doped diodes with various band gaps. One obvious feature of this figure is that the *I–V* characteristics appear more "square" than exponential; that is, the current is very small until a critical forward bias is reached, and then the current increases rapidly. Furthermore, there is a limiting value of forward-bias voltage for each device, and in each case this limiting voltage is slightly less than the value of the band gap in electron volts.

The reason for the small current at low voltages for these devices can be understood from a simple rearrangement of the diode equation. If we rewrite Eq. (5-36) for a forward-biased p⁺-n diode (with $V \gg kT/q$) and include the exponential form for the minority carrier density p_n, we obtain

$$(5\text{-}80) \qquad I = \frac{qAD_p}{L_p} p_n e^{qV/kT} = \frac{qAD_p}{L_p} N_v e^{[qV-(E_{Fn}-E_{vn})]/kT}$$

Hole injection into the n material is small if the forward bias V is much less than $(E_{Fn} - E_{vn})/q$. For a p⁺-n diode, this quantity is essentially the con-

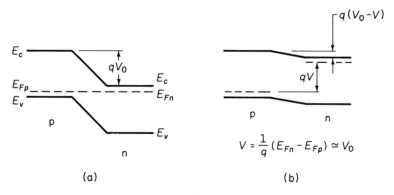

FIGURE 5-25. Examples of contact potential for a heavily doped p-n junction: (a) at equilibrium; (b) approaching the maximum forward bias $V = V_0$.

tact potential, since the Fermi level is near the valence band on the p side. If the n region is also heavily doped, the contact potential is almost equal to the band gap (Fig. 5-25). From another point of view, the minority carrier density $p_n = n_i^2/N_d$ is very small at low temperature (n_i small) and with heavy doping (N_d large). In fact, the minority carrier density can become almost negligible for materials with large band gaps (Prob. 5.18). Thus for small or moderate forward-bias voltages, the usual diode equation predicts a very small current, and this accounts for the fact that in Fig. 5-24 there is no observable current in the low voltage range.

It is somewhat more difficult to explain why the forward current becomes large at a limiting voltage near $V = E_g/q$ for the devices of Fig. 5-24. This effect is not predicted by the simple diode equation, for which the current increases exponentially with applied voltage. The reason this important result is excluded in the simple theory is that in Eq. (5-28) we neglect changes in the majority carrier densities on either side of the junction. This assumption is valid only for low injection levels; for large injected carrier densities, the concentration of excess majority carriers becomes important compared with the majority doping density. For example, at low injection $\Delta n_p = \Delta p_p$ is important compared with the equilibrium minority electron density n_p, but is negligible compared with the majority hole density p_p; this was the basis for neglecting Δp_p in Eq. (5-28). For high injection levels, however, Δp_p can be comparable to p_p and we must write Eq. (5-27) in the form

$$(5\text{-}81) \qquad \frac{p(-x_{p0})}{p(x_{n0})} = \frac{p_p + \Delta p_p}{p_n + \Delta p_n} = e^{q(V_0-V)/kT} = \frac{n_n + \Delta n_n}{n_p + \Delta n_p}$$

To find the more complete diode equation from this expression, we can follow essentially the same steps used in obtaining the low-level relation, Eq. (5-36). The algebra is more complicated, but the basic method is straightforward (Prob. 5.20). The resulting equation for the forward current through the diode is

$$(5\text{-}82) \qquad I = qA\left[\frac{e^{qV/kT} - 1}{1 - e^{-2q(V_0 - V)/kT}}\right]\left[\frac{D_p p_n}{L_p}\left(1 + \frac{n_i^2}{p_p^2}e^{qV/kT}\right)\right.$$
$$\left. + \frac{D_n n_p}{L_n}\left(1 + \frac{n_i^2}{n_n^2}e^{qV/kT}\right)\right]$$

For $V \ll V_0$ (low-level injection) and $p_p, n_n \gg n_i$, this equation reduces to the simple diode equation, Eq. (5-36). However, as the applied voltage V approaches the contact potential V_0, the denominator of the first bracketed term decreases toward zero. Thus the diode forward current increases rapidly near $V = V_0$, and *the contact potential is the limiting forward voltage for the junction.*† For the heavily doped diodes of Fig. 5-24, V_0 is only slightly less than E_g/q (Fig. 5-25), with the result that the forward voltage approaches asymptotically a value slightly less than the band gap expressed in electron volts.

Summarizing the effects of contact potential and carrier injection, we must distinguish between low-level and high-level injection in a junction. When the injected excess carriers are negligible compared with the equilibrium majority carrier densities, the simple diode equation is valid. However, the resulting current may be very small, particularly in heavily doped junctions and at low temperatures. As the injection level is increased to the point where the majority carrier densities deviate significantly from their equilibrium values, the more complete expression for current, Eq. (5-82), must be used. As the applied voltage approaches the contact potential, the current increases rapidly and is limited only by the external circuit and the power rating of the device. The contact potential limits the forward bias which can be applied to the junction. Of course, V may not approach V_0 in devices for which the exponential current described by the simple diode equation is significant (e.g., if p_n or n_p is fairly large). In such a diode, the current may become larger than the maximum rating of the device before a forward bias of V_0 is reached.

In effect, the contact potential serves as a barrier to the net injection of excess carriers across a junction. If V_0 is small, the diode current can be significant at fairly low voltage; however, if V_0 is essentially the full band gap voltage, the current is negligible until the forward bias approaches E_g/q. These effects are illustrated clearly in Fig. 5-24 because the devices were

---

†The forward voltage applied to the diode can exceed V_0 if ohmic effects are significant (Section 5.6.3).

purposely chosen with heavy doping and observed at low temperature. For lightly doped samples and higher temperatures, the *I–V* characteristics include ohmic losses (Section 5.6.3) and other effects.

5.6.2 Recombination and Generation in the Transition Region.

In analyzing the p-n junction, we have assumed that recombination and thermal generation of carriers occur primarily in the neutral p and n regions, outside the transition region. In this model, forward current in the diode is carried by recombination of excess minority carriers injected into each neutral region by the junction. Similarly, the reverse saturation current is due to the thermal generation of EHP's in the neutral regions and the subsequent diffusion of the generated minority carriers to the transition region, where they are swept to the other side by the field. In many devices this model is adequate; however, a more complete description of junction operation should include recombination and generation within the transition region itself.

When a junction is forward biased, the transition region contains excess carriers of both types, which are in transit from one side of the junction to the other. Unless the width of the transition region W is very small compared with the carrier diffusion lengths L_n and L_p, significant recombination can take place within W. This provides an additional forward current, since holes must be supplied from the p side and electrons from the n side to feed the recombination. Therefore, the actual current may be somewhat larger than that predicted in the simple theory. An accurate calculation of the additional recombination current is complicated by the fact that the recombination rate, which depends on the carrier densities [Eq. (4-5)], varies with position within the transition region. Analysis of the recombination kinetics shows that the current due to recombination within W is proportional to n_i and increases with forward bias according to approximately $\exp(qV/2kT)$. On the other hand, current due to recombination in the neutral regions is proportional to p_n and n_p [Eq. (5-36)] and therefore to n_i^2/N_d and n_i^2/N_a, and increases according to $\exp(qV/kT)$. Therefore, the ratio of the two currents is

$$(5\text{-}83) \qquad \frac{I \text{ (recombination in neutral regions)}}{I \text{ (recombination in transition region)}} \propto \frac{n_i^2 \, e^{qV/kT}}{n_i \, e^{qV/2kT}} \propto n_i e^{qV/2kT}$$

This ratio becomes small for wide band gap materials and low temperatures (small n_i). Thus the forward current for low injection in a Si diode is likely to be dominated by recombination in the transition region, while a Ge diode may follow the usual diode equation. In either case, injection through W into the neutral regions becomes more important with increased voltage, until finally the current begins to increase appreciably as the forward bias approaches the contact potential.

Just as recombination within W can affect the forward characteristics, the reverse current through a junction can be influenced by carrier generation in the transition region. We found in Section 5.3.2 (and Prob. 5.7) that the reverse saturation current can be accounted for by the thermal generation of EHP's within a diffusion length of either side of the transition region. The generated minority carriers diffuse to the transition region, where they are swept to the other side of the junction by the electric field (Fig. 5-26).

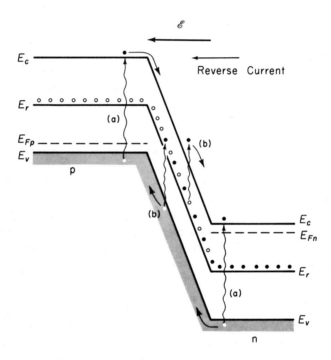

FIGURE 5-26. Current in a reverse-biased p-n junction due to thermal generation of carriers by (a) band-to-band EHP generation, and (b) generation from a recombination level.

However, carrier generation can take place within the transition region itself. If W is small compared with L_n or L_p, band-to-band generation of EHP's within the transition region is not important compared with generation in the neutral regions. However, the lack of free carriers within the space charge of the transition region can create a current due to the net generation of carriers by *emission from recombination centers*. Of the four generation–recombination processes depicted in Fig. 5-27, the two capture

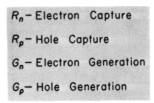

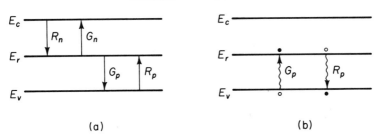

FIGURE 5-27. Capture and generation of carriers at a recombination center: (a) capture and generation of electrons and holes; (b) hole capture and generation processes redrawn in terms of valence band electron excitation to E_r (hole generation) and electron de-excitation from E_r to E_v (hole capture by E_r).

rates R_n and R_p are negligible within W because of the very small carrier densities in the reverse-bias space charge region. Therefore, a recombination level E_r near the center of the band gap can provide carriers through the thermal generation rates G_n and G_p. Each recombination center alternately emits an electron and a hole; physically, this means that an electron at E_r is thermally excited to the conduction band (G_n) and a valence band electron is subsequently excited thermally to the empty state on the recombination level, leaving a hole behind in the valence band (G_p). The process can then be repeated over and over, providing electrons for the conduction band and holes for the valence band. At equilibrium, these emission processes are exactly balanced by the corresponding capture processes R_n and R_p. However, in the reverse-bias transition region, generated carriers are swept out before recombination can occur, and net generation results.

Of course, the importance of thermal generation within W depends on the temperature and the nature of the recombination centers. A level near the middle of the band gap is most effective, since for such centers neither G_n nor G_p requires thermal excitation of an electron over more than about half the band gap. If no recombination level is available near the middle of the band gap, this type of generation is negligible. However, in most materials recombination centers exist near the middle of the gap due to trace impurities or lattice defects. Generation from centers within W is most important in materials with large band gaps, for which band-to-band generation in the neutral regions is small. Thus for Si, generation within W is generally more important than for a narrower band gap material such as Ge.

The saturation current due to generation in the neutral regions was found to be essentially independent of reverse bias. However, generation within W naturally increases as W increases with reverse bias. As a result, the reverse current can increase almost linearly with W, or with the square root of reverse-bias voltage (Fig. 5-28).

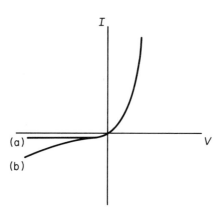

FIGURE 5-28. Comparison of reverse current mechanisms: (a) carrier generation in the neutral regions; (b) carrier generation within the transition region.

5.6.3 Ohmic Losses. In deriving the diode equation we assumed that the voltage applied to the device appears entirely across the junction. Thus we neglected any voltage drop in the neutral regions or at the external contacts. For most devices this is a valid assumption; the doping is usually fairly high, so that the resistivity of each neutral region is low, and the area of a typical diode is large compared with its length. However, some devices do exhibit ohmic effects, which cause significant deviation from the expected *I–V* characteristic.

We can seldom represent ohmic losses in a diode accurately by including a simple resistance in series with the junction. The effects of voltage drops outside the transition region are complicated by the fact that the voltage drop depends on the current, which in turn is dictated by the voltage across the junction. For example, if we represent the series resistance of the p and n regions by R_p and R_n, respectively, we can write the junction voltage V as

(5-84)
$$V = V_a - I[R_p(I) + R_n(I)]$$

where V_a is the external voltage applied to the device. As the current increases, there is an increasing voltage drop in R_p and R_n, and the junction voltage V decreases. This reduction in V lowers the level of injection so that the current increases more slowly with increased bias (Fig. 5-29). A further

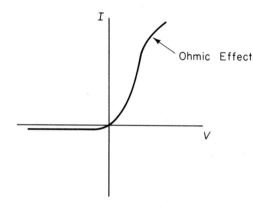

FIGURE 5-29. Effects of ohmic losses at high injection in a p-n junction diode.

complication in calculating the ohmic loss is that the conductivity of each neutral region increases with increasing carrier injection. Since the effects of Eq. (5-84) are most pronounced at high injection levels, this *conductivity modulation* by the injected excess carriers can reduce R_p and R_n significantly.

Ohmic losses are purposely avoided in properly designed devices by appropriate choices of doping and geometry. Therefore, deviations of the current as shown in Fig. 5-29 generally appear only for very high currents, outside the normal operating range of the device.

5.6.4 Graded Junctions. While the abrupt junction approximation accurately describes the properties of alloyed junctions and many epitaxial structures, it is often inadequate in analyzing diffused junction devices. For shallow diffusions, in which the diffused impurity profile is very steep (Fig. 5-30a), the abrupt approximation is usually acceptable. If the impurity profile is spread out into the sample, however, a graded junction can result (Fig. 5-30b). Several of the expressions we have derived for the abrupt junction must be modified for this case.

The graded junction problem can be solved analytically if, for example, we make a linear approximation of the net impurity distribution near the junction (Fig. 5-30c). We assume that the graded region can be described approximately by

$$(5\text{-}85) \qquad\qquad N_d - N_a = Gx$$

where G is a grade constant giving the slope of the net impurity distribution.

In Gauss's law [Eq. (5-14)], we can use our linear approximation to write

$$(5\text{-}86) \qquad\qquad \frac{d\mathscr{E}}{dx} = \frac{q}{\epsilon}(p - n + N_d^+ - N_a^-) \simeq \frac{q}{\epsilon}Gx$$

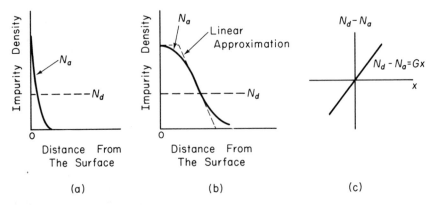

FIGURE 5-30. Approximations to diffused junctions: (a) shallow diffusion (abrupt); (b) deep drive-in diffusion with source removed (graded); (c) linear approximation to the graded junction.

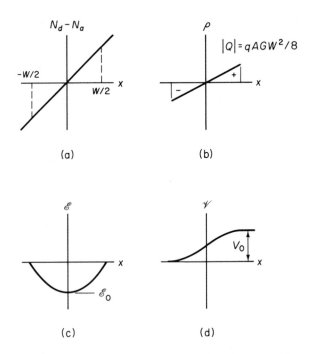

FIGURE 5-31. Properties of the graded junction transition region: (a) net impurity profile; (b) net charge distribution; (c) electric field; (d) electrostatic potential.

within the transition region. In this approximation we assume complete ionization of the impurities and neglect the carrier densities in the transition region, as before. The net space charge varies linearly over W, and the electric field distribution is therefore parabolic. The expressions for contact potential and junction capacitance are different from the abrupt junction case (Fig. 5-31 and Prob. 5.21), since the electric field is no longer linear on each side of the junction.

In a graded junction the usual depletion approximation is often inaccurate. If the grade constant G is small, the carrier densities $(p - n)$ can be important in Eq. (5-86). Similarly, the usual assumption of negligible space charge outside the transition region is questionable for small G. It would be more accurate to refer to the regions just outside the transition region as quasi-neutral rather than neutral. Thus, the edges of the transition region are not sharp as Fig. 5-31 implies but are spread out in x. These effects complicate calculations of junction properties, and a computer must be used in solving the problem accurately.

Most of the conclusions we have made regarding carrier injection, recombination and generation currents, and other properties are qualitatively applicable to graded junctions, with some alterations in the functional form of the resulting equations. Therefore, we can apply most of our basic concepts of junction theory to reasonably graded junctions as long as we remember that certain modifications should be made in accurate computations.

READING LIST

P. E. GRAY et al., *Physical Electronics and Circuit Models of Transistors.* (SEEC series, vol. 2). New York: John Wiley & Sons, Inc., 1964, pp. 1–98.

L. ERIKSSON, J. A. DAVIS, and J. W. MAYER, "Ion Implantation Studies in Silicon," *Science*, vol. 163, pp. 627–633, 14 February 1969.

J. L. MOLL, "The Evolution of the Theory for the Voltage–Current Characteristic of p-n Junctions," *Proc. IRE*, vol. 46, pp. 1076–1082, June 1958. There are also numerous other articles of interest in this special issue on transistors and other semiconductor devices.

A. S. GROVE, *Physics and Technology of Semiconductor Devices.* New York: John Wiley & Sons, Inc., 1967, pp. 149–207.

S. K. GHANDHI, *The Theory and Practice of Microelectronics.* New York: John Wiley & Sons, Inc., 1968, pp. 258–317.

A. VAN DER ZIEL, *Solid State Physical Electronics* (2nd ed.). Englewood Cliffs, N.J.: Prentice-Hall, Inc., 1968, pp. 299–336.

PROBLEMS

5.1 (a) A Si sample is doped with 10^{21} cm^{-3} phosphorus atoms. Is it possible to form a p-n junction in this sample by alloying or diffusion? If so, what impurity would you recommend for the counterdoping?

(b) A Ge sample contains 10^{19} cm^{-3} Sb atoms. Is In an appropriate metal for forming an alloy junction in this material?

5.2 The mathematics used in deriving Eq. (5-7) can be avoided if we accept *a priori* the proposition that the Fermi level must be invariant across the junction $(E_{F_n} = E_{F_p})$ at equilibrium. Draw a band diagram such as in Fig. 5-8b, and show that Eq. (5-7) follows automatically from this condition.

5.3 An n-type sample of Ge contains $N_d = 10^{16}$ cm^{-3}. A junction is formed by alloying with In at 160°C. Assume the acceptor concentration in the regrown region equals the solid solubility at the alloying temperature.

(a) Calculate the Fermi level positions at 300°K in the p and n regions.

(b) Draw an equilibrium band diagram for the junction and determine the contact potential V_0 from the diagram.

(c) Compare the results of part (b) with V_0 as calculated from Eq (5-8).

5.4 Aluminum is alloyed into an n-type Si sample $(N_d = 10^{16}$ cm$^{-3})$, forming an abrupt junction of circular cross section, with a diameter of 20 mils. Assume the acceptor density in the alloyed regrown region is $N_a = 4 \times 10^{18}$ cm^{-3}. Calculate V_0, x_{n0}, x_{p0}, Q_+ and $\mathscr{E}_0$ for this junction at equilibrium (300°K). Sketch $\mathscr{E}(x)$ and charge density to scale, as in Fig. 5-9.

5.5 Use the balance of drift and diffusion, Eq. (5-2), to derive Eq. (5-27) for a junction with external bias V.

5.6 Draw a band diagram for a forward-biased p-n junction, including a sketch of the quasi-Fermi levels $F_n(x)$ and $F_p(x)$ throughout the junction and for several diffusion lengths on either side of the junction. Explain qualitatively the variations in F_n and F_p.

5.7 Consider a volume of n-type material of area A, with a length of one hole diffusion length L_p.

(a) What is the rate of thermal generation of holes within this volume (holes per second) in terms of p_n, τ_p, and the dimensions?

(b) Assume each thermally generated hole diffuses out of the volume before it can recombine. Compare the resulting hole current with

(Continued 5.7)

the reverse saturation current of a p^+-n junction. What does this tell you physically about saturation current?

5.8 In a p^+-n junction the hole diffusion current in the neutral n material is given by Eq. (5-32). What are the electron diffusion and electron drift components of current at point x_n in the n region?

5.9 Assume the doping density N_a on the p side of an abrupt junction is the same as N_d on the n side. Each side is many diffusion lengths long. Show that the expression for the electron current I_n in the n-type material is

$$I_n = qA\left[\frac{D_p}{L_p}(1 - e^{-x_n/L_p}) + \frac{D_n}{L_n}\right]\frac{n_i^2}{N_d}(e^{qV/kT} - 1)$$

5.10 The hole injection efficiency of a junction is I_p/I at $x_n = 0$.

(a) Assuming the junction follows the simple diode equation, express I_p/I in terms of the diffusion constants, diffusion lengths, and equilibrium minority carrier concentrations.

(b) Show that I_p/I can be written as $[1 + (\sigma_n/L_n)/(\sigma_p/L_p)]^{-1}$, where σ_n and σ_p are the conductivities of the n and p regions, respectively. What should be done to increase the hole injection efficiency of a junction?

5.11 Assume both sides of an abrupt p-n junction can be doped at will. What is the maximum reverse voltage required to cross the bands of any junction as illustrated in Fig. 5-19b?

5.12 Use sketches of band diagrams to illustrate that the tunneling distance d in Fig. 5-19b becomes smaller with an increase in reverse bias. Assume the doping is sufficiently heavy that changes in W are negligible.

5.13 Assuming the critical field for Zener breakdown is 10^6 V/cm, calculate the reverse bias required for this type of breakdown in the following abrupt Si junction: N_a on the p side $= N_d$ on the n side $= 2 \times 10^{18}$ cm^{-3}. Assume breakdown can occur if the peak electric field in the junction reaches 10^6 V/cm.

5.14 Holes are injected into the long n region of a forward-biased p^+-n diode. The current is I until $t = 0$; after $t = 0$ the current decays according to $i(t) = Ie^{-t/\tau_p}$.

(a) Solve for the stored charge in the n region as a function of time, $Q_p(t)$.

(b) Assuming the hole distribution is always exponential, find the voltage across the junction $v(t)$ for $v \gg kT/q$.

5.15 A long p^+-n junction diode is forward biased with current I flowing. The current is suddenly tripled at $t = 0$.

(a) The junction voltage does not change at $t = 0$. Why not? What is the slope of the hole distribution in the n material at the edge of the transition region just after the current is tripled?

(b) How is the final junction voltage (at $t = \infty$) related to the original voltage (before $t = 0$)? Assume the voltage is always much larger than kT/q.

(c) Assume that the stored charge in the n region can be represented by an exponential at each instant, for simplicity. Write the expression for the instantaneous current as a sum of recombination current and current due to changes in the stored charge. Using proper boundary conditions, solve this equation for the instantaneous hole distribution and find the expression for the instantaneous junction voltage.

5.16 Assume a p^+-n diode is biased in the forward direction, with a current I_f. At time $t = 0$ the current is switched to $-I_r$. Use the appropriate boundary conditions to solve Eq. (5-52) for $Q_p(t)$. Apply the quasi-steady state approximation, Eq. (5-57), and show that the storage delay time can be written as $t_{sd} = \tau_p \ln (I_r + I_f)/I_r$ for this approximation.

5.17 Explain physically why the charge storage capacitance is unimportant for reverse-biased junctions.

5.18 In this problem we calculate the reverse saturation current and the forward current for typical diodes. Assume a junction area of 10^{-3} cm². The junctions are p^+-n, with N_d on the n side $= 10^{16}$ cm⁻³.

(a) Calculate the reverse saturation current for Ge ($D_p = 50$ cm²/sec, $\tau_p = 100$ μsec) and for Si ($D_p = 12.5$ cm²/sec, $\tau_p = 1$ μsec).

(b) Calculate the diode currents from the simple diode equation for a forward bias equal to one-half the band gap voltage, $E_g(eV)/2$.

5.19 In this problem we wish to check the approximation in Eq. (5-28) that the majority carrier densities do not change with bias; that is, we wish to determine what constitutes "high-level injection," calling for the application of the more general current expression, Eq. (5-82). Assume an abrupt p-n⁺ Ge junction, with $N_d = 10^{20}$ cm⁻³ on the n side and $N_a = 10^{16}$ cm⁻³ on the p side. The electron lifetime on the p side is 100 μsec, and the area of the junction A is 10^{-3} cm². Use Eq. (5-34) to calculate Δn_p for a current of 1 μA. Assuming space charge neutrality at $x_p = 0$, is Eq. (5-28) valid? Repeat for a current of 1 mA.

5.20 In this problem we wish to derive the general diode equation, Eq. (5-82). Assume space charge neutrality in the n and p regions, $\Delta n_p = \Delta p_p$ and $\Delta p_n = \Delta n_n$.

(Continued 5.20)

    (a) Use Eq. (5-81) to solve for Δp_n and Δn_p in terms of equilibrium quantities and the applied voltage.

    (b) Show that Eq. (5-82) follows from

$$I = qA(D_p \Delta p_n/L_p + D_n \Delta n_p/L_n).$$

5.21 Assume a linearly graded junction as in Fig. (5-31), with a doping distribution described by Eq. (5-85). The doping is symmetrical, so that $x_{p0} = x_{n0} = W/2$.

    (a) Integrate Eq. (5-86) to show that

$$\mathscr{E}(x) = \frac{q}{2\epsilon} G\left[x^2 - \left(\frac{W}{2}\right)^2\right]$$

    (b) Show that the width of the depletion region is

$$W = \left[\frac{12\epsilon(V_0 - V)}{qG}\right]^{1/3}$$

    (c) Show that the junction capacitance is

$$C_j = A\left[\frac{qG\epsilon^2}{12(V_0 - V)}\right]^{1/3}$$

5.22 When impurities are diffused into a sample from an unlimited source such that the surface concentration N_0 is held constant, the impurity distribution (profile) is given by

$$N(x, t) = N_0 \, \text{erfc}\left(\frac{x}{2\sqrt{Dt}}\right)$$

where D is the diffusion constant for the impurity and t is the diffusion time.

    If a certain number of impurities are placed in a thin layer on the surface before diffusion, and if no impurities are added and none escape during diffusion, a gaussian distribution is obtained:

$$N(x, t) = \frac{N_s}{\sqrt{\pi Dt}} e^{-(x/2\sqrt{Dt})^2}$$

where N_s is the quantity of impurity placed on the surface (atoms/cm²) prior to $t = 0$. Notice that this expression differs from Eq. (4-47) by a factor of two. Why?

    Figure P5-22 gives curves of the complementary error function and gaussian factors for the variable u, which in our case is $x/2\sqrt{Dt}$. Assume boron is diffused into n-type Si (uniform $N_d = 5 \times 10^{16}$ cm^{-3})

(Continued)

(Continued 5.22)

    at 1000°C for 30 min. The diffusion constant for B in Si at this temperature is $D = 3 \times 10^{-14}$ cm²/sec.

    (a) Plot $N_a(x)$ after the diffusion, assuming the surface concentration is held constant at $N_0 = 5 \times 10^{20}$ cm⁻³. Locate the position of the junction below the surface.

    (b) Plot $N_a(x)$ after diffusion, assuming that a B concentration of 5×10^{20} cm⁻³ is deposited in a thickness of 10 Å on the surface prior to diffusion ($N_s = 5 \times 10^{13}$ cm⁻²), and no additional B atoms are available during the diffusion. Locate the junction for this case.

    Hint: Plot the curves on five-cycle semilog paper, with an abscissa varying from zero to one-half micron. In plotting $N_a(x)$, choose values of x which are simple multiples of $2\sqrt{Dt}$.

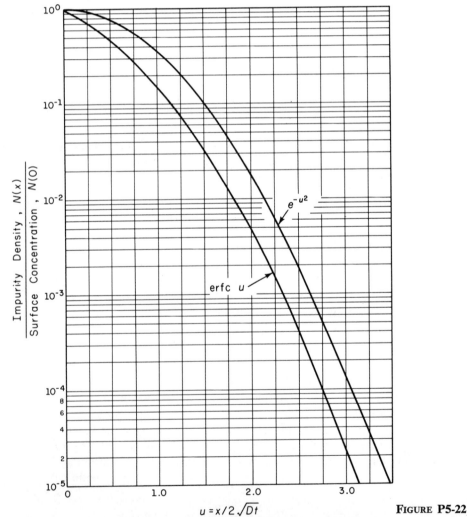

FIGURE P5-22

p-n JUNCTION DEVICES **6**

In this chapter we shall investigate some of the important electronic devices built with p-n junctions. The devices discussed here are primarily single-junction structures (diodes), with transistors and other multijunction devices left for later chapters. The devices discussed in this chapter illustrate many of the fundamental applications of junction properties such as rectification, variable capacitance, tunneling, and light emission and detection. In addition, we shall examine the use of diode arrays in optical detection and image processing. Since detailed descriptions of applications can be found in various texts on electronic circuits, we shall limit discussion of circuit applications to examples which illustrate particular device properties.

6.1 The Junction Diode

We have investigated many of the properties of junction diodes in Chapter 5. In this section we discuss the use of these diodes for rectification and other circuit applications. In some cases, diodes are designed to exploit specific junction properties, such as capacitance or charge storage. Therefore, we shall extend the treatment beyond the usual applications of diodes in rectification and switching to include those devices which depend on "secondary" junction properties.

6.1.1 Rectifiers. The most obvious property of a p-n junction is its *unilateral* nature; that is, to a good approximation it conducts current in only one direction. We can think of an *ideal diode* as a short circuit when forward biased and as an open circuit when reverse biased (Fig. 6-1a). The p-n junction diode does not quite fit this description, but the $I-V$ character-

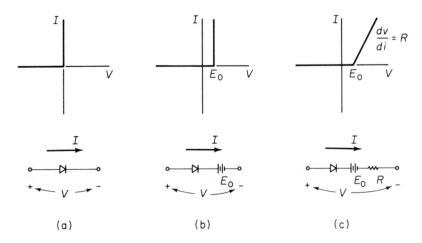

FIGURE **6-1.** Piecewise-linear approximations of junction diode characteristics: (a) the ideal diode; (b) ideal diode with an offset voltage; (c) ideal diode with an offset voltage and a resistance to account for slope in the forward characteristic.

istics of many junctions can be approximated by the ideal diode in series with other circuit elements to form an equivalent circuit. For example, most forward-biased diodes exhibit an *offset voltage* E_o which can be approximated in a circuit model by a battery in series with the ideal diode (Fig. 6-1b). The series battery in the model keeps the ideal diode turned off for applied voltages less than E_o. From Section 5.6.1 we expect E_o to be approximately the contact potential of the junction. In some cases the approximation to the actual diode characteristic is improved by adding a series resistor R to the circuit equivalent (Fig. 6-1c). The equivalent circuit approximations illustrated in Fig. 6-1 are called *piecewise-linear equivalents*, since the approximate characteristics are linear over specific ranges of voltage and current.

An ideal diode can be placed in series with an a-c voltage source to provide *rectification* of the signal (Fig. 6-2a). Since current can flow only in the forward direction through the diode, only the positive half-cycles of the input sine wave are passed to the resistor. During the negative half-cycles the diode is nonconducting, and the resulting voltage across the resistor is zero. The output voltage v_o is a *half-rectified sine wave*. Whereas the input sinusoid has no average value, the rectified signal has a positive average value and therefore contains a d-c component. By appropriate filtering this d-c level can be extracted from the rectified signal.

Both half-cycles of the input signal contribute to the *full-rectified sine wave* in Fig. 6-2b. On the positive half-cycle, current flows through diode D_2, the resistor, and D_4. On the negative half-cycle, D_3 and D_1 conduct, and

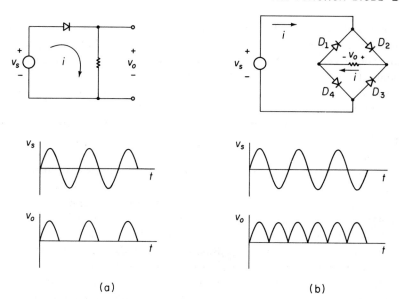

FIGURE 6-2. Rectifying circuits: (a) half-wave rectifier; (b) full-wave rectifier.

current flows through the resistor in the same direction as before. Thus in each half-cycle the voltage v_o has the polarity indicated, and the output signal consists of a series of positive sinusoidal half-cycles.

The unilateral nature of diodes is useful for many other circuit applications which require *waveshaping*. This involves alteration of a-c signals by passing only certain portions of the signal while blocking other portions (Prob. 6.1).

Junction diodes designed for use as rectifiers should have I–V characteristics as close as possible to that of the ideal diode. The reverse current should be negligible, and the forward current should exhibit little voltage dependence (negligible *forward resistance R*). The reverse breakdown voltage should be large, and the offset voltage E_o in the forward direction should be small. Unfortunately, not all of these requirements can be met by a single device; compromises must be made in the design of the junction to provide the best diode for the intended application.

From the theory derived in Chapter 5 we can easily list the various requirements for good rectifier junctions. *Band gap* is obviously an important consideration in choosing a material for rectifier diodes. Since n_i is small for large band gap materials, the reverse saturation current (which depends on thermally generated carriers) decreases with increasing E_g. A rectifier made with a wide band gap material can be operated at higher temperatures, because thermal excitation of EHP's is reduced by the increased band gap.

Such temperature effects are critically important in rectifiers which must carry large currents in the forward direction and are thereby subjected to appreciable heating. On the other hand, the contact potential and offset voltage E_o generally increase with E_g. This drawback is usually outweighed by the advantages of low n_i; for example, Si is generally preferred over Ge for power rectifiers because of its wider band gap, lower leakage current, and higher breakdown voltage, as well as its more convenient fabrication properties.

The *doping density* on each side of the junction influences the avalanche breakdown voltage, the contact potential, and the series resistance of the diode. If the junction has one highly doped side and one lightly doped side (such as a p^+-n junction), the lightly doped region determines many of the properties of the junction. From Eq. (5-48) we see that a high resistivity region should be used for at least one side of the junction to increase the breakdown voltage V_{br}. However, this approach tends to increase the forward resistance R of Fig. 6-1c, and therefore contributes to the problems of thermal effects due to I^2R heating. To reduce the resistance of the lightly doped region, it is necessary to make its area large and reduce its length. Therefore, the physical *geometry* of the diode is another important design variable. Limitations on the practical area for a diode include problems of obtaining uniform starting material and junction processing over large areas. Localized flaws in junction uniformity can cause premature reverse breakdown in a small region of the device. Similarly, the lightly doped region of the junction cannot be made arbitrarily short. One of the primary problems with a short, lightly doped region is an effect called *punch-through*. Since the transition region width W increases with reverse bias and extends primarily into the lightly doped region, it is possible for W to increase until it fills the entire length of this region (Prob. 6.4). The result of punch-through is a breakdown below the value of V_{br} expected from Eq. (5-48).

In devices designed for use at high reverse bias, care must be taken to avoid premature breakdown across the edge of the sample. This effect can be reduced by *beveling* the edge or by diffusing a *guard ring* to isolate the junction from the edge of the sample (Fig. 6-3). The electric field is lower at the beveled edge of the sample in Fig. 6-3b than it is in the main body of the device. The depletion layer on the n side must be wide enough to "uncover" a positive charge due to ionized donors just equal to the exposed negative charge on the p^+ side. Therefore, the depletion width must decrease near the edge where the width of the p^+ region decreases. Since the resulting electric field is small near the edge, avalanche breakdown must occur in the main body of the sample. Similarly, the junction at the lightly doped p guard ring of Fig. 6-3c breaks down at higher voltage than the p^+-n junction, according to Eq. (5-48). Since the depletion region is wider in the p ring than in the p^+ region, the average electric field is smaller at the ring for a given diode reverse voltage.

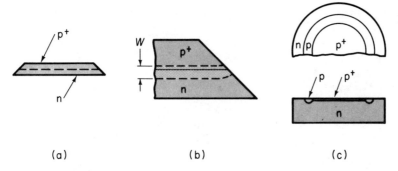

(a)   **(b)**   **(c)**

FIGURE 6-3. Beveled edge and guard ring to prevent edge breakdown under reverse bias: (a) p^+-n diode with beveled edge; (b)
closeup view of edge, showing reduction of depletion region
near the bevel; (c) guard ring.

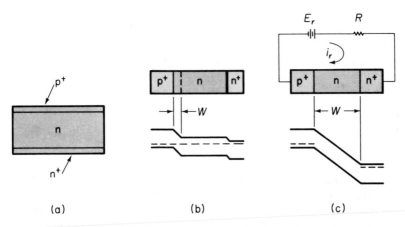

(a)   **(b)**   **(c)**

FIGURE 6-4. A p^+-n-n^+ junction diode: (a) device configuration;
(b) zero-bias condition; (c) reverse-biased to punch-through.

In fabricating a p^+-n or a p-n^+ junction, it is common to terminate the
lightly doped region with a heavily doped layer of the same type (Fig. 6-4a),
to ease the problem of making ohmic contact to the device. The result is a
p^+-n-n^+ structure with the p^+-n layer serving as the active junction, or a
p^+-p-n^+ device with an active p-n^+ junction. The lightly doped center region
determines the avalanche breakdown voltage. If this region is short compared
with the minority carrier diffusion length, the excess carrier injection for
large forward currents can increase the conductivity of the region significantly. This type of *conductivity modulation*, which reduces the forward
resistance R, can be very useful for high-current devices. On the other hand,

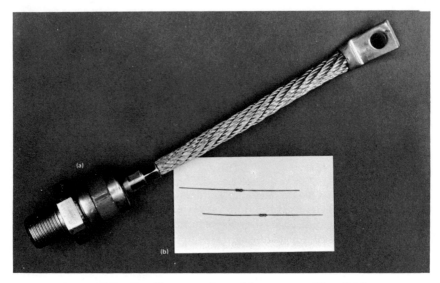

FIGURE 6-5. Rectifier mountings: (a) power rectifier; (b) low-current diodes. The power rectifier is rated at 250 *A*, with 2800 *V* maximum reverse voltage; the small diodes are rated at less than one ampere with several hundred volts reverse bias breakdown. Photograph courtesy of Delco Electronics Division, General Motors Corp.

a short, lightly doped center region can also lead to punch-through under reverse bias, as in Fig. 6-4c.

The mounting of a rectifier junction is critical to its ability to handle power. For diodes used in low-power circuits, glass or plastic encapsulation or a simple header mounting is adequate. However, high-current devices which must dissipate large amounts of heat require special mountings to transfer thermal energy away from the junction. A typical Si power rectifier is mounted on a molybdenum or tungsten disk to match the thermal expansion properties of the Si. This disk is fastened to a large stud of copper or other thermally conductive material which can be bolted to a heat sink with an appropriate cooling facility. The contrast in size and type of mounting between a power rectifier and low-current diodes is shown in Fig. 6-5.

6.1.2 Switching Diodes; the Step-Recovery Diode. In discussing rectifiers we emphasized the importance of minimizing the reverse-bias current and the power losses under forward bias. In many applications, time response can be important as well. If a junction diode is to be used to switch rapidly from the conducting to the nonconducting state and back again, special consideration must be given to its charge control properties. We have discussed the equations governing the turn-on time and the reverse recovery time of a junction in Section 5.5. From Eqs. (5-52) and (5-59) it is clear that a diode with fast switching properties must either store very little charge in the neutral regions for steady forward currents, or have a very short carrier lifetime, or both.

A common technique for improving the switching speed of a diode is adding efficient recombination centers to the bulk material. For Si diodes, Au doping is useful for this purpose. To a good approximation the carrier lifetime varies with the reciprocal of the recombination center density. Thus, for example, a p^+-n Si diode may have $\tau_p = 1$ μsec and a reverse recovery time of 0.1 μsec before Au doping. If the addition of 10^{14} Au atoms/cm³ reduces the lifetime to 0.1 μsec and t_{sd} to 0.01 μsec, 10^{15} cm⁻³ Au atoms could reduce τ_p to 0.01 μsec and t_{sd} to 1 nsec (10^{-9} sec). This process cannot be continued indefinitely, however. The reverse current due to generation of carriers from the Au centers in the depletion region becomes appreciable with large Au density (Section 5.6.2). In addition, as the Au density approaches the lightest doping of the junction, the equilibrium carrier density of that region can be affected.

A second approach to improving the diode switching time is to make the lightly doped neutral region shorter than a minority carrier diffusion length. This is the *narrow base diode* (Prob. 6.5). In this case the stored charge for forward conduction is very small, since most of the injected carriers diffuse through the lightly doped region to the end contact. When such a diode is switched to reverse conduction, very little time is required to eliminate the stored charge in the narrow neutral region. The mathematics involved in Prob. 6.5 is particularly interesting, because it closely resembles the calculations we shall make in analyzing the junction transistor in Chapter 9.

One interesting type of diode is designed to store charge under forward conduction but to have a very fast turn-off time once the stored charge is withdrawn below a certain level. This type of device is called a *step-recovery diode* or *snap-off diode*. The geometry of a step-recovery device is typically a p^+-p-n^+ structure in which the charge is stored in a narrow p region. With proper design of the lightly doped center region, this device yields fast transitions in the picosecond (10^{-12} sec) range. We recall from Fig. 5-22 that a reverse recovery transient takes place in essentially two steps: (1) The stored charge is withdrawn near the edge of the transition region as the excess carrier density approaches zero at x_{n0} (for a p^+-n junction); the voltage during this phase is positive and the current is negative. (2) Once V reaches zero and begins to go negative, the reverse current decreases to its reverse saturation value; during this phase the remaining charge is withdrawn and the minority carrier distribution approaches its steady state value for reverse saturation.

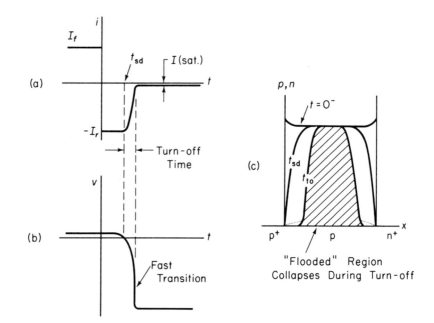

FIGURE 6-6. Reverse recovery transient for a step recovery diode: (a) current transient; (b) voltage transient; (c) carrier densities in the lightly doped p region before switching ($t = 0^-$), when V reaches zero ($t = t_{sd}$), and during the turn-off transient (t_{to}) while the stored charge is swept out. Mobility differences for electrons and holes are neglected in this figure.

The p^+-p-n^+ structure provides the important feature of *field-aided withdrawal* of stored charge in the second phase of the recovery transient (Fig. 6-6b). As the excess carrier density reaches zero at each edge of the center p region, the voltage across the device goes to zero and then becomes negative. This reverse bias appears primarily across the center region, in the direction to sweep holes to the p^+ side and electrons to the n^+ side. The combination of the narrow p region and the field-aided withdrawal of carriers provides a turn-off time much shorter than the carrier lifetime. This is in contrast to the simple p-n junction which depends on diffusion of excess carriers in the neutral regions.

Under forward bias, holes are injected into the lightly doped p region from p^+ and electrons from n^+. The center region is filled with an essentially homogeneous mixture of electrons and holes in transit across the region ($t = 0^-$ of Fig. 6-6c). At $t = t_{sd}$ the carrier densities reach zero at the two edges of the center region and $V = 0$. Actually, a single t_{sd} is hard to define, since the difference in μ_n and μ_p causes the left- and right-hand sides of the carrier distribution to collapse at slightly different rates. This effect is neglected in Fig. 6-6c for simplicity. During the field-aided turn-off time, the carrier distribution collapses from both sides as electrons and holes are swept to n^+ and p^+, respectively, by the growing electric field in the center region. The dotted lines in Fig. 6-6c indicate the small hole density on the left and small electron density on the right of the "flooded" region, due to carriers in transit to their respective junctions.

Most of the turn-off time is required for the field-aided withdrawal of the remaining charge in the center region after the voltage goes through zero. Once the stored charge is reduced to zero, the remaining applied voltage appears across the device in an extremely short time. This "fast transition" occurs in approximately ten picoseconds per micron of center region thickness. The fast transition can be used as a source for high-frequency harmonics, for generation of sharp pulses, and in various microwave applications. In many ways the step-recovery diode can be considered as a nonlinear capacitance, as is the varactor diode to be described in Section 6.1.4.

6.1.3 The Breakdown Diode.
As we discussed in Section 5.4, the reverse-bias breakdown voltage of a junction can be varied by choice of junction doping densities. The breakdown mechanism is the Zener effect (tunneling) for abrupt junctions with extremely heavy doping; however, the more common breakdown is avalanche (impact ionization), typical of more lightly doped or graded junctions. By varying the doping density we can fabricate diodes with specific breakdown voltages ranging from less than one volt to several hundred volts. If the junction is well designed, the breakdown will be sharp and the current after breakdown will be essentially independent of voltage (Fig. 6-7a). When a diode is designed for a specific breakdown voltage, it is called a *breakdown diode*. Such diodes are also called *Zener diodes*, despite the fact that the actual breakdown mechanism is usually the avalanche effect. This error in terminology is due to an early mistake in identifying the first observations of breakdown in p-n junctions.

Breakdown diodes can be used as *voltage regulators* in circuits with varying inputs. The 15-V breakdown diode of Fig. 6-7 holds the circuit output voltage v_o constant at 15 V, while the input varies at voltages greater than 15 V. For example, if v_s is a rectified and filtered signal composed of a 17-V d-c component and a 1-V ripple variation above and below 17 V, the output v_o will remain constant at 15 V. More complicated voltage regulator circuits can be designed using breakdown diodes, depending on the type of signal being regulated and the nature of the output load. In a similar application, such a device can be used as a *reference diode*; since the breakdown voltage of a particular diode is known, the voltage across it during breakdown can be used as a reference in circuits which require a known value of voltage.

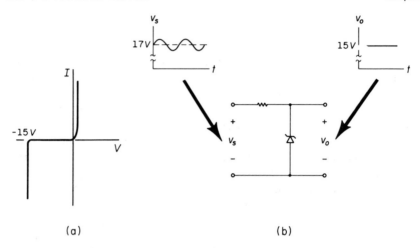

(a)   (b)

FIGURE 6-7. A breakdown diode: (a) I-V characteristic;
(b) application as a voltage regulator.

6.1.4 The Varactor Diode.

The term *varactor* is a shortened form of *var*iable re*actor*, referring to the voltage-variable capacitance of a reverse-biased p-n junction. The equations derived in Section 5.5.3 indicate that junction capacitance depends on the applied voltage and the design of the junction. In some cases a junction with fixed reverse bias may be used as a capacitance of a set value. More commonly, the varactor diode is designed to exploit the voltage-variable properties of the junction capacitance. For example, a varactor (or a set of varactors) may be used in the tuning stage of a radio receiver to replace the bulky variable plate capacitor. The size of the resulting circuit can be greatly reduced, and its dependability is often improved. Other applications of varactors include use in harmonic generation, microwave frequency multiplication, and active filters.

If the p-n junction is abrupt, the capacitance varies as the square root of the reverse bias V_r [Eq. (5-66)]. In a graded junction, however, the capacitance can usually be written in the form

$$(6\text{-}1) \qquad\qquad C_j \propto V_r^{-n}, \qquad \text{for } V_r \gg V_0$$

For example, in a linearly graded junction the exponent **n** is one-third. Thus the voltage sensitivity of C_j is greater for an abrupt junction than for a linearly graded junction. For this reason, varactor diodes are commonly made by alloying or epitaxial growth techniques. The epitaxial layer and the substrate doping profile can be designed to obtain junctions for which the

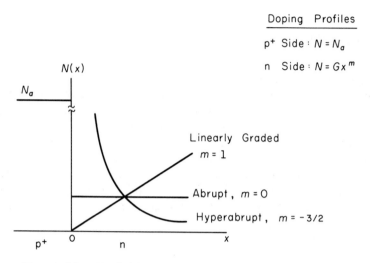

FIGURE 6-8. Graded junction profiles: linearly graded, abrupt, hyperabrupt.

exponent **n** in Eq. (6-1) is greater than one-half. Such junctions are called *hyperabrupt junctions.*

In the set of doping profiles shown in Fig. 6-8, the junction is assumed p^+-n so that the depletion layer width W extends primarily into the n side. Three types of doping profiles on the n side are illustrated, with the donor density $N_d(x)$ given by Gx^m, where G is a constant and the exponent **m** is 0, 1, or $-\frac{3}{2}$. We can show (Prob. 6.7) that the exponent **n** in Eq. (6-1) is $1/(m+2)$ for the p^+-n junction. Thus for the profiles of Fig. 6-8, **n** is $\frac{1}{2}$ for the abrupt junction and $\frac{1}{3}$ for the linearly graded junction. The hyperabrupt junction with $\mathbf{m} = -\frac{3}{2}$ is particularly interesting for certain varactor applications, since for this case $\mathbf{n} = 2$ and the capacitance is proportional to V_r^{-2}. When such a capacitor is used with an inductor L in a resonant circuit, the resonant frequency varies linearly with the voltage applied to the varactor

$$(6\text{-}2) \qquad \omega_r = \frac{1}{\sqrt{LC}} \propto \frac{1}{\sqrt{V_r^{-n}}} \propto V_r, \qquad \text{for } \mathbf{n} = 2$$

Because of the wide variety of C_j vs. V_r dependencies available by choosing doping profiles, varactor diodes can be designed for specific applications. For some high-frequency applications, varactors can be designed to exploit the forward-bias charge storage capacitance. The step-recovery diode is an example of this type of device.

6.2 Tunnel Diodes

The tunnel diode is a p-n junction device which operates in certain regions of its *I–V* characteristic by the quantum mechanical tunneling of electrons through the potential barrier of the junction (see Sections 2.4.4 and 5.4.1). The tunneling process for reverse current is essentially the Zener effect, although negligible reverse bias is needed to initiate the process in tunnel diodes. This device is useful in many applications, including high-speed switching and logic circuits. As we shall see in this section, the tunnel diode (often called the *Esaki diode* after L. Esaki, who first demonstrated the effect) exhibits the important feature of *negative resistance* over a portion of its *I–V* characteristic.

6.2.1 Degenerate Semiconductors. Thus far we have discussed the properties of relatively pure semiconductors; any impurity doping represented a small fraction of the total atomic density of the material. Since the few impurity atoms were so widely spaced throughout the sample, we could be confident that no charge transport could take place within the donor or acceptor levels themselves. What happens, however, if we continue to dope a semiconductor with impurities of either type? As might be expected, a point is reached at which the impurities become so closely packed within the lattice that interactions between them cannot be ignored. For example, donors present in high densities (e.g., 10^{20} donors/cm^3) are so close together that their wave functions (and therefore their position probability functions) begin to overlap. When this happens we can no longer consider the donor level as being composed of discrete, noninteracting energy states. Instead, the donor states form a band which may overlap the bottom of the conduction band. If the conduction band electron density n exceeds the effective density of states N_c, the Fermi level is no longer within the band gap but lies within the conduction band. When this occurs, the material is called *degenerate* n-type. The analogous case of degenerate p-type material occurs when the acceptor density is very high and the Fermi level lies in the valence band. In Fig. 6-9 the region between the Fermi level and the edge of the impurity band is shaded to illustrate overlapping with the valence and conduction bands. Generally, this overlapping is implied if E_F lies within one of the bands, and it is common practice to omit such shading from the band pictures. We recall that the energy states below E_F are mostly filled and states above E_F are empty, except for a small distribution dictated by the Fermi statistics. Thus in a degenerate n-type sample the region between E_c and E_F is for the most part filled with electrons, and in degenerate p-type the region between E_v and E_F is almost completely filled with holes.

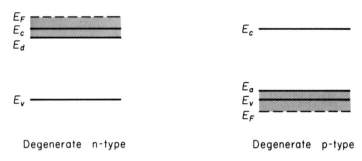

Degenerate n-type   Degenerate p-type

FIGURE 6-9. Band pictures for degenerate semiconductors, illustrating the spreading of doping levels into impurity bands.

6.2.2 Tunnel Diode Operation. A p-n junction between two degenerate semiconductors is illustrated in terms of energy bands in Fig. 6-10a. This is the equilibrium condition, for which the Fermi level is constant throughout the junction. We notice that E_{F_p} lies below the valence band edge on the p side and E_{F_n} is above the conduction band edge on the n side. Thus the bands must overlap on the energy scale in order for E_F to be constant. This overlapping of bands is very important; it means that with a small forward or reverse bias, filled states and empty states appear opposite each other, separated by essentially the width of the depletion region. If the metallurgical junction is sharp, as in an alloyed junction, the depletion region will be very narrow for such high doping densities, and the electric field at the junction will be quite large. Thus the conditions for electron tunneling are met—filled and empty states separated by a narrow potential barrier of finite height.

As mentioned above, the filled and empty states are distributed about E_F according to the Fermi distribution function; thus there are some filled states above E_{F_p} and some empty states below E_{F_n}. In Fig. 6-10 the bands are shown filled to the Fermi level for convenience of illustration, with the understanding that a distribution is implied.

Since the bands overlap under equilibrium conditions, a small reverse bias (Fig. 6-10b) allows electron tunneling from the filled valence band states below E_{F_p} to the empty conduction band states above E_{F_n}. This condition is similar to the Zener effect except that no bias is required to create the condition of overlapping bands. As the reverse bias is increased, E_{F_n} continues to move down the energy scale with respect to E_{F_p}, placing more filled states on the p side opposite empty states on the n side. Thus the tunneling of electrons from p to n increases with increasing reverse bias. The resulting conventional current is opposite to the electron flow, that is, from n to p. At equilibrium (Fig. 6-10a) there is equal tunneling from n to p and from p to n, giving a zero net current.

When a small forward bias is applied (Fig. 6-10c), E_{F_n} moves up in

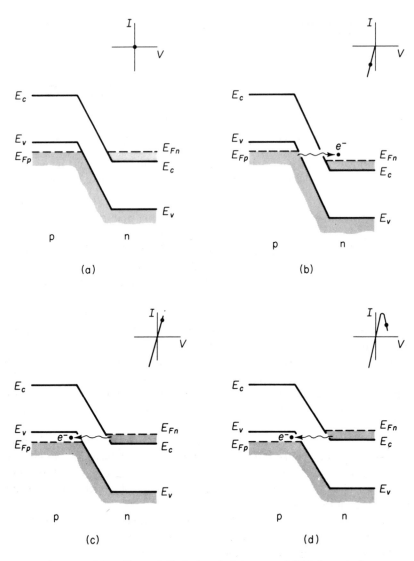

FIGURE 6-10. Tunnel diode band pictures and *I-V* characteristics for various biasing conditions: (a) equilibrium (zero bias) condition, no net tunneling; (b) small reverse bias, electron tunneling from p to n; (c) small forward bias, electron tunneling from n to p; (d) increased forward bias, electron tunneling from n to p decreases as bands pass by each other.

energy with respect to E_{F_p} by the amount qV. Thus electrons below E_{F_n} on the n side are placed opposite empty states above E_{F_p} on the p side. Electron tunneling occurs from n to p as shown, with the resulting conventional current from p to n. This forward tunneling current continues to increase with increased bias as more filled states are placed opposite empty states. However, as E_{F_n} continues to move up with respect to E_{F_p}, a point is reached at which the bands begin to pass by each other. When this occurs, the number of filled states opposite empty states decreases. The resulting decrease in tunneling current is illustrated in Fig. 6-10d. This region of the *I–V* characteristic is important in that the *decrease* of tunneling current with *increased* bias produces a region of negative slope; that is, the *dynamic resistance dV/dI* is negative. This negative resistance region is useful in a number of applications, as we shall see in the following section.

If the forward bias is increased beyond the negative resistance region, the current begins to increase again (Fig. 6-11). Once the bands have passed each other, the characteristic resembles that of a conventional diode. The forward current is now dominated by the diffusion current—electrons surmounting the potential barrier from n to p and holes surmounting their potential barrier from p to n. Of course, the diffusion current is present in the forward tunneling region, but it is negligible compared to the tunneling current.

The total tunnel diode characteristic (Fig. 6-12) has the general shape of an *N* (if a little imagination is applied); therefore, it is common to refer to this characteristic as exhibiting a *type N negative resistance*. It is also called a *voltage-controlled negative resistance*, meaning that the current

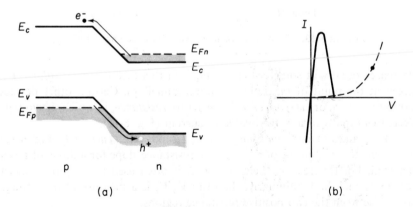

(a)   (b)

FIGURE 6-11. Band picture (a) and *I-V* characteristic (b) for the tunnel diode beyond the tunnel current region. In (b) the tunneling component of current is shown by the solid curve and the diffusion current component is dashed.

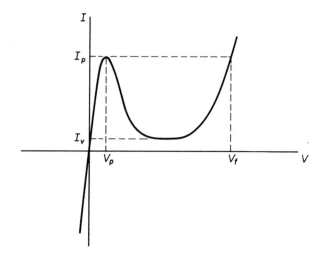

FIGURE 6-12. Total tunnel diode characteristic.

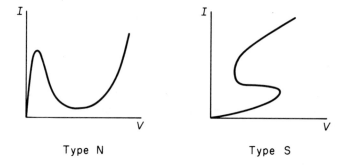

Type N   Type S

FIGURE 6-13. Two types of negative resistance characteristics.

decreases rapidly at some critical voltage (in this case the *peak voltage* V_p, taken at the point of maximum forward tunneling). Certain other devices exhibit a *type S*, or *current-controlled negative resistance*, in which the voltage decreases rapidly at a critical value of current (Fig. 6-13).

The values of *peak tunneling current* I_p and *valley current* I_v (Fig. 6-12) determine the magnitude of the negative resistance slope for a diode of given material. For this reason, their ratio I_p/I_v is often used as a figure of merit for the tunnel diode. Similarly, the ratio V_p/V_f is a measure of the voltage spread between the two positive resistance regions.

6.2.3 Circuit Applications.

The type N negative resistance of the tunnel diode can be used in a number of ways to achieve switching, oscilla-

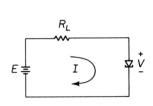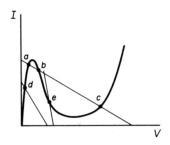

FIGURE **6-14.** Three types of load line intersections with the tunnel diode characteristic.

tion, amplification, and other circuit functions. This wide range of applications, coupled with the fact that the tunneling process does not present the time delays of drift and diffusion, makes the tunnel diode a natural choice for many high-speed circuits.

A simple bias circuit can result in several types of load lines on the tunnel diode characteristic (Fig. 6-14).† When the slope of the load line $(-1/R_L)$ is less than the slope of the negative resistance region, there can be three points of intersection with the characteristic (a, b, and c). Points a and c are stable operating points, but the circuit is unable to maintain the condition represented by point b. Since the slope of the characteristic is negative at b, any small increase in I will cause V to decrease. This will cause a further increase in I until the stable point a is reached. Similarly, a small decrease in current at b causes an eventual shift to the stable point c. Since there are two stable operating points with this type of load line, the circuit is called *bistable*. There are obvious switching and logic applications of the bistable circuit, since its operation can be shifted at will between points a and c by the application of pulses to the d-c-biased circuit. For example, if the circuit is in state a, a positive pulse (sufficient to overcome the peak current of the characteristic) can shift the operating point to c. It will remain at c until a negative pulse shifts it back to a. Thus the condition of the circuit depends on which type of pulse was last applied, indicating applications such as a high-speed computer memory storage element.

Other load lines can be drawn which cross the characteristic in only one point. If the crossing is in a positive resistance region (point d in Fig. 6-14), the circuit is called *monostable*. A crossing in the negative resistance region (point e) indicates an *astable* circuit. The astable condition is useful for amplifiers and oscillators. A simple illustration of a monostable circuit is the *threshold detector* (Fig. 6-15), which gives a very small output for any input voltage below a certain threshold value and a significantly larger

---

†For a discussion of load lines, see Section 8.1.1.

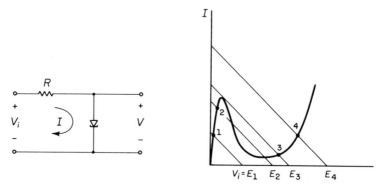

FIGURE 6-15. Threshold detector.

output above threshold. Suppose a series of voltage pulses of varying size appear at the input of the circuit. A small pulse such as $v_i = E_1$ produces an even smaller output at 1. A larger pulse (E_2) still registers a small output at 2, but when v_i is as large as E_3, the output shifts all the way to the diffusion current region of the characteristic (point 3). The bias E_3 represents the threshold value, and all pulses larger than E_3 will be registered at the output with a large voltage (e.g., point 4). The load line for E_2 has more than one crossing in this example, but since switching occurs only above the threshold voltage, the circuit is basically monostable in its operation.

If the tunnel diode circuit is biased in the negative resistance region (e.g., point *e* in Fig. 6-14), a-c gain can be achieved. For example, let us consider the simple a-c circuits of Fig. 6-16. We assume here that the tunnel diode is biased in its negative resistance region and omit the d-c portion of the circuit for a-c analysis. The tunnel diode is represented by a single negative resistance in these simplified models.

In the parallel circuit biased in the negative resistance region, the a-c current gain is simply

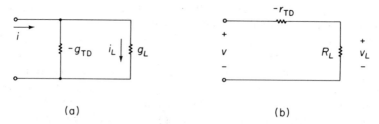

(a)   (b)

FIGURE 6-16. Equivalent a-c circuits with the tunnel diode biased in the negative resistance region: (a) parallel circuit; (b) series circuit.

(6-3) $$A_i = \frac{i_L}{i} = \frac{g_L}{g_L - g_{TD}} = \frac{1}{1 - g_{TD}/g_L}$$

Current amplification is obtained when the load conductance g_L is greater than the magnitude of the tunnel diode conductance g_{TD}. When $g_L \leq g_{TD}$, the device cannot be maintained in a stable condition. Similarly, in the series circuit, the voltage gain is

(6-4) $$A_v = \frac{v_L}{v} = \frac{R_L}{R_L - r_{TD}} = \frac{1}{1 - r_{TD}/R_L}$$

To maintain the device in the negative resistance region without switching, we must choose $R_L < r_{TD}$ (Fig. 6-14).

Actually, the circuits of Fig. 6-16 are oversimplified and do not reflect the true nature of the device, particularly at high frequencies. A better a-c equivalent circuit for the tunnel diode is shown in Fig. 6-17. The device

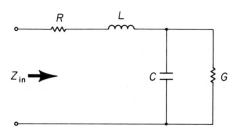

FIGURE 6-17. An a-c equivalent circuit for the tunnel diode which includes the junction capacitance and the lead resistance and inductance.

conductance G (either positive or negative, depending on the Q-point) is shunted by C to account for the junction capacitance and whatever capacitance may be associated with the physical mounting of the device. The elements R and L account for resistance and inductance external to the actual junction, such as in the leads. Now the tunnel diode has a complex input impedance given by

$$Z_{in} = R + j\omega L + \frac{1}{G + j\omega C} = R + j\omega L + \frac{G - j\omega C}{G^2 + \omega^2 C^2}$$

(6-5) $$Z_{in} = R + \frac{G}{G^2 + \omega^2 C^2} + j\left(\omega L - \frac{\omega C}{G^2 + \omega^2 C^2}\right)$$

In the negative resistance region, we can represent G by $-g$; in this case the input resistance R_{in} (the real part of Z_{in}) is given by

(6-6) $$R_{in} = R - \frac{g}{g^2 + \omega^2 C^2}$$

This is negative only when $R < g/(g^2 + \omega^2 C^2)$. Thus, a high series resistance can eliminate the negative resistance character of the device. We notice the real part of the diode impedance is zero when $Rg^2 + \omega^2 C^2 R = g$, that is, when

$$\omega^2 = \frac{1}{C^2 R}(g - Rg^2)$$

(6-7)
$$f_{RCO} = \frac{g}{2\pi C}\sqrt{\frac{1}{gR} - 1}$$

This is the *resistive cutoff* frequency, the upper frequency for which the diode exhibits negative resistance. There is also a *self-resonant* frequency f_{SR}, at which the reactive part of the device impedance is zero. This occurs when the imaginary terms in Eq. (6-5) are zero

$$\omega L - \frac{\omega C}{G^2 + \omega^2 C^2} = 0$$

$$\frac{C}{L} = G^2 + \omega^2 C^2$$

(6-8)
$$f_{SR} = \frac{1}{2\pi}\sqrt{\frac{1}{LC} - \frac{G^2}{C^2}}$$

This is also called the *reactive cutoff frequency*. In typical circuits $f_{SR} > f_{RCO} > f$, where f is the operating frequency.

6.2.4 Relation to Other Diode Forms.
As pointed out previously, the tunnel diode operates by principles introduced with the Zener diode and simply represents a greater shift in the energy bands due to higher doping. Actually, there is a continuous variation of device properties with doping level from the tunnel diode to the conventional diode, as Fig. 6-18 illustrates. This example, first demonstrated by Lesk et al.,† serves as an excellent illustration of the effects of Fermi level position on device behavior. For simplicity, we have chosen a constant degenerate doping density in the p region and let the density of donors vary on the n side of the junction. The variations of n in Fig. 6-18 can be obtained experimentally by forming a diffused n region (Fig. 6-19a). The donor density N_d decreases from a degenerate concentration at the surface to a small value deep in the material (Fig. 6-19b). This region is contacted by an alloyed p region of uniform degenerate doping. By successively etching the sample to smaller values of N_d, the characteristics of Fig. 6-18 are obtained.

---

†I. A. Lesk et al., "Germanium and Silicon Tunnel Diodes—Design, Operation and Application," 1959 IRE *WESCON*. Convention Record, Part 3.

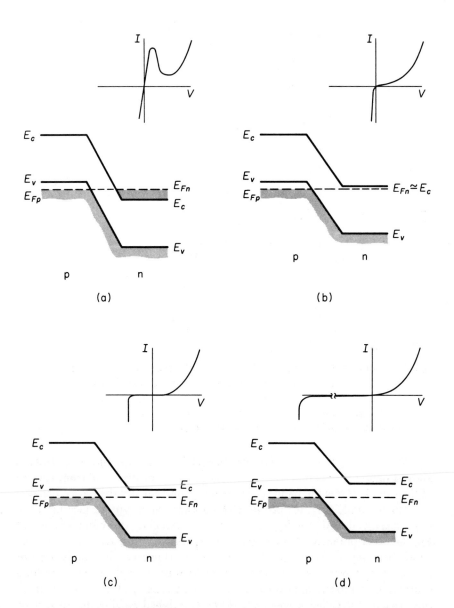

FIGURE 6-18. Relation of diode forms to Fermi level positions: (a) tunnel diode; (b) backward diode; (c) Zener diode; (d) conventional diode.

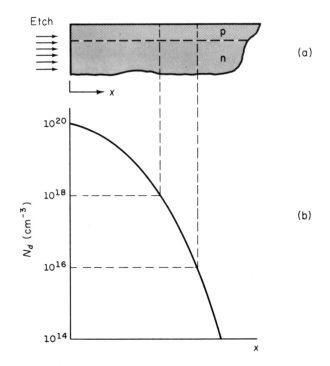

FIGURE 6-19. A device which allows observation of the characteristics of Fig. 6-18 by successive etching: (a) device with degenerate alloyed p region and diffused n region; etching of n region exposes successively lower doping densities; (b) distribution of donor densities in the n region.

When the junction in this example is first formed, the degenerate p alloyed region is in contact with the degenerate n region at the front surface of the sample. Thus the bands overlap at equilibrium and the device behaves as a tunnel diode (Fig. 6-18a). By etching away the n-type surface until the most heavily doped nondegenerate layer is exposed, the band picture of Fig. 6-18b is obtained. Notice that any reverse bias will serve to cross the bands, leading to an immediate Zener effect. However, the bands are not crossed for forward bias, so there is no forward tunneling. The resulting characteristic appears opposite to that normally expected for a diode, and for this reason it is called a *backward diode*. Continued etching of the surface moves E_{F_n} further below E_c on the n-side (Fig. 6-18c), so that a small reverse voltage is required to cross the bands. This is the condition for Zener breakdown. As the etching proceeds, E_{F_n} moves further below E_c until finally the conventional diode characteristic is obtained (Fig. 6-18d), for which avalanche breakdown occurs at higher reverse biases.

The results of Fig. 6-18 illustrate dramatically that these four types of junction diodes should be considered in terms of related mechanisms.

6.2.5 Fabrication. The basic requirements for tunneling (a sharp metallurgical junction and overlapping degenerate bands) can be met by forming an alloyed junction of one type on degenerate material of the opposite type. One of the simplest tunnel diode configurations is shown in Fig. 6-20. This device can be made easily on a strip heater for research or demon-

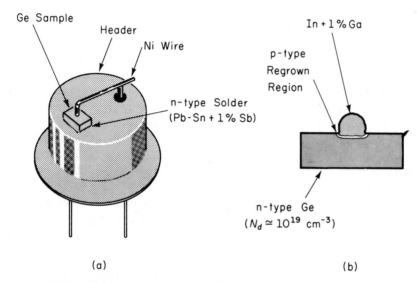

FIGURE 6-20. Germanium tunnel diode configuration for demonstration purposes: (a) mounting of sample on a header; (b) sample cross section.

stration purposes. The usual alloys for Ge devices have conveniently low melting points, and the devices are generally easy to handle. An n-type sample of degenerate Ge is polished or etched to obtain a smooth surface. It is then cut or cleaved into small dice for alloying. The header is cleaned and placed on the strip heater. At this point, a thin (e.g., 5–10 mil) Ni wire can be soldered to one of the header posts with a high melting point solder (e.g., commercial silver solder), or it can be spot-welded to the post before cleaning. Next the Ge sample is soldered to the header with Pb–Sn solder containing a small percentage of Sb. This forms an ohmic contact to the n-type Ge sample. Then a sphere of In containing a small amount of Ga is alloyed to the top surface of the sample while the Ni wire is pressed into the sphere.

Each of these alloying processes is done in a reducing (hydrogen) or an inert (e.g., argon) atmosphere.

The p-type regrown region obtained by this method is sharp and degenerate, as required. However, the I–V characteristic of the device is usually poor because of leakage currents. A common problem is that in small regions around the junction the In + Ga metal may touch the n-type surface, thereby offering paths for current to bypass the junction. The characteristic can be improved greatly by etching away part of the sample material, leaving an isolated junction as shown in Fig. 6-21a. This etching can be done in a diluted HNO_3–HF mixture or in an electrolytic NaOH etching apparatus. Figure 6-21b shows the characteristic of a device prepared by an undergraduate student using the method described here.

Although the procedures illustrated by Figs. 6-20 and 6-21 are convenient for fabricating individual units, commercial tunnel diodes are usually built by techniques more suitable for automation. Alloying is generally required, since diffusion processes do not result in sufficiently sharp junctions. An interesting variation on the alloying process is the *pulsed bond* method, in which a wire of the alloying metal is held against the semiconductor while short pulses of current are passed between the wire and the sample. Localized heating at the junction results in alloying. An advantage of this method is that the diode I–V characteristic can be monitored between pulses until the desired junction is obtained. Liquid epitaxy can be used to grow sharp junctions in some materials, and then individual diodes can be separated by selective masking and etching, leaving mesas of p-n junctions.

There is considerable variation in tunnel diode characteristics due to influences such as choice of materials and the structure of the junction. For

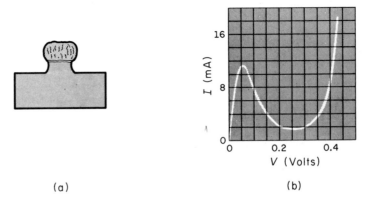

(a)   (b)

FIGURE 6-21. An example of tunnel diode fabrication: (a) cross section of an etched junction; (b) typical Ge tunnel diode I-V characteristic.

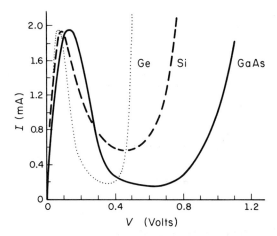

FIGURE **6-22.** Variation of tunnel diode characteristics for several semiconductors. From N. Holonyak, Jr. and I. A. Lesk, *Proc. IRE*, vol. 48, pp. 1405–1409, August 1960.

example, the forward voltage V_f is a strong function of the band gap E_g of the material (Fig. 6-22).

The peak current I_p and the valley current I_v are, of course, strongly dependent on the cross-sectional area of the junction. But in addition, the *peak-to-valley current ratio* I_p/I_v is influenced by the junction area and whatever defects are present in the crystal or surrounding the junction. Etching a junction after fabrication may improve the ratio I_p/I_v by a factor of ten or more by reducing the number of leakage current paths. However, the valley current I_v always has a value higher than expected from the contributions of band-to-band tunneling and the diffusion current. This *excess current* may be caused not only by leakage around the junction but also by processes within the junction itself. For example, one important contribution to I_v can be the tunneling of electrons through trapping levels within the band gap. After the bands have uncrossed there is no current due to band-to-band tunneling. However, if a large density of trapping centers is present (Fig. 6-23), tunneling can occur from the n-side conduction band to the trapping level (A–B). Then the electrons may drop to the valence band on the p side (B–C), thereby completing a two-step process of charge transport across the junction. Thus there can be an appreciable forward current in the valley of the characteristic, where a small current is normally desired. In fact, if the density of trapping centers is large, it is possible to observe an increase in current as the states below E_{Fn} pass by the trapping level with increased bias. This may result in a second hump in the characteristic located between V_p and V_f. Such traps can arise from impurities or from crystal defects. For example, extensive nuclear irradiation of tunneling junctions can introduce a large number of traps which may lower the value of I_p/I_v and render the

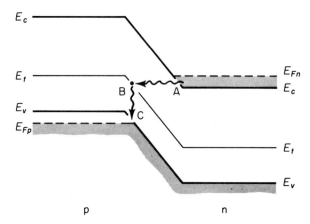

Figure 6-23. Tunneling via trapping centers.

device useless. Therefore, even though the tunnel diode is not as sensitive to irradiation as are devices which depend on long minority carrier lifetime, the effects of defect introduction cannot be ignored.

6.3 Photodiodes

In Section 4.3.4 we saw that bulk semiconductor samples can be used as photoconductors by providing a change in conductivity proportional to an optical generation rate. Often, junction devices can be used to improve the speed of response and sensitivity of detectors of optical or high-energy radiation. Single-junction devices designed to respond to photon absorption are called *photodiodes*. Some photodiodes have extremely high sensitivity and response speed. Since modern electronics often involves optical as well as electrical signals, photodiodes serve important functions as electronic devices. In this section we shall investigate the response of p-n junctions to optical generation of EHP's and discuss a few typical *photodiode detector* structures. We shall also consider the very important use of junctions as *solar cells*, which convert absorbed optical energy into useful electrical power.

6.3.1 Current and Voltage in an Illuminated Junction. In Chapter 5 we identified the current due to drift of minority carriers across a junction as a generation current. In particular, minority carriers generated thermally within a diffusion length of each side of the junction diffuse to the

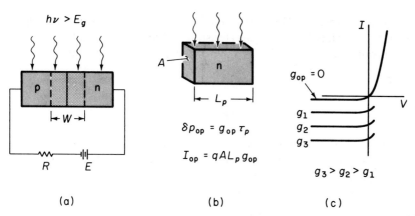

FIGURE 6-24. Optical generation of carriers in a p-n junction: (a) absorption of light by the device; (b) current I_{op} resulting from EHP generation within a diffusion length of the junction on the n side; (c) I-V characteristics of an illuminated junction.

transition region and are swept to the other side by the electric field (Fig. 6-24). If the junction is illuminated by photons with $hv > E_g$, an added generation rate g_{op} (EHP/cm³-sec) participates in this current. The number of holes created per second within a diffusion length of the transition region on the n side is $AL_p g_{op}$. Similarly $AL_n g_{op}$ electrons are generated per second within L_n of x_{p0}. The resulting current due to collection of these optically generated carriers by the junction is

$$(6\text{-}9) \qquad I_{op} = qAg_{op}(L_p + L_n)$$

We arrived at Eq. (6-9) by a rather qualitative argument; however, this expression can be derived easily from the diffusion equation if the generation rate g_{op} is included (Prob. 6.10).

Since the current of Eq. (6-9) is directed from n to p, it subtracts from the total current from p to n, and the diode equation [Eq. (5-36)] must be modified accordingly. The I-V relationship of an illuminated diode is then

$$(6\text{-}10) \qquad I = qA\left(\frac{L_p}{\tau_p}p_n + \frac{L_n}{\tau_n}n_p\right)(e^{qV/kT} - 1) - qAg_{op}(L_p + L_n)$$

Thus the I-V curve is lowered by an amount proportional to the generation rate (Fig. 6-24c). This equation can be considered in two parts—the current described by the usual diode equation, and the current due to optical generation.

When the device is short circuited ($V = 0$), the thermal generation, injection, and recombination currents (the normal diode terms) cancel in

Eq. (6-10), as expected. However, there is a short-circuit current from n to p equal to I_{op}. Thus the *I–V* characteristics of Fig. 6-24c cross the *I*-axis at negative values proportional to g_{op}. When there is an open circuit across the device, $I = 0$ and the voltage $V = V_{oc}$ is

$$(6\text{-}11) \qquad V_{oc} = \frac{kT}{q} \ln\left[\frac{L_p + L_n}{(L_p/\tau_p)p_n + (L_n/\tau_n)n_p} \cdot g_{op} + 1 \right]$$

For the special case of a symmetrical junction, $p_n = n_p$ and $\tau_p = \tau_n$, we can rewrite Eq. (6-11) in terms of the thermal generation rate $p_n/\tau_n = g_{th}$ and the optical generation rate g_{op}

$$(6\text{-}12) \qquad V_{oc} = \frac{kT}{q} \ln \frac{g_{op}}{g_{th}}, \qquad \text{for } g_{op} \gg g_{th}$$

Actually, the term $g_{th} = p_n/\tau_n$ represents the *equilibrium* thermal generation–recombination rate. As the minority carrier density is increased by optical generation of EHP's, the lifetime τ_n becomes shorter, and p_n/τ_n becomes larger (p_n is fixed, for a given N_d and T). Therefore, V_{oc} cannot increase indefinitely with increased generation rate; in fact, the limit on V_{oc} is the equilibrium contact potential V_0 (Fig. 6-25). This result is to be expected, since the contact potential is the maximum forward bias which can appear across a junction. The appearance of a forward voltage across an illuminated junction is known as the *photovoltaic effect*.

Depending on the intended application, the photodiode of Fig. 6-24 can be operated in either the third or fourth quadrants of its *I–V* characteristic. As Fig. 6-26 illustrates, power is delivered to the device from the external circuit when the current and junction voltage are both positive or both

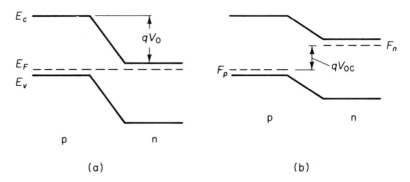

FIGURE 6-25. Effects of illumination on the open circuit voltage of a junction: (a) junction at equilibrium; (b) appearance of a voltage V_{oc} with illumination.

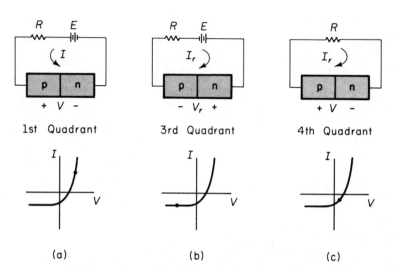

FIGURE 6-26. Operation of an illuminated junction in the various quadrants of its *I-V* characteristic: in (a) and (b) power is delivered to the device by the external circuit; in (c) the device delivers power to the load.

negative (first or third quadrants). In the fourth quadrant, however, the junction voltage is positive and the current is negative. In this case power is delivered from the junction to the external circuit (notice that in the fourth quadrant the current flows from the negative side of V to the positive side, as in a battery).

If power is to be extracted from the device, the fourth quadrant is used; on the other hand, in applications as a photodetector we reverse bias the junction and operate it in the third quadrant. We shall investigate these applications more closely in the discussion to follow. Thus far, we have neglected generation of carriers within the transition region; as we shall see, this can be the dominant generation effect in some photodiodes.

6.3.2 Solar Cells. Since power can be delivered to an external circuit by an illuminated junction, it is natural to ask about the practicality of converting solar energy into electrical energy. If we consider the fourth quadrant of Fig. 6-26c, it appears doubtful that much power can be delivered. The voltage is restricted to values less than the contact potential, which in turn is generally less than the band gap voltage E_g/q. For Si the voltage V_{oc} is less than about one volt. The current generated depends on the illuminated area, but typically I_{op} is in the milliampere range for a large area junction. However, if many such devices are used, the resulting power can be significant. In fact,

FIGURE 6-27. Solar cell arrays attached to the 1969 Mars fly-by satellite Mariner. NASA Photograph courtesy of Jet Propulsion Laboratory.

arrays of p-n junction solar cells are currently used to supply electrical power for many space satellites. Solar cells can supply power for the electronic equipment aboard a satellite over a long period of time, which is a distinct advantage over batteries. The array of junctions can be distributed over the surface of the satellite or can be contained in solar cell "paddles" attached to the main body of the satellite (Fig. 6-27).

To utilize a maximum amount of available optical energy, it is necessary to design a solar cell with a large area junction located near the surface of the device (Fig. 6-28). The planar junction is formed by diffusion, and the surface is coated with appropriate materials to reduce reflection and to decrease surface recombination. Many compromises must be made in solar cell design. For example, the junction depth d must be less than L_p in the n material to allow holes generated near the surface to diffuse to the junction before they recombine; similarly, the thickness of the p region must be such that electrons generated in this region can diffuse to the junction before recombination takes place. This requirement implies a proper match between the electron diffusion length L_n, the thickness of the p region, and the mean optical penetration depth $1/\alpha$ [see Eq. (4-2)]. It is desirable to have a large contact potential V_0 to obtain a large photovoltage, and therefore heavy

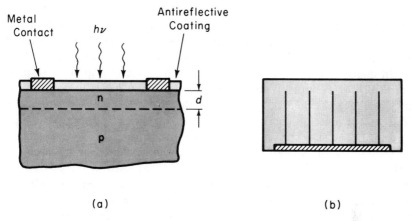

FIGURE 6-28. Configuration of a solar cell: (a) enlarged view of the planar junction; (b) top view, showing metal contact "fingers."

doping is indicated; on the other hand, long lifetimes are desirable and these are reduced by doping too heavily. It is important that the series resistance of the device be very small so that power is not lost to heat due to ohmic losses in the device itself. A series resistance of only a few ohms can seriously reduce the output power of a solar cell. Since the area is large, the resistance of the p-type body of the device can be made small. However, contacts to the thin n region require special design. If this region is contacted at the edge, current must flow along the thin n region to the contact, resulting in a large series resistance. To prevent this effect, the contact can be distributed over the n surface by providing small contact "fingers" as in Fig. 6-28b. These thin contacts serve to reduce the series resistance without interfering appreciably with the incoming light.

The maximum power delivered by a solar cell to a load resistance occurs when the product $I_r V$ is maximum (Fig. 6-26c). It is easy to show (Prob. 6.11) that the voltage for maximum power V_{mp} can be found from the equation

$$(6\text{-}13) \qquad \left(1 + \frac{q}{kT} V_{mp}\right) e^{qV_{mp}/kT} = 1 + \frac{I_{op}}{I_{th}}$$

where I_{th} is the thermally induced reverse saturation current. When the voltage and current for maximum power are found for a given excitation level, the appropriate load resistance R can be chosen as $R = V_{mp}/I_{mp}$.

6.3.3 Photodetectors. When the photodiode is operated in the third quadrant of its I–V characteristic (Fig. 6-26b), the current is essentially independent of voltage but is proportional to the optical generation rate. Such a device provides a useful means of measuring illumination levels or of converting time-varying optical signals into electrical signals.

In many optical detection applications the detector's speed of response is critical. For example, if the photodiode of Fig. 6-26b is to respond to a series of light pulses 1 μsec apart, the photogenerated minority carriers must diffuse to the junction and be swept across to the other side in a time much less than 1 μsec. The carrier diffusion step in this process is time consuming and should be eliminated if possible. Therefore, it is desirable that the width of the depletion region W be large enough so that most of the photons are absorbed within W rather than in the neutral p and n regions. When an EHP is created in the depletion region, the electric field sweeps the electron to the n side and the hole to the p side. Since this carrier drift occurs in a very short time, the response of the photodiode can be quite fast. When the carriers are generated primarily within the depletion layer W, the detector is called a *depletion layer photodiode*. Obviously, it is desirable to dope at least one side of the junction lightly so that W can be made large. The appropriate width for W is chosen as a compromise between sensitivity and speed of response. If W is wide, most of the incident carriers will be absorbed in the depletion region. Also, a wide W results in a small junction capacitance [see Eq. (5-67)], thereby reducing the RC time constant of the detector circuit. On the other hand, W must not be so wide that the time required for drift of photogenerated carriers out of the depletion region is excessive.

One convenient method of controlling the width of the depletion region is to build a *p-i-n photodetector* (Fig. 6-29). The "i" region need not be truly intrinsic, as long as the resistivity is high. It can be grown epitaxially on the n-type substrate, and the p region can be obtained by diffusion. When this device is reverse biased, the applied voltage appears almost entirely across the i region. If the carrier lifetime within the i region is long compared with

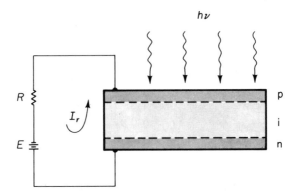

FIGURE 6-29. Schematic representation of a p-i-n photodiode.

the drift time, most of the photogenerated carriers will be collected by the n and p regions.

The type of photodiode described here is sensitive to photons with energies near the band gap energy. If $h\nu$ is less than E_g, the photons will not be absorbed; on the other hand, if the photons are much more energetic than E_g, they will be absorbed very near the surface, where the recombination rate is high. Therefore, it is necessary to choose a photodiode material which is sensitive to a particular region of the spectrum. This generally means that semiconductor diodes such as GaAs, Si, and Ge respond best to band gap light. Detectors sensitive to longer wavelengths result if photons can excite electrons into or out of impurity levels.

Mixed compounds can be utilized to vary the band gap of the detector to match a particular region of the spectrum. For example, the mixed compound HgCdTe can be grown with a band gap chosen from about 1.6 eV (CdTe) to near zero (HgTe), depending on the composition. Even a relatively narrow-gap material such as Ge is useful in *nuclear particle detectors*. When energetic particles (electrons, protons, etc.) are incident on a semiconductor, absorption takes place over a considerable distance within the lattice. Much of the energy of each particle is given up in a cascade of EHP's as the particle slows down. For example, a 1-MeV electron creates about 0.3×10^6 EHP in absorption by Si (i.e., about one-third of the absorbed energy results in EHP generation, the rest going to heat and other losses, including the creation of lattice disorder). Energetic particle absorption in a reverse-biased p-i-n diode induces current which can be analyzed to measure the particle energy and flux density.

If very low-level optical signals are to be detected, it is often desirable to operate the photodiode in the avalanche region of its characteristic. In this mode each photogenerated carrier results in a significant change in the current because of avalanche multiplication. In the *avalanche photodiode* the junction must be very uniform, and a guard ring or bevel (Fig. 6-3) is generally used to ensure that the avalanche current is truly proportional to the optical signal. With proper design a Si avalanche photodiode can have high sensitivity to low-level optical signals, and the response speed is in the neighborhood of one nanosecond.

6.4 Light-Emitting Junctions

When carriers are injected across a forward-biased junction, the current is usually accounted for by recombination in the transition region and in the neutral regions near the junction. In a semiconductor with an indirect band gap, such as Si or Ge, the recombination releases heat to the lattice. On the other hand, in a material characterized by direct recombination, considerable light may be given off from the junction under forward bias. This effect, called *injection electroluminescence* (Section 4.2.3), provides an important application of diodes as generators of light.

The III–V and II–VI compound semiconductors offer direct and indirect band gaps with a wide variety of energies from the infrared through the visible regions of the spectrum. Although junctions cannot be formed in all of these materials, diodes of many semiconductors are possible. If the electron–hole recombination occurs directly across the band gap, photons are given off with energies restricted to a narrow band near E_g. In other materials transitions via impurity levels dominate, but in certain of these materials the energy difference is given up primarily as photons. In either case, the light output of the diode is an essentially pure color characteristic of the dominant transition. We shall discuss methods of obtaining truly monochromatic light output in the description of lasers (Chapter 7).

6.4.1 Applications of Diode Lamps.

Diode lamps are useful in many applications for which small incandescent or gas discharge lamps are unsuitable. Among the advantages of diode light emitters are the following: unlike incandescent bulbs, they operate best when cold; they require little power; they can be switched on and off in nanoseconds; they are inherently very small and amenable to integration into low-power electronic circuitry; they have good durability and long operating life. For example, diodes are used as indicator lamps in applications where reliability, low power consumption, and small size are important.

A light-emitting diode (LED) can be used in conjunction with a photodiode or other photosensitive device to transmit information optically between locations. By varying the current through a diode, the light output can be modulated such that the information to be transmitted appears in the optical signal directed at the detector (Fig. 6-30). The information trans-

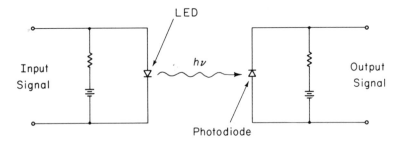

FIGURE 6-30. Optical signal transmission using a light-emitting diode and a photodiode.

ferred from the light emitter to the detector may be introduced ahead of the diode lamp, as in information transmission between locations in a computer or other system. On the other hand, the information may be introduced between the source and detector. For example, the arrangement illustrated by Fig. 6-30 can be used in a system for reading punched computer cards, in which the cards pass between a row of diode lamps and a row of detectors. A light emitter and a photodiode form an *optoelectronic pair* which provides complete electrical isolation between input and output, since the only link between the two devices is optical. In fabricating an isolator, both devices may be mounted on a ceramic substrate and packaged in a single header.

There are many other potential applications for LED's. Lamps and detectors can be used to perform logic functions and information storage and retrieval. Arrays of diode lamps can be used as color displays of printed information. Tiny, low-power lamps have many applications in medicine. Depending on whether the optical information is sent to an electronic detector or the human eye, infrared or visible light may be chosen for a particular application. The infrared light from GaAs diodes is useful for data transmission or other applications where electronic detectors are used. Applications requiring human observation call for a variety of colors in the visible part of the spectrum.

6.4.2 Materials.
The band gaps of various compound semiconductors are illustrated in Fig. 4-4 relative to the spectrum. There is a wide variation in band gaps and, therefore, in available photon energies, extending from the ultraviolet (ZnS, 3.6 eV) into the infrared (InSb, 0.18 eV). In fact, by utilizing mixed compounds the number of available energies can be increased significantly. A good example of the variation in photon energy obtainable from the compound semiconductors is the material gallium arsenide–phosphide, which is a combination of GaAs (infrared) and GaP (green). As the percentage of As is reduced and P is increased in the material, the resulting band gap varies from the direct 1.43-eV gap of GaAs to the indirect 2.26-eV gap of GaP. The symbol for the resulting material is often written as $GaAs_{1-x}P_x$, where the x represents the fraction of P atoms serving as the column V impurity. For example, a crystal designated as $GaAs_{.8}P_{.2}$ has 80 per cent As atoms and 20 per cent P atoms on the non-gallium sites of the lattice. The band gap of the resulting material varies almost linearly with x until the 44 per cent composition is reached, and electron–hole recombination is direct over this range. For compositions with P concentrations above 44 per cent, the band gap becomes indirect. This mixed compound offers wavelength variations from the infrared at $x = 0$ to the red part of the visible spectrum at $x = 0.44$.

Color variation can be accomplished by the proper choice of impurity levels as well as by variation of band gap. For example, recombination in GaP proceeds by indirect transitions via impurity states, with photons given off in many of the transitions. By choosing proper impurity doping in GaP, we can achieve considerable variation in color of the radiation. Shallow donors (Te, S, Se) introduce levels less than 0.1 eV below the conduction band in GaP ($E_g = 2.26$ eV), and acceptors such as Zn and Cd introduce levels close to the valence band. A typical recombination event occurs as an electron is captured at a shallow sulphur level, a hole is captured at a Zn level, and the electron drops from the donor to the empty state in the acceptor level (Fig. 6-31a). In this case the two events in which free carriers are captured at

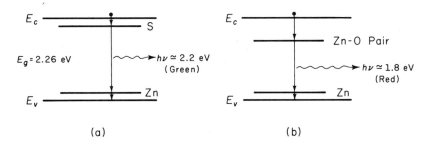

(a)   (b)

FIGURE **6-31.** Luminescent transitions between impurity levels in GaP: (a) 2.2 eV (green) transition from a S donor level to a Zn acceptor level; (b) 1.8 eV (red) transition from a zinc–oxygen pair level to a Zn acceptor level.

impurity levels result in thermal energy given up to the lattice, while the transition between the impurity levels is radiative, with emission of a photon in the green part of the spectrum. A more important mechanism is electron–hole recombination at nitrogen impurities. A neutral N atom traps an electron about 8 meV below the conduction band. The resulting negative center then binds a hole by coulombic attraction. When the bound electron and hole recombine, a photon is emitted with energy only slightly less than E_g, and green luminescence results. This is called *bound exciton* recombination.

Deeper levels can be introduced in GaP to achieve radiation of lower energy photons; for example, a complex composed of a zinc–oxygen pair can trap an electron about 0.4 eV below the conduction band. In an oxygen-doped GaP sample counterdoped with Zn, electron transitions from the zinc–oxygen pair level to the Zn acceptor level result in photons in the red part of the spectrum (Fig. 6-31b). Bound exciton recombination at the Zn–O trap is also important. The *external quantum efficiency* (optical power out divided by electrical power in) of LED's of this type can be as high as seven per cent.

6.4.3 Fabrication. Generally, in forming a p-n junction for a light-emitting diode, the junction should be near the surface to minimize reabsorption of the emitted light. Shallow diffusions can be made to keep the junction near the surface. Another problem exists, however, in transmitting the light through the surface of the device. Unless the photons are emitted almost perpendicular to the plane of the surface, it is likely that the light will be reflected back into the sample (Fig. 6-32a). The critical angle θ_c for internal

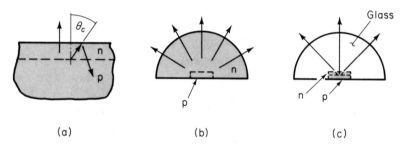

(a)   (b)   (c)

FIGURE **6-32.** Transmission and reflection at the surface of a diode: (a) internal reflection from a semiconductor–air boundary; (b) transmission through a hemispherical semiconductor surface; (c) transmission through a glass or plastic dome.

reflection is only about 16° for a GaAs-to-air boundary. More light can be extracted by polishing the diode surface into a hemispherical shape so that more light rays strike the surface at small angles to the perpendicular (Fig. 6-32b). This technique requires a junction deep in the material, but the gain in reducing surface reflection is greater than the loss in absorption between the junction and the surface. Unfortunately, forming the semiconductor material into a hemispherical shape is difficult. Alternatively, a plastic or glass dome can be molded onto the surface of the diode. The critical angle θ_c is larger for boundaries between materials with similar dielectric constants. Thus if the dome has a dielectric constant between that of the diode and air, more light can be emitted.

Some of the II–VI compound semiconductors cannot be doped p-type or n-type at will (Section 4.2.1). In these materials, crystal defects are formed in the doping process, such that self-compensation occurs. For example, CdS and CdSe can be made n-type but cannot be converted to p-type by counterdoping with acceptor atoms. In the process of counterdoping, a lattice defect with donor properties is created for each acceptor atom introduced. Thus each acceptor state is automatically compensated by a donor state, and the Fermi level remains essentially unchanged by the doping process. This difficulty makes it impossible to form the usual p-n junction structure necessary for a luminescent diode. In some materials which do not allow p-n junction formation, a *heterojunction* can be formed between the sample and a metal or another semiconductor such that avalanche breakdown or tunneling of carriers through a thin insulating barrier can occur. With such methods enough carriers can be injected into the semiconductor to cause luminescence, although the efficiency of the resulting device is generally poor compared with a p-n junction.

6.5 Diode Arrays

We have already discussed applications which call for large arrays of p-n junction diodes. For example, the solar cell arrays of Fig. 6-27 are useful in multiplying the power conversion of individual devices. Similarly, arrays of photodetectors can be used in image processing, and arrays of LED's can be used to display letters or numbers. As a further example of diode array applications, we shall discuss the *Picturephone*† vidicon camera tube developed by Bell Telephone Laboratories for use in audiovisual telephone communications. The target for this camera tube is an array of more than three-quarters of a million p-n junctions.

To understand the operation of the Si diode vidicon, let us first consider the more conventional vidicon structure (Fig. 6-33a). The purpose of this device is to translate an optical image into electronic information, which can be transmitted to a television receiver. The essential features of the vidicon tube are a heated cathode which emits electrons, an acceleration and deflection system to scan the electron beam periodically across the face of the tube, and a photoconductive target. The photoconductor has a high dark resistance and is capable of storing negative charge when scanned by the electron beam. The front surface of the photoconductor is terminated in a transparent conductive layer which is connected to a positive potential by the bias circuit. After a scan of the beam across the screen, the photoconductor is charged to the potential of the cathode. However, if photons enter the photoconductor at a given spot, the generated EHP's increase the conductivity at that location, and the stored negative charge is drained off to the positive bias. As a result, when the electron beam passes that spot on the next scan ($\frac{1}{30}$ sec later), more electrons are deposited to bring the spot back to cathode potential. The resulting pulse of current can be synchronized with the position of the beam to relate light intensity to position on the target. This information can then be translated into a spot of proportional intensity on the CRT screen of the receiver.

---

†Registered trademark of the American Telephone and Telegraph Co.

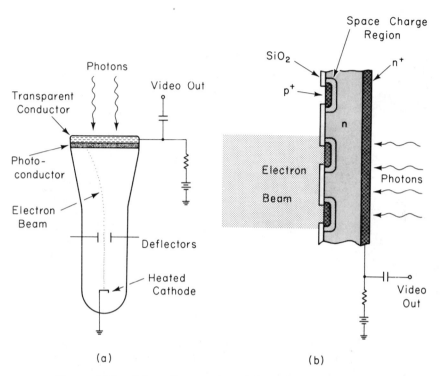

(a)   (b)

FIGURE 6-33. Schematic operation of the vidicon tube: (a) conventional vidicon with photoconductor target; (b) diode array target.

There are several disadvantages inherent in the photoconductive target, including limitations on time response and the fact that the photoconductor is easily damaged by intense light. Furthermore, desirable heat treatments of the evacuated tube during fabrication are difficult in the presence of the photoconductor. These limitations are overcome by replacing the photoconductor with an array of p^+-n diodes (Fig. 6-33b). In the diode target the junctions are reverse biased by the negative charge deposited onto the p^+ regions by the electron beam. As a result, a space charge surrounds each junction, extending primarily into the more lightly doped n region. If photons create EHP's between scans, this depletion region is relaxed and the reverse bias is reduced. Therefore, more electrons are deposited by the beam during the next scan, and the resulting current pulse is used in the video information. An interesting feature of this diode array is the fact that electrical contacts to the p^+ regions are not required, since these regions are charged and sampled periodically by the electron beam.

Fabrication of the diode array for the Picturephone image tube is an

(a)   (b)

FIGURE 6-34. Diode array targets for the Picturephone vidicon:
(a) microscope view of a portion of the diode array, compared
with a human hair; (b) Si targets and vidicon tubes. The three
diode arrays are compared with a Si ingot; the completed diode
array vidicon tube (background) is compared with a conventional
photoconductor vidicon. Photographs courtesy of Bell Telephone Laboratories, Incorporated.

excellent example of the capabilities of Si planar device technology. The
target is a wafer of n-type Si approximately 1 in. in diameter and 15 μm
thick. An array of almost 8×10^5 diodes is fabricated in a square about
$\frac{1}{2}$ in. on a side in this thin wafer. The p^+ regions of these diffused diodes are
about eight microns in diameter. As an illustration of the size of these diodes,
Fig. 6-34a shows a microscope view of a portion of the array compared with
a human hair. To produce arrays of sufficient quality and uniformity, extreme
cleanliness and careful handling are required in each fabrication step. The
completed junction array vidicon tube illustrated in Fig. 6-34b is a much
more hardy and reliable device than the older forms of vidicons.

READING LIST

A. A. BERGH and P. J. DEAN, "Light-Emitting Diodes," *Proceedings of the
IEEE*, vol. 60, pp. 156–223, February 1972.

L. P. HUNTER, *Handbook of Semiconductor Electronics* (3rd ed.). New York: McGraw-Hill, Inc., 1970. See particularly Sec. 3, "Rectification, Diodes, and Photocells."

H. A. WATSON, *Microwave Semiconductor Devices and Their Circuit Applications*. New York: McGraw-Hill, Inc., 1969. See particularly Ch. 7, "Varactor Diodes," Ch. 8, "Varactor Applications," Ch. 9, "p-i-n Diodes," Ch. 13, "Tunnel Diodes," and Ch. 14, "Tunnel-Diode Circuits."

S. M. SZE, *Physics of Semiconductor Devices*. New York: John Wiley & Sons, 1969. See particularly Ch. 4, "Tunnel Diode and Backward Diode," and Ch. 12, "Optoelectronic Devices."

H. BENDA and E. SPENKE, "Reverse Recovery Processes in Silicon Power Rectifiers," *Proc. IEEE*, vol. 55, pp. 1331–1354, August 1967.

J. L. MOLL and S. A. HAMILTON, "Physical Modeling of the Step Recovery Diode for Pulse and Harmonic Generation Circuits," *Proc. IEEE*, vol. 57, pp. 1250–1259, July 1969.

H. MELCHIOR, M. B. FISHER, and F. R. ARAMS, "Photodetectors for Optical Communication Systems," *Proc. IEEE*, vol. 58, pp. 1466–1486, October 1970.

H. C. CASEY, Jr., and F. A. TRUMBORE, "Single Crystal Electroluminescent Materials," *Materials Science and Engineering*, vol. 6, pp. 69-109, 1970.

J. I. PANKOVE, *Optical Processes in Semiconductors*. Englewood Cliffs, N. J.: Prentice-Hall, Inc., 1971.

PROBLEMS

6.1 Assume $v_s = 3 \sin \omega t$ in the circuits shown in Fig. P6-1. The diodes are ideal, with $I-V$ characteristics given by Fig. 6-1a. Sketch $v_o(t)$ over a full cycle for each circuit.

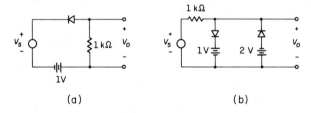

(a) (b)

FIGURE P6-1

6.2 Assume $v_s = \sqrt{2} \sin \omega t$ for the circuit of Fig. P6-1a. For what fraction of a cycle is the diode conducting?

6.3 The diode in the simple half-wave rectifier circuit of Fig. 6-2a is described by the piecewise linear model of Fig. 6-1c. The offset voltage E_o is 0.5 V, and $R = dv/di = 500\,\Omega$; in the circuit, $v_s = 2 \sin \omega t$ and the load resistor is 1 kΩ. Sketch v_o over two cycles of v_s.

6.4 It is found experimentally that the avalanche breakdown voltage V_{br} is 60 V for an abrupt Si p$^+$-n junction with $N_d = 10^{16}$ cm^{-3}. What is the minimum thickness of the n region (between the metallurgical junction and the ohmic contact) required to ensure avalanche breakdown rather than punch-through?

6.5 Assume a p$^+$-n diode is built with an n region width l smaller than a hole diffusion length ($l < L_p$). This is the so-called *narrow base diode*. Since for this case holes are injected into a short n region under forward bias, we cannot use the assumption $\delta p(x_n = \infty) = 0$ in Eq. (4-38). Instead, we must use as a boundary condition the fact that $\delta p = 0$ at $x_n = l$.

(a) Solve the diffusion equation to obtain

$$\delta p(x_n) = \frac{\Delta p_n [e^{(l-x_n)/L_p} - e^{(x_n-l)/L_p}]}{e^{l/L_p} - e^{-l/L_p}}$$

(b) Show that the current in the diode is

$$I = \left(\frac{qAD_p p_n}{L_p} \operatorname{ctnh}\frac{l}{L_p}\right)(e^{qV/kT} - 1)$$

6.6 Given the narrow base diode result (Prob. 6.5), (a) calculate the current due to recombination in the n region, and (b) show that the current due to recombination at the ohmic contact is

$$I \text{ (ohmic contact)} = \left(\frac{qAD_p p_n}{L_p} \operatorname{csch}\frac{l}{L_p}\right)(e^{qV/kT} - 1)$$

6.7 Assume a p$^+$-n junction is built with a graded n region in which the doping is described by $N_d(x) = Gx^m$. The depletion region ($W \cong x_{n0}$) extends from essentially the junction at $x = 0$ to a point W within the n region.

(a) Integrate Gauss's law across the depletion region to obtain the maximum value of the electric field $\mathscr{E}_0 = -qGW^{(m+1)}/\epsilon(m + 1)$.

(b) Find the expression for $\mathscr{E}(x)$ and use the result to obtain $V_0 - V = qGW^{(m+2)}/\epsilon(m + 2)$.

(c) Find the charge Q due to ionized donors in the depletion region; write Q explicitly in terms of $(V_0 - V)$.

(d) Using the results of (c), take the derivative $dQ/d(V_0 - V)$ to show that the capacitance is

$$C_j = A\left[\frac{qG\epsilon^{(m+1)}}{(m + 2)(V_0 - V)}\right]^{1/(m+2)}$$

6.8 Estimate the minimum density of donor atoms required to dope Ge to the degenerate condition. A common acceptor alloying metal used in making tunnel diodes in degenerate n-type Ge is In containing a small percentage of Ga (Fig. 6-20b). Is the Ga necessary? What function does it serve?

6.9 The tunnel diode of Fig. 6-23 has a trapping level E_t located 0.3 eV above the valence band. Assume $E_g = 1$ eV, and $E_{Fn} - E_c$ on the n side equals $E_v - E_{Fp}$ on the p side, equals 0.1 eV.
(a) Calculate the minimum forward bias at which tunneling through E_t occurs.
(b) Calculate the maximum forward bias for tunneling via E_t.
(c) Sketch the I–V curve for this tunnel diode. Assume the maximum tunneling current via E_t is about one-third of the peak band-to-band tunneling current.

6.10 For steady state optical excitation, we can write the hole diffusion equation as

$$D_p \frac{d^2 \delta p}{dx^2} = \frac{\delta p}{\tau_p} - g_{op}$$

Assume a long p^+-n diode is uniformly illuminated by an optical signal resulting in g_{op} EHP/cm³-sec.
(a) Show that the excess hole distribution in the n region is

$$\delta p(x_n) = \left[p_n(e^{qV/kT} - 1) - g_{op} \frac{L_p^2}{D_p} \right] e^{-x_n/L_p} + \frac{g_{op} L_p^2}{D_p}$$

(b) Calculate the hole diffusion current $I_p(x_n)$ and evaluate it at $x_n = 0$. Compare the result with Eq. (6-10) evaluated for a p^+-n junction.

6.11 (a) Derive Eq. (6-13) from Eq. (6-10) by finding the condition for a maximum in the product IV.
(b) Write this equation in the form $\ln x = C - x$ for the case $I_{op} \gg I_{th}$, and $V_{mp} \gg kT/q$.
(c) Assume a Si solar cell with a dark saturation current of 1.5 nA is illuminated such that the short-circuit current is $I_{op} = 100$ mA. Use a graphical solution to obtain the voltage V_{mp} at maximum delivered power.
(d) What is the maximum power output of the cell at this illumination?

6.12 This problem is composed of several short qualitative questions.
(a) Is it possible for the contact potential V_0 of a junction to exceed the band gap E_g(eV)?
(b) Sketch the band diagram, including the quasi-Fermi levels F_n and F_p, throughout a uniformly illuminated diode under open-circuit conditions.

(Continued)

(Continued 6.12)

    (c) Why does the forward voltage V_f vary for the diodes of Fig. 6-22?

    (d) The backward diode (Fig. 6-18b) can be operated as a high-frequency diode; the frequency characteristics of this device in the "forward-bias" (easy current flow) direction are better than those of a conventional diode. Explain.

LASERS 7

The word LASER is an acronym for *light amplification by stimulated emission of radiation,* which sums up the operation of an important optical and electronic device. The laser is a source of highly directional, monochromatic, coherent light, and as such it has revolutionized some long-standing optical problems and has created some new fields of basic and applied optics. The light from a laser, depending on the type, can be a continuous beam of low or medium power, or it can be a short burst of intense light delivering millions of watts. Light has always been a primary communications link between man and his environment, but until the invention of the laser, the light sources available for transmitting information and performing experiments were generally neither monochromatic nor coherent, and were of relatively low intensity. Thus the laser is of great interest to people working with optics; but it is equally important in several areas of electronics. The laser is basically an electronic device in that the light emisson occurs when excited electrons in an atomic medium drop to lower energy levels and give off the energy difference as photons. The effort of creating a useful laser system involves the techniques of solid state (or gaseous) electronics, and the transmission and detection of information with the laser makes use of a broad portion of the electronics field.

7.1 Stimulated Emission

The last three letters in the word *laser* are intended to imply how the device operates: by the *stimulated emission* of radiation. In Chapter 2 we discussed the emission of radiation when excited electrons fall to lower energy states; but generally, these processes occur randomly and can therefore be classed as *spontaneous emission.* This means that the rate at which electrons fall from some upper level of energy E_2 to a lower level E_1 is at every instant proportional to the number of electrons remaining in E_2 (the *population* of E_2). Thus if an initial electron population in E_2 were allowed to decay, we would expect an exponential emptying of the electrons to the lower energy level, with a mean decay time describing how much time an average electron spends in the upper level. An electron in a higher or excited state need not wait for spontaneous emission to occur, however; if conditions are right, it can be *stimulated* to fall to the lower level and emit its photon in a time much shorter than its mean spontaneous decay time. The stimulus is provided by the presence of photons of the proper wavelength. Let us visualize an electron in state E_2 waiting to drop spontaneously to E_1 with the emission of a photon of energy $h\nu_{12} = E_2 - E_1$ (Fig. 7-1). Now we

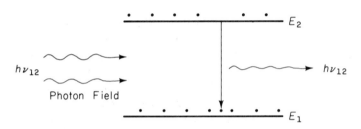

FIGURE 7-1. Stimulated transition of an electron from an upper state to a lower state, with accompanying photon emission.

assume this electron in the upper state is immersed in an intense field of photons, each having an energy $h\nu_{12} = E_2 - E_1$, and in phase with the other photons. The electron is induced to drop in energy from E_2 to E_1, contributing a photon whose wave is *in phase* with the radiation field. If this process continues and other electrons are stimulated to emit photons in the same fashion, a large radiation field can build up. This radiation will be *monochromatic* since each photon will have an energy of precisely $h\nu_{12} = E_2 - E_1$ and will be *coherent*, because all the photons released will be in phase and reinforcing. This process of stimulated emission can be described quantum mechanically to relate the probability of emission to the intensity of the radiation field. Without quantum mechanics we can make a few observations here about the relative rates at which the absorption and emission processes occur. Let us assume the instantaneous populations of E_1 and E_2 to be n_1 and n_2, respectively. We know from earlier discussions of distributions and the Boltzmann factor that at *thermal equilibrium* the relative population will be

(7-1) $$\frac{n_2}{n_1} = e^{-(E_2 - E_1)/kT} = e^{-h\nu_{12}/kT}$$

if the two levels contain an equal number of available states.

The negative exponent in this equation indicates that $n_2 \ll n_1$ at equilibrium; that is, most electrons are in the lower energy level as expected. If the atoms exist in a radiation field of photons with energy hv_{12}, such that the energy density of the field is $\rho(v_{12})$,† then stimulated emission can occur along with absorption and spontaneous emission. The rate of stimulated emission is proportional to the instantaneous number of electrons in the upper level n_2 and to the energy density of the stimulating field $\rho(v_{12})$. Thus we can write the stimulated emission rate as $B_{21}n_2\,\rho(v_{12})$, where B_{21} is a proportionality factor. The rate at which the electrons in E_1 absorb photons should also be proportional to $\rho(v_{12})$, and to the electron population in E_1. Therefore, the absorption rate is $B_{12}n_1\,\rho(v_{12})$, where B_{12} is a proportionality factor for absorption. Finally, the rate of spontaneous emission is proportional only to the population of the upper level. Introducing still another coefficient, we can write the rate of spontaneous emission as $A_{21}n_2$. For steady state the two emission rates must balance the rate of absorption to maintain constant populations n_1 and n_2 (Fig. 7-2)

$$(7\text{-}2) \qquad B_{12}n_1\,\rho(v_{12}) = \qquad A_{21}n_2 \qquad + B_{21}n_2\,\rho(v_{12})$$

$$\text{Absorption} = \text{spontaneous} + \text{stimulated}$$
$$\text{emission} \qquad\quad \text{emission}$$

This relation was described by Einstein, and the coefficients B_{12}, A_{21}, B_{21} are called the *Einstein coefficients*. We notice from Eq. (7-2) that no energy density ρ is required to cause a transition from an upper to a lower state;

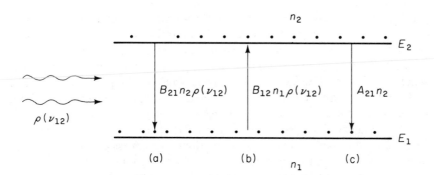

FIGURE 7-2.   Balance of absorption and emission in steady state: (a) stimulated emission; (b) absorption; (c) spontaneous emission.

---

†The energy density $\rho(v_{12})$ indicates the total energy in the radiation field per unit volume and per unit frequency, due to photons with $hv_{12} = E_2 - E_1$.

spontaneous emission occurs without an energy density to drive it. The reverse is not true, however; exciting an electron to a higher state (absorption) requires the application of energy, as we would expect thermodynamically.

At thermal equilibrium, the ratio of the stimulated to spontaneous emission rates is generally very small, and the contribution of stimulated emission is negligible.

$$(7\text{-}3) \qquad \frac{\text{Stimulated emission rate}}{\text{Spontaneous emission rate}} = \frac{B_{21} n_2 \, \rho(\nu_{12})}{A_{21} n_2} = \frac{B_{21}}{A_{21}} \rho(\nu_{12})$$

As Eq. (7-3) indicates, the way to enhance the stimulated emission over spontaneous emission is to have a very large photon field energy density $\rho(\nu_{12})$. In the laser, this is encouraged by providing an *optical resonant cavity* in which the photon density can build up to a large value through multiple internal reflections at certain frequencies (ν).

Similarly, to obtain more stimulated emission than absorption we must have $n_2 > n_1$

$$(7\text{-}4) \qquad \frac{\text{Stimulated emission rate}}{\text{Absorption rate}} = \frac{B_{21} n_2 \, \rho(\nu_{12})}{B_{12} n_1 \, \rho(\nu_{12})} = \frac{B_{21}}{B_{12}} \frac{n_2}{n_1}$$

Thus if stimulated emission is to dominate over absorption of photons from the radiation field, we must have a way of maintaining more electrons in the upper level than in the lower level. This condition is quite unnatural, since Eq. (7-1) indicates that n_2/n_1 is less than unity for any equilibrium case. Because of its unusual nature, the condition $n_2 > n_1$ is called *population inversion*. It is also referred to as a condition of *negative temperature*. This rather startling terminology emphasizes the nonequilibrium nature of population inversion, and refers to the fact that the ratio n_2/n_1 in Eq. (7-1) could be larger than unity only if the temperature were negative. Of course, this manner of speaking does not imply anything about temperature in the usual sense of that word. The fact is that Eq. (7-1) is a thermal equilibrium equation and cannot be applied to the situation of population inversion without invoking the concept of negative temperature.

In summary, Eqs. (7-3) and (7-4) indicate that if the photon density is to build up through a predominance of stimulated emission over both spontaneous emission and absorption, two requirements must be met. We must provide (1) an optical resonant cavity to encourage the photon field to build up and (2) a means of obtaining population inversion. A large part of our discussion in the following sections will be devoted to examining the second requirement for various energy level structures. There are ways of obtaining population inversion in the atomic levels of many solids, liquids, and gases, and in the energy bands of semiconductors. Thus the possibilities for laser systems with various materials are quite extensive.

7.2 The Ruby Laser

The first working laser was built in 1960 by Maiman,† using a ruby crystal. Ruby belongs to the family of gems consisting of Al_2O_3 with various types of impurities. For example, pink ruby contains about 0.05 per cent Cr atoms. Similarly, Al_2O_3 doped with Ti, Fe, or Mn results in variously colored sapphire. Most of these materials can be grown by modern techniques as single crystals.

7.2.1 The Resonant Cavity. Ruby crystals are available in rods several inches long, convenient for forming an optical cavity (Fig. 7-3). The crystal is cut and polished so that the ends are flat and parallel, with the

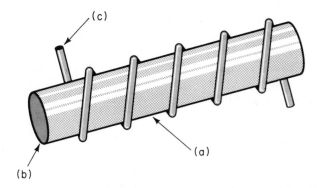

FIGURE 7-3.   Schematic diagram of a ruby laser: (a) ruby rod; (b) end mirror; (c) flash tube.

end planes perpendicular to the axis of the rod. These ends are coated with a highly reflective material, such as Al or Ag, producing a resonant cavity in which light intensity can build up by multiple reflections. One of the end mirrors is constructed to be partially transmitting so that a fraction of the light will "leak out" of the resonant system. This transmitted light is the output of the laser. Of course, in designing such a laser one must choose the amount of transmission to be a small perturbation on the resonant system. The gain in photons per pass between the end plates must be larger than the transmission at the ends, scattering from impurities, absorption, and other losses. The arrangement of parallel plates providing multiple internal reflec-

---

†T. H. Maiman, *Nature*, vol. 187, pp. 493–494, 6 August 1960.

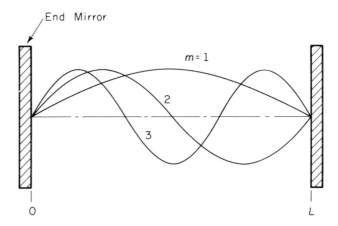

FIGURE 7-4. Resonant modes within a laser cavity.

tions is similar to that used in the Fabry–Perot interferometer†; thus the silvered ends of the laser cavity are often referred to as Fabry–Perot faces. As Fig. 7-4 indicates, light of a particular frequency can be reflected back and forth within the resonant cavity in a reinforcing (coherent) manner if an integral number of half-wavelengths fit between the end mirrors. Thus the length of the cavity for stimulated emission must be

(7-5)
$$L = \frac{m\lambda}{2}$$

where **m** is an integer. In this equation λ is the photon wavelength within the laser material. If we wish to use the wavelength λ_0 of the output light in the atmosphere (often taken as the vacuum value), the index of refraction **n** of the laser material must be considered

(7-6)
$$\lambda_0 = \lambda \mathbf{n}$$

In practice, the distance in a half-wavelength of visible or near infrared light is so small that Eq. (7-5) is automatically satisfied over some portion of a real mirror spacing. Mechanical adjustment of L to some $\mathbf{m}\lambda/2$ is not necessary, since many values of **m** and $\lambda/2$ will fit the resonant condition.

7.2.2 Population Inversion in Ruby. In the case of ruby, Cr atoms in the crystal have energy levels as shown in Fig. 7-5. Without attempting to indicate fine points of the energy level scheme, this figure portrays those

---

†Interferometers are discussed in many sophomore physics texts.

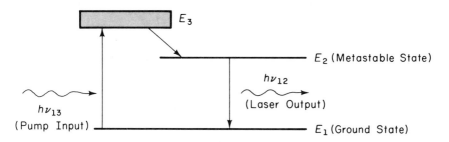

FIGURE 7-5. Energy levels for chromium ions in ruby.

ruby levels important for stimulated emission. This is basically a three-level system. Absorption occurs in a rather broad range in the green part of the spectrum, raising electrons from the ground state E_1 to the band of levels designated E_3 in the figure. These excited levels are highly unstable, and the electrons decay rapidly to the level E_2. This transition occurs with the energy difference $E_3 - E_2$ given up as heat. The level E_2 is very important for the stimulated emission process since electrons in this level have a mean lifetime of about 5 msec before they fall to the ground state. Because this lifetime is relatively long, E_2 is called a *metastable* state. If electrons are excited from E_1 to E_3 at a rate faster than the decay rate from E_2 back to E_1, the population of the metastable state E_2 becomes larger than that of the ground state E_1 (we assume that electrons fall from E_3 to E_2 in a negligibly short time). This buildup of electrons in the metastable level can be compared to the filling of a sink with an open drain. When water is poured into the sink faster than it drains out, the water level must increase. Similarly, population inversion can be established between E_2 and E_1 if electrons arrive at E_2 faster than they leave.

Population inversion is obtained by *optical pumping* of the ruby rod with a flash lamp such as the one shown in Fig. 7-3c. A common type of flash lamp is a glass tube wrapped around the ruby rod and filled with xenon gas. A capacitor can be discharged through the xenon-filled tube, creating a pulse of very intense light over a broad spectral range.

If the light pulse from the flash tube is several milliseconds in duration, we might expect an output from the ruby laser over a large fraction of that time. However, the laser does not operate continuously during the light pulse but instead emits a series of very short spikes (Fig. 7-6). When the flash lamp intensity becomes large enough to create population inversion (the *threshold pumping level*), stimulated emission from the metastable level to the ground level occurs, with a resulting laser output. Once the stimulated emission begins, however, the metastable level is depopulated very quickly. Thus the laser output consists of an intense spike lasting from a few nanoseconds to microseconds. After the stimulated emission spike, population

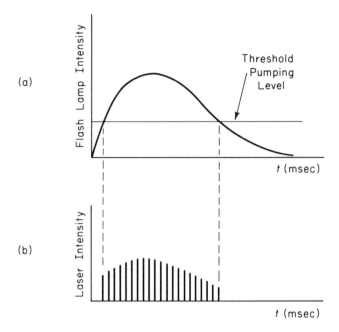

FIGURE 7-6. Laser spikes in the output of a ruby laser: (a) typical variation of flash lamp intensity with time; (b) laser spikes occurring while the flash lamp intensity is above the threshold pumping level.

inversion builds up again and a second spike results. This process continues as long as the flash lamp intensity is above the threshold pumping level.

7.2.3 Giant Pulse Lasers. We conclude from Fig. 7-6 that the metastable level never receives a highly inverted population of electrons. Whenever the population of E_2 reaches the minimum required for stimulated emission, these electrons are depleted quickly in one of the spikes. To prevent this, we must somehow keep the coherent photon field in the ruby rod from building up (and thus prevent stimulated emission) until after a larger population inversion is obtained. This can be accomplished if we temporarily interrupt the resonant character of the optical cavity. This process is called *Q-spoiling* or *Q-switching*, where *Q* is the quality factor of the resonant structure. A straightforward method for doing this is illustrated in Fig. 7-7. The front face of the ruby rod is silvered to be partially reflecting, but the back face is left unsilvered. The back reflector of the optical cavity is provided by an external mirror, which can be rotated at high speeds. When the mirror plane is aligned exactly perpendicular to the laser axis, a resonant structure exists; but as the mirror rotates away from this position, there is

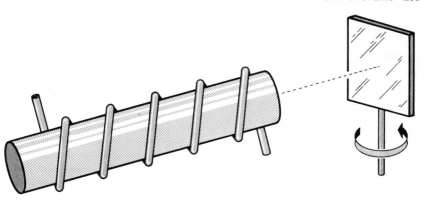

Figure 7-7. Q-switched ruby laser in which one face of the resonant cavity is an external rotating mirror.

no buildup of photons by multiple reflection, and no laser action can occur. Thus during a flash from the xenon lamp, a very large inverted population builds up as the mirror rotates off-axis. When the mirror finally returns to the position at which light reflects back into the rod, stimulated emission can occur, and the large population of the metastable level is given up in one intense laser pulse. This structure is called a *giant pulse laser* or a *Q-switched laser*. By saving the electron population for a single pulse, a large amount of energy is given up in a very short time. For example, if the total energy in the pulse is 1 J and the pulse width is 100 nsec (10^{-7} sec), the pulse power is 10^7 J/sec $= 10$ MW.

A number of alternative methods of *Q*-spoiling are available. For example, in the *Kerr cell* a transverse electric field is used to control the transmission of light longitudinally through a material in the laser path. By controlling the polarization of light transmitted through the cell, the Q of the laser cavity can be reduced until the cell is pulsed to the "on" state by the applied electric field. Another example is the use of bleachable dyes placed between the rod and one end mirror. The dye solution is opaque until the population inversion reaches a point at which the spontaneous emission is intense enough to optically bleach the dye. Then the path to the mirror is clear and the resonant cavity is established. The reading list at the end of this chapter includes discussions of other giant pulse techniques.

7.3 Other Laser Systems

One important disadvantage of the three-level laser system, such as the ruby laser, is that no significant stimulated emission occurs until at least half the ground state electrons have been excited to the metastable state. That is, population inversion exists only when the metastable state is more densely populated than the ground state. This makes for a high threshold for optical pumping and a considerable waste of energy. Better performance may be expected from a system with an extra level above the ground state (Fig. 7-8). In such a four-level laser system, population inversion begins between the metastable state (E_3) and the laser terminal state (E_2) with very little pumping. Of course, the emission rate from E_2 to the ground state must be faster than the rate from E_3 to E_2; otherwise, the electron population of E_2 could build up and exceed that of E_3, destroying the population inversion. In addition, E_2 must be substantially empty at equilibrium; that is, it must be several kT above E_1 to prevent its being populated by thermal excitation of ground state electrons.

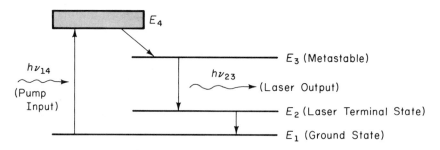

FIGURE 7-8. Typical four-level laser system.

7.3.1 Rare Earth Systems. The most common four-level solid state lasers utilize host materials doped with atoms from the *rare earth* section of the periodic table. For example, neodymium (Nd) is used as the active atom in a number of host materials. The literature abounds with descriptions of four-level laser systems, since the active spectroscopic levels of each rare earth atom are somewhat different when placed in various host lattices. For example, the laser wavelengths of Nd in the $CaWO_4$ (calcium tungstate) lattice are different from those obtained for the same dopant in Yb_2O_3 or various glasses. The number of useful doping atoms is continually being expanded, and laser host materials now include not only exotic crystals and glasses but also plastics and even liquid solutions.

7.3.2 Gas Lasers. Even the four-level solid state laser is quite inefficient, in that a large amount of optical energy is required to create population inversion. Most of the light from the flash lamp is of the wrong wavelength for the desired pumping and is therefore wasted. Generally, it is necessary to operate these lasers on a pulsed basis, allowing the system to cool between pulses. Many laser applications, however, require a steady, long-term output of light (often called *continuous wave*, or *cw* operation). A laser system which can be excited by means other than optical pumping is needed. A good alternative is provided by the gas laser systems, in which electric discharges can be created to excite atoms. In a discharge, energy is transferred to the atoms of the gas by electron impact and collisions between atoms. The resulting mixture of electrons and gas ions (*plasma*) is a good source of spectral emission and, in many cases, can be maintained continuously as long as power is supplied.

The most commonly used gas laser system employs a mixture of helium and neon gases. This mixture uses a process of resonant energy transfer between colliding He and Ne atoms. Energy exchange between atoms becomes highly probable when they possess a pair of similar energy levels. For example, He has metastable levels at almost the same energies as the $2s$ and $3s$ levels of Ne. Since the lifetimes for electrons in these He levels are relatively long, they will be populated for many of the He atoms at a given time. The close coincidence of the He metastable levels to the $2s$ and $3s$ levels of Ne allows a resonant exchange of energy and a resultant increase in population of the Ne levels. The laser emissions of interest are due to electronic transitions in the Ne atoms; thus the primary purpose of He in the mixture is to enhance the excitation process.

The Ne levels important for laser action are shown in Fig. 7-9. The strongest emission line occurs between a $2s$ level and a $2p$ level, with a wavelength of 1.1 μm (in the infrared). Another important transition occurs between a $3s$ level and a $2p$ level, with photons given off at 0.6328 μm (red). This is the most popular operating mode of the He–Ne laser, since photons are created in the red part of the visible spectrum. Thus the He–Ne laser in the 6328 Å mode is a natural choice in many applications which require visible light and cw operation.

A typical He–Ne laser consists of a glass tube to contain the gas mixture, a means of creating a discharge in the gas, and end mirrors for forming the

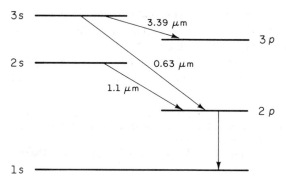

FIGURE 7-9. Schematic of laser transitions in Ne.

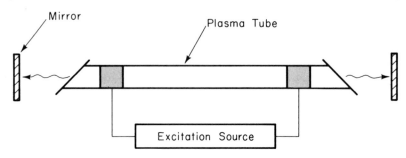

FIGURE 7-10. Gas laser structure.

resonant cavity (Fig. 7-10). It is common now to place the mirrors outside the glass envelope, or *plasma tube,* to make alignment easier. However, when this is done, it is necessary to arrange for the light to be transmitted through the end of the tube with a minimum of reflection. This is accomplished by terminating the plasma tube with an optically flat, glass plate oriented at an angle to the tube axis. There is a particular angle, called the *Brewster angle,* for which internal reflection within the plasma tube is zero for light with a given polarization. Excitation of the system is commonly accomplished by a d-c discharge between two electrodes within the plasma tube; alternatively, rf power can be applied to generate the plasma. Although it is capable of cw operation, the He–Ne gas laser is rather inefficient (typically less than 1 per cent). Thus for a He–Ne laser with a cw light output of 10 mW, we should expect to supply at least several watts of power to the gas discharge. Other gas systems are better suited for generating large amounts of optical power output; for example, CO_2 lasers can generate thousands of watts of continuous power at rather high conversion efficiencies.

7.4 Semiconductor Lasers

The laser became an important part of semiconductor device technology in 1962 when the first p-n junction lasers were built in GaAs (infrared)[†] and GaAsP (visible).[‡] We have already discussed the incoherent light emission from p-n junctions, generated by the spontaneous recombination of electrons and holes injected across the junction. In this section we shall

---

[†]R. N. Hall et al., *Phys. Rev. Letters,* vol. 9, pp. 366–368, 1 November 1962; M. I. Nathan et al., *Appl. Phys. Letters,* vol. 1, pp. 62–64, 1 November 1962; T. M. Quist et al., *Appl. Phys. Letters,* vol. 1, pp. 91–92, 1 December 1962.

[‡]N. Holonyak, Jr., and S. F. Bevacqua, *Appl. Phys. Letters,* vol. 1, pp. 82–83 ,1 December 1962.

concentrate on the requirements for population inversion due to these injected carriers and the nature of the coherent light from p-n junction lasers. These devices differ from the solid, gas, and liquid lasers discussed previously in several important respects. Junction lasers are remarkably small (typically on the order of $5 \times 5 \times 20$ mils), they exhibit high efficiency, and the laser output is easily modulated by controlling the junction current. Semiconductor lasers operate at low power compared, for example, with ruby or CO_2 lasers; on the other hand, these junction lasers compete easily with He–Ne lasers in power output. Thus the function of the semiconductor laser is to provide a portable and easily controlled source of low-power coherent radiation.

7.4.1 Population Inversion at a Junction. If a p-n junction is formed between degenerate materials, the bands under forward bias appear as shown in Fig. 7-11. If the bias (and thus the current) is large enough, electrons and holes are injected into and across the transition region in

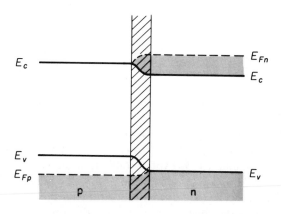

FIGURE 7-11. Band diagram of a p-n junction laser under forward bias. The cross-hatched region indicates the inversion region at the junction.

considerable concentrations. As a result, the region about the junction is far from being depleted of carriers. This region contains a large concentration of electrons within the conduction band and a large concentration of holes within the valence band. If these population densities are high enough, a condition of population inversion results, and the region about the junction over which it occurs is called an *inversion region*.

Population inversion at a junction is best described by the use of the concept of *quasi-Fermi levels* (Section 4.3.3). Since the forward-biased condition of Fig. 7-11 is a distinctly nonequilibrium state, the equilibrium equations defining the Fermi level are not applicable. In particular, the density of electrons in the inversion region (and for several diffusion lengths into the p material) is larger than equilibrium statistics would imply; the same is also true for the injected holes in the n material. We can still use the same *form* of equations to describe the carrier concentrations, if we define new quantities F_n and F_p as the quasi-Fermi levels for electrons and holes, respectively. Thus

$$(7\text{-}7a) \qquad n = N_c e^{-(E_c - F_n)/kT} = n_i e^{(F_n - E_i)/kT}$$

$$(7\text{-}7b) \qquad p = N_v e^{-(F_p - E_v)/kT} = n_i e^{(E_i - F_p)/kT}$$

If Eqs. (7-7a) and (7-7b) are considered definitions of the quasi-Fermi levels for electrons and holes, we can draw F_n and F_p on any band diagram for which we know the electron and hole distributions. For example, in Fig. 7-12, F_n in the neutral n region is essentially the same as the equilibrium Fermi

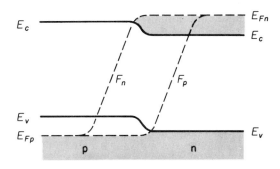

FIGURE 7-12. Quasi-Fermi levels in a laser junction under forward bias.

level E_{Fn}. This is true to the extent that the electron concentration on the n side is equal to its equilibrium value. However, since large numbers of electrons are injected across the junction, the electron concentration begins at a high value near the junction and decays exponentially to its equilibrium value n_p deep in the p material. Therefore, F_n drops from E_{Fn} to E_{Fp} as shown in Fig. 7-12. We notice that, deep in the neutral regions, the quasi-Fermi levels are essentially equal. Thus the separation of F_n and F_p at any point is a measure of the departure from equilibrium at that point. Obviously, this departure is considerable in the inversion region, since F_n and F_p are separated by an energy greater than the band gap (Fig. 7-13).

Unlike the case of the two-level system discussed in Section 7.1, the condition for population inversion in semiconductors must take into account the distribution of energies available for transitions between the bands. The basic definition of population inversion holds—for dominance of stimulated emission between two energy levels separated by energy $h\nu$, the electron population of the upper level must be greater than that of the lower level. The unusual aspect of a semiconductor is that bands of levels are available for such transitions. Population inversion obviously exists for transitions

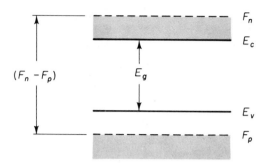

FIGURE 7-13. Expanded
view of the inversion region.

between the bottom of the conduction band E_c and the top of the valence
band E_v in Fig. 7-13. In fact, transitions between levels in the conduction band
up to F_n and levels in the valence band down to F_p take place under conditions
of population inversion. For any given transition energy $h\nu$ in a semiconductor, population inversion exists when

$$(7\text{-}8a) \qquad\qquad (F_n - F_p) > h\nu$$

For band-to-band transitions, the minimum requirement for population
inversion occurs for photons with $h\nu = E_c - E_v = E_g$

$$(7\text{-}8b) \qquad\qquad (F_n - F_p) > E_g$$

When F_n and F_p lie within their respective bands (as in Fig. 7-13), stimulated
emission can dominate over a range of transitions, from $h\nu = (F_n - F_p)$ to
$h\nu = E_g$. As we shall see below, the dominant transitions for laser action are
determined largely by the resonant cavity and the strong recombination
radiation occurring near $h\nu = E_g$.

In choosing a material for junction laser fabrication, it is necessary that
electron–hole recombination occur directly, rather than through trapping
processes such as are dominant in Si or Ge. Gallium arsenide is an example
of such a "direct" semiconductor. Furthermore, we must be able to dope
the material n-type or p-type to form a junction. In the following section we
shall consider some of the fabrication steps in making a p-n junction laser.
If an appropriate resonant cavity can be constructed in the junction region,
a laser results in which population inversion is accomplished by the bias
current applied to the junction (Fig. 7-14).

7.4.2 Fabrication. To build a p-n junction laser, we need to form a
junction in a highly doped, direct semiconductor (GaAs, for example), construct a resonant cavity in the proper geometrical relationship to the junction,

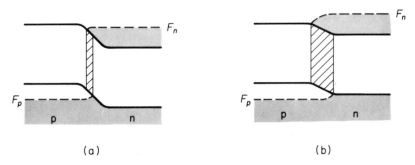

FIGURE 7-14. Variation of inversion region width with forward bias: $V(a) < V(b)$.

and make contact to the junction in a mounting which allows for efficient heat transfer. One common fabrication technique is outlined in Fig. 7-15. Beginning with a degenerate n-type sample, a p region is diffused into one side, to a thickness of 1 mil or less. In practice this step may be accomplished by diffusing Zn into the n-type GaAs sample and by then removing the diffused layer from one side by lapping. Since Zn is in column II of the periodic table and is introduced substitutionally on Ga sites, it serves as an acceptor in GaAs; therefore, the heavily doped Zn diffused layer forms a p region (Fig. 7-15b). At this point we have a large-area planar p-n junction. Next, grooves are cut or etched along the length of the sample as in Fig. 7-15c, leaving a series of long p regions isolated from each other. These long

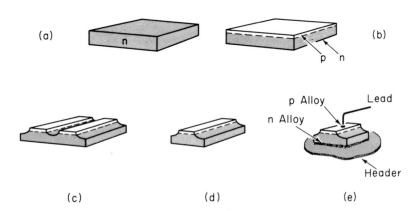

FIGURE 7-15. Fabrication of a junction laser: (a) degenerate n-type sample; (b) diffused p layer; (c) isolation of junctions by cutting or etching; (d) individual junction to be cut or cleaved into devices; (e) mounted laser structure.

p-n junctions can be cut or broken apart (Fig. 7-15d) and then sawed or cleaved into devices of the desired length.

At this point in the fabrication process, the very important requirement of a resonant cavity must be considered. It is necessary that the front and back faces (Fig. 7-15e) be flat and parallel. This can be accomplished by very careful polishing of the faces, or by cleaving. If the sample has been oriented so that the long junctions of Fig. 7-15d are perpendicular to a crystal plane of the material, it is possible to cleave the sample along this plane into laser devices, letting the crystal structure itself provide the parallel faces. The device is then mounted on a suitable header, and contact is made to the p region. Various techniques are used to provide adequate heat sinking of the device for large forward current levels.

To operate junction lasers at room temperature, it is necessary to design very efficient structures and provide good heat sinking. The most effective approach to lowering the threshold current for laser operation is a heterostructure such as those illustrated in Fig. 7-16. The object of these structures is to *confine the injected carriers* to a narrow region so that population inversion can build up at lower current levels. In GaAs the laser action occurs primarily on the p side of the junction due to a higher efficiency for electron injection than for hole injection. In a normal p-n junction the injected electrons diffuse into the p material such that population inversion occurs for only part of the electron distribution near the junction. However, if the p material is narrow and terminated in a barrier, the injected electrons can be confined near the junction. In Fig. 7-16a a layer of p-type AlGaAs ($E_g \simeq$ 2eV) is grown by liquid epitaxy on top of the thin p-type GaAs region. The wider band gap of AlGaAs effectively terminates the p-type GaAs layer, since injected electrons do not surmount the barrier at the GaAs–AlGaAs heterojunction (Fig. 7-16b). As a result of the confinement of injected electrons, laser action begins at a substantially lower current than for simple p-n junctions. In addition to the effects of carrier confinement, the change of refractive index at the heterojunction provides a waveguide effect for optical confinement of the photons.

The multilayer heterostructure shown in Fig. 7-16c has been used to obtain continuous laser operation at room temperature.[†] In this structure AlGaAs layers are grown on both sides of the thin p-type GaAs region, in which laser action takes place. The n-type AlGaAs layer provides confinement for holes in a manner analogous to the electron confinement provided by the p-type AlGaAs layer. The top p-type GaAs layer provides improved

---

[†] I. Hayashi, M. B. Panish, P. W. Foy, and S. Sumski, *Appl. Phys. Letters*, vol. 17, pp. 109–111, 1 August 1970. See also M. B. Panish and I. Hayashi, *Scientific American*, vol. 225, no. 1, pp. 32–40, July 1971.

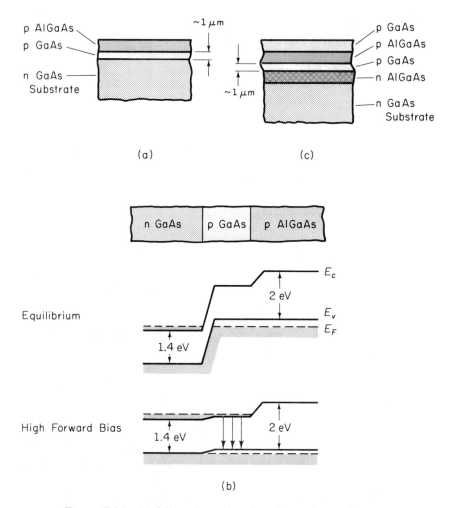

FIGURE 7-16. Multilayer heterojunctions for carrier confinement in laser diodes: (a) AlGaAs heterojunction grown on the thin p-type GaAs layer; (b) band diagrams for the structure of (a), showing confinement of electrons to the thin p region under bias; (c) heterostructure for room temperature cw laser operation.

contact to the p-type AlGaAs. In these structures AlGaAs is chosen as the wide band gap material, because its lattice parameters closely match those of GaAs. Thus high-quality heterojunctions can be grown with little mismatch of the lattice and therefore with little mechanical strain at the junctions. The room-temperature cw lasers described by Hayashi et al. were mounted on diamond heat sinks to provide efficient transfer of heat away from the junctions.

7.4.3 *Emission Spectra for p-n Junction Lasers.* Under forward bias, an inversion layer can be obtained along the plane of the junction, where a large population of electrons exists at the same location as a large hole population. A second look at Fig. 7-13 indicates that spontaneous emission of photons can occur due to direct recombination of electrons and holes, releasing energies ranging from approximately $F_n - F_p$ to E_g. That is, an electron can recombine over an energy from F_n to F_p, yielding a photon of energy $hv = F_n - F_p$, or an electron can recombine from the bottom of the conduction band to the top of the valence band, releasing a photon with $hv = E_c - E_v = E_g$. These two energies serve as the approximate outside limits of the laser spectra (with exceptions noted below).

The photon wavelengths which participate in stimulated emission are determined by the length of the resonant cavity as in Eq. (7-5). Figure 7-17 illustrates a typical plot of emission intensity vs. photon energy for a semiconductor laser. At low current levels (Fig. 7-17a), a spontaneous emission spectrum containing energies in the range $E_g < hv < (F_n - F_p)$ is obtained. As the current is increased to the point that significant population inversion exists, stimulated emission occurs at frequencies corresponding to the cavity modes as shown in Fig. 7-17b. These modes correspond to successive numbers of integral half-wavelengths fitted within the cavity, as described by Eq. (7-5). Finally, at a still higher current level, a most preferred mode or set of modes will dominate the spectral output (Fig. 7-17c). This very intense mode represents the main laser output of the device; the output light will be com-

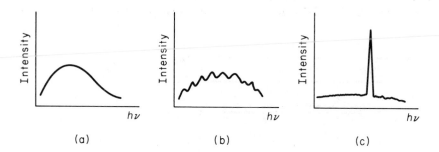

(a)   (b)   (c)

FIGURE **7-17.** Light intensity vs. photon energy hv for a junction laser: (a) incoherent emission below threshold; (b) laser modes at threshold; (c) dominant laser mode above threshold. The intensity scales are greatly compressed from (a) to (b) to (c).

posed of almost monochromatic radiation superimposed on a relatively weak radiation background, due primarily to spontaneous emission.

The separation of the modes in Fig. 7-17b is complicated by the fact that the index of refraction **n** for GaAs depends on wavelength λ. From Eq. (7-5) we have

$$(7\text{-}9) \qquad \mathbf{m} = \frac{2L\mathbf{n}}{\lambda_0}$$

If **m** (the number of half-wavelengths in L) is large, we can use the derivative to find its rate of change with λ_0.

$$(7\text{-}10) \qquad \frac{d\mathbf{m}}{d\lambda_0} = -\frac{2L\mathbf{n}}{\lambda_0{}^2} + \frac{2L}{\lambda_0}\frac{d\mathbf{n}}{d\lambda_0}$$

Now reverting to discrete changes in **m** and λ_0, we can write

$$(7\text{-}11) \qquad -\Delta\lambda_0 = \frac{\lambda_0^2}{2L\mathbf{n}}\left(1 - \frac{\lambda_0}{\mathbf{n}}\frac{d\mathbf{n}}{d\lambda_0}\right)^{-1}\Delta\mathbf{m}$$

If we let $\Delta\mathbf{m} = -1$, we can calculate the change in wavelength $\Delta\lambda_0$ between adjacent modes (i.e., between modes **m** and **m** $-$ 1).

It was mentioned above that the photon energies are not restricted exactly to the most obvious limits of E_g to $(F_n - F_p)$. One complicating factor is that the inversion layer may contain donor and acceptor bands, as in Fig. 7-18, since the junction is usually obtained by a diffusion of large numbers of acceptors into a sample already containing a high donor density. The impurity bands can serve as intermediate steps in the recombination processes; in fact, the impurity bands may overlap the valence and conduction bands. As Fig. 7-18 suggests, several transitions involving the release

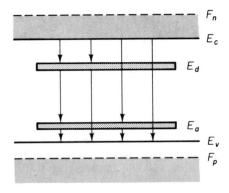

FIGURE 7-18. Several possible transitions within the inversion region of a junction laser with donor and acceptor bands.

of photons with energies less than E_g are possible. In addition, there are more complicated processes, including tunneling via the impurity bands, which may give rise to photons of lower energy.

7.4.4 Materials for Semiconductor Lasers. We have discussed
the properties of the junction laser largely in terms of GaAs, since this material has been used most widely in laser studies. Gallium arsenide crystals can be grown with high purity and fairly low defect density and are therefore suited for reliable device fabrication. On the other hand, the light emitted from GaAs is in the infrared part of the spectrum. This is an advantage in many communications applications, where infrared detectors can be used for receiving the information. For applications in which the eye is the detector, however, coherent light in the visible spectrum is needed.

Considerable variation in wavelength is available for the compound semiconductors, as Appendix III suggests. The choices, however, are restricted by the requirements of direct band gap and availability of both doping types. From Appendix III it is clear that most III–V compounds have direct band gaps, as do all II–VI compounds listed. All of the III–V semiconductors can be grown n-type or p-type, and junctions can be formed. Furthermore, mixed compounds such as GaAsP, AlGaAs, InGaP, etc., can be employed to vary further the available band gap energies. In some mixtures, one of the constituent compounds is direct and the other is indirect; for example, GaAsP can be thought of as a mixture of GaAs (direct, $E_g =$ 1.43 eV) and GaP (indirect, $E_g = 2.26$ eV). As mentioned in Section 6.4.2, this mixed compound is direct over a wide range of its composition, including transitions in the red part of the visible spectrum. Infrared lasers can be built in the mixed compound PbSnTe, with output wavelengths from about seven microns to more than thirty microns at low temperatures, depending on the composition of the material.

The problem of self-compensation in the II–VI compounds is a serious limitation in designing laser diodes. If a compound cannot be doped both n-type and p-type, a junction laser cannot be formed. It is possible that a heterojunction could result in a laser device in these materials, but the junction would have to be of high quality, with little mismatch of the lattices. In the absence of an appropriate heterostructure, lasers in these materials must utilize optical or electron–beam pumping techniques. As Fig. 7-19 indicates, a sample containing Fabry–Perot faces can be excited optically by a laser of shorter wavelength to obtain laser operation of the sample, and hence, a frequency conversion of coherent light. Generally, samples pumped in this fashion must be quite thin, however, since the optical absorption coefficient α is often high for these materials. Thus a thick sample

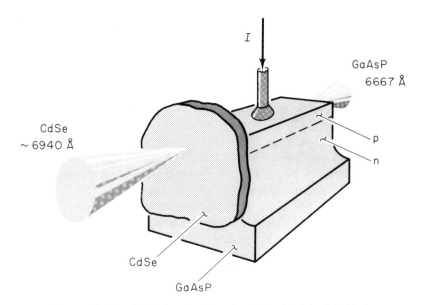

FIGURE 7-19. Optical excitation of a thin platelet of CdSe by the laser output of a GaAsP diode. The parallel faces of the platelet form the resonant structure for the CdSe. From G. E. Stillman et al., *Appl. Phys. Letters*, vol. 9, pp. 268–269, 1 October 1966.

would absorb most of the pumping radiation in a thin layer near the surface, and the population inversion within the cavity would be insufficient to cause laser emission.

7.5 Laser Applications

The application of lasers to scientific, engineering, and medical problems is a fast-changing subject. Like any device, the laser can solve many problems in unique ways, but it may fail to produce results in other applications which seemed promising at first. Among those laser applications which have proved effective, we should include

1. Experiments and measurements which take advantage of the monochromaticity and coherence of laser light.
2. Communications, from simple on–off signaling to modulation and detection of multichannel signals.
3. Mechanical and medical applications, which utilize the short-term power of the pulsed laser.

We shall mention a few representative applications from these categories, without attempting to choose the most current or most dramatic. The reading list for this chapter contains several references to recent developments, and there is no shortage of application notes in the current literature.

7.5.1 Experiments.

Lasers were used first in the laboratory and the classroom. The obvious advantages of having a source of monochromatic and coherent light available make many of the classical experiments and demonstrations much more effective. For example, with lasers students may now perform experiments involving diffraction, interference, and other studies of the properties of light with much improved convenience and accuracy.

An example of a classical measurement which can be performed now with greatly improved accuracy is the Michelson–Morley experiment. This experiment, which was one of the important factors in the development of the theory of relativity, demonstrated for the first time that the speed of light is independent of the motion of the Earth. The experiment required an interferometer which could be oriented in various directions. The measurements were extremely delicate and difficult to perform with the standard sources of monochromatic light. With lasers as the monochromatic sources, however, the experiment can be performed with considerably less difficulty and more precision. It is still necessary to use a room with minimal vibrations, but now the laser source is monochromatic and is already built in the form of an interferometer. Thus two cw lasers (such as He–Ne gas lasers) can be mounted at right angles on a turntable and positioned with respect to the rotation of the Earth. When the axis of one laser is moving with the Earth's rotation, the other is at right angles to this motion. Thus any change in the speed of light within the cavity of the first laser due to the rotation of the Earth would shift the operating frequency relative to the second laser. With this arrangement, one can demonstrate with accuracies several times greater than before that there is no such change in the speed of light with motion.

In *holography*, lasers can be used to photograph three-dimensional objects by storing both amplitude and phase information on a film (*hologram*). Then the hologram can be illuminated by a laser to reproduce the object in three dimensions. This interesting and useful application of lasers is discussed in the articles by Herriot and Pennington in the reading list for this chapter.

7.5.2 Communications.

The ultimate application of the laser in the communications field would be the realization of a system by which thousands of telephone conversations or television programs could be carried simultaneously on a single laser beam. The high frequencies involved in the optical spectrum make such large numbers of information channels theoretically possible. However, the problems of mixing modulated optical signals for transmission and the subsequent detection, separation, and demodulation of the individual signals at the receiving end of such a system are very difficult. Intensive efforts are being made toward the development of this type of optical communications system, but the direction of the ultimate solution is not clear. Therefore, we shall leave investigations of progress in this area to surveys of the current literature and discuss here a few of the more straightforward communications applications.

An obvious use of the pulsed laser is in *range-finding*, or the measurement of distances. If a pulse from a *Q*-switched laser is aimed at an object, the time delay between sending the pulse and detection of reflected light from the object gives the time for a pulse traveling at the speed of light to make the round trip. A counting circuit can be actuated by a signal from the sending pulse, so that very accurate delay times can be measured between pulses. By such techniques, accurate distance measurements can be made in space or military applications and in surveying. Applications in space are particularly promising, since laser range-finding can be used without the atmospheric scattering which may disturb Earth-based optical signals.

Another interesting proposal is the use of lasers for the transmission of information within and between computers. This also is a controlled environment application, so that the disadvantages of scattering are not so important. Complete electrical isolation between circuits is possible with optical signaling, and the digital nature of computer information allows for simple laser pulsing instead of the much more complicated modulation–demodulation of telephone or television signals. For the low powers necessary in such applications, the small semiconductor diode lasers seem most appropriate.

7.5.3 Mechanical and Medical Applications. These two apparently different categories are considered together here, since in both cases the applications depend on the sudden release of energy which the laser provides, as opposed to its information-carrying ability. Mechanical applications include the use of pulsed lasers in the cutting and welding of small parts and in various small-scale melting and vaporizing tasks. Since a typical ruby laser pulse is delivered in microseconds, for example, the absorbed energy is concentrated in the small illuminated area, with little heat carried away by thermal conduction. The exact focusing of the laser beam allows for the formation of circular holes a few microns in diameter in thin metal parts, with minimal damage to the surrounding material. Other applications include forming holes in diamonds and other hard materials, trimming resistors, and

balancing turbines and watch movements. The formation of tiny holes in diamonds has been particularly useful in making dies for the drawing of fine wire. In the semiconductor industry, lasers are used to scribe between integrated circuits on a Si wafer prior to separation into individual chips. In many cases lasers have replaced the diamond stylus for this operation.

Similar applications can be made in the medical field, especially in microsurgery. A typical medical application is the detached retina operation, in which a laser pulse is focused, by external optics and the lens of the patient's eye, onto the retina. If the pulse is applied properly, a sudden coagulation of the retinal material occurs, and the retina is reattached by a very small lesion which produces a negligible area of damage compared to the total surface of the retina. This "spot-welding" operation is so quick that no anesthetic is needed. Another medical application is the selective absorption of optical energy by opaque tissue, which may lead to the successful destruction of cancer cells, particularly those on the skin.

We have mentioned only a few of the fields in which laser applications are being investigated. The laser is still relatively new, and much work still must be done in its development and its uses. There is no doubt, however, that this device will play an increasing role in the future of electronics and in other fields.

READING LIST

Special Issue on Lasers, *Applied Optics*, vol. 5, October 1966.

A. L. SCHAWLOW, "Advances in Optical Lasers," *Scientific American*, vol. 209, no. 1, pp. 34–45, July 1963.

A. L. SCHAWLOW, "Laser Light," *Scientific American*, vol. 219, no. 3, pp. 120–126, September 1968.

C. K. N. PATEL, "High-Power Carbon Dioxide Lasers," *Scientific American*, vol. 219, no. 2, pp. 22–23, August 1968.

N. HOLONYAK, JR., "Semiconductor Lasers," *The Bridge of Eta Kappa Nu*, vol. 61, Winter, 1965, pp. 3–7, 12.

F. J. MOREHEAD, JR., "Light-Emitting Semiconductors," *Scientific American*, vol. 216, no. 5, pp. 108–122, May 1967.

M. R. LORENZ and M. H. PILKUHN, "Semiconductor-Diode Light Sources," *IEEE Spectrum*, vol. 4, no. 4, pp. 87–96, April 1967.

G. BURNS and M. I. NATHAN, "p-n Junction Lasers," *Proc. IEEE*, vol. 52, pp. 770–794, July 1964.

D. R. HERRIOTT, "Applications of Laser Light," *Scientific American*, vol. 219, no. 3, pp. 141–156, September 1968.

K. S. PENNINGTON, "Advances in Holography," *Scientific American*, vol. 218, no. 2, pp. 40–48, February 1968.

Special Issue on Optical Communication, *Proc. IEEE*, vol. 58, no. 10, October 1970. See particularly the following review articles: "Optical Communications—A Decade of Preparations," by N. LINDGREN, pp. 1410–1418; and "Coherent Optical Sources for Communications," by J. E. GEUSIC et al., pp. 1419–1439.

R. GLICKSMAN, "Technology and Design of GaAs Laser and Non-Coherent IR-Emitting Diodes," *Solid State Technology*, vol. 13, no. 9, pp. 29–35, September 1970, and no. 10, pp. 39–44, October 1970.

B. A. LENGYEL, *Introduction to Laser Physics*. New York: John Wiley & Sons, 1966.

P. R. THORNTON, *The Physics of Electroluminescent Devices*. London: E. and F. N. Spon, Ltd., 1967.

PROBLEMS

7.1 Assume the system described by Eq. (7-2) is in thermal equilibrium at an extremely high temperature such that the energy density $\rho(\nu_{12})$ is essentially infinite. Show that $B_{12} = B_{21}$.

7.2 The system described by Eq. (7-2) interacts with a blackbody radiation field whose energy density per unit frequency at ν_{12} is

$$\rho(\nu_{12}) = \frac{8\pi h \nu_{12}{}^3}{c^3}[e^{h\nu_{12}/kT} - 1]^{-1}$$

from Planck's radiation law. Given the result of Prob. 7-1, find the value of the ratio A_{21}/B_{12}.

7.3 Assuming equal electron and hole densities and band-to-band transitions, calculate the minimum carrier density $n = p$ for population inversion in GaAs at $300°K$. The intrinsic carrier density in GaAs is about 10^7 cm^{-3}.

7.4 Read the article by Pennington in the reading list (or a more recent article) and write a brief description of holography.

AMPLIFYING DEVICES 8

Although junction diodes are quite versatile and useful in many circuits, the nature of modern electronics is due primarily to the transistor. The invention of the transistor in 1949 by Shockley, Bardeen, and Brattain provided the impetus for serious study of semiconductors on a large scale; it is questionable that we would have any of the present-day solid state devices if the transistor had not inspired the considerable research and development of semiconductor materials and device processing. The transistor occupies a dominant role in modern electronics largely because it is an excellent *amplifying device* which is small, reliable, and requires little power. As an amplifying device the transistor converts weak time-varying signals into strong signals. There are other important functions which the transistor can perform in electronic circuits, but certainly its ability to amplify is central to its usefulness.

In this chapter we shall discuss the transistor as an amplifier. There are two basic types of transistor to be considered. In one type the current is carried by the majority carriers and is controlled by an applied electric field. This device is called a *unipolar* (one type of carrier) or *field-effect transistor*. The second type involves the flow of both minority and majority carriers and is called a *bipolar junction transistor*. We shall begin the discussion by investigating some of the basic principles of amplification, using the *vacuum tube triode* as an example. Then we shall discuss the operation of the two transistor types, with the major emphasis on basic principles of operation and amplification. Most of the details of the conduction processes in bipolar transistors will be left for Chapter 9. We shall not consider in detail here the very important subject of creating circuit models for amplifying devices. A few simple models will be discussed to illustrate device operation and amplification fundamentals, but a full discussion of circuit models is left for texts on electronic circuits.

8.1 Principles of Amplification: Triodes

In amplifying circuits we are usually interested in increasing the amplitude and power of a-c signals. Since we cannot extract more energy from a circuit than we put in, what really happens in an amplifier is the conversion of d-c power into a-c power proportional to an input signal. A d-c source is used to obtain steady state voltage and current at the terminals of an amplifying device, and an a-c signal is superimposed on the d-c level in response to the input. The function of the device is to cause a large variation in the output of the amplifier proportional to the a-c variation at the input. In this section we shall consider amplification in terms of the vacuum tube triode. The basic operation of this device is relatively straightforward, and it illustrates the principles of amplification to be used later in the discussion of transistors. The triode served as the primary amplifying device before the invention of the transistor, and it is still used in some applications. Various vacuum tube devices related to the triode are useful in applications requiring a combination of high frequency and high power.

8.1.1 Triode Operation and Characteristics. The *thermionic triode* is a vacuum tube device which operates by controlling the flow of electrons through a vacuum from a heated *cathode* to an unheated *anode*. We can understand triodes best by first considering the thermionic diode of Fig. 8-1.

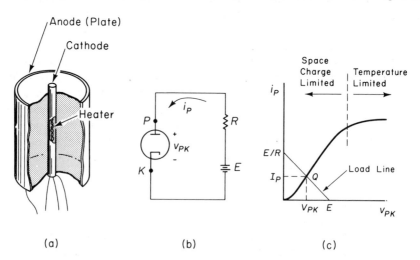

(a) (b) (c)

FIGURE 8-1. A thermionic diode: (a) typical diode construction; (b) schematic representation in a circuit; (c) *I-V* characteristic and load line.

The cathode is typically a metal cylinder surrounding a spiral heating element (Fig. 8-1a). When the proper current is passed through the heater, the cathode becomes hot enough to emit electrons from its surface into the vacuum; this effect is called *thermionic emission*. If the surrounding anode (commonly called the *plate*) is biased positive with respect to the cathode, an electric field is set up which sweeps the thermally emitted electrons from the cathode K to the plate P. The magnitude of the resulting current i_p depends on the plate-to-cathode voltage v_{PK}. If v_{PK} and the resulting electric field are sufficiently large to sweep every emitted electron to the plate, i_p will be *temperature limited*; that is, the number of electrons collected per second at P will depend only on the rate of electron emission from K. However, for smaller values of v_{PK} the field is insufficient to sweep every emitted electron from K to P, and the resulting current is less than the temperature-limited value. In this range of operation, uncollected electrons form a cloud of negative charge around the cathode, and this negative space charge reduces the electric field between the electrodes. Some emitted electrons are repelled by the negative space charge and return to the cathode. The result is a *space charge limited* current which reaches almost zero when v_{PK} is zero. A very small current flows with no voltage applied, since some emitted electrons wander to the plate even at zero bias. It is clear, however, that only a small reverse bias is needed to repel any emitted electron, so that no electrons flow from K to P for reverse bias. Since the plate is not heated, no electrons move from P to K, and no reverse current flows.

To find the steady state current and voltage we need two simultaneous equations. The diode characteristic $i_P = f(v_{PK})$ is a graphical equation (Fig. 8-1c) relating the device voltage and current. We can obtain a second relation by writing a loop equation around the circuit of Fig. 8-1b

$$(8\text{-}1) \qquad\qquad E = i_P R + v_{PK}$$

This linear equation can be plotted on the diode characteristic diagram by locating intersections with the ordinate (E/R) and the abscissa (E). The intersection of this *load line* with the diode characteristic is the simultaneous solution, representing the steady state or *quiescent operating point Q*. This point locates the d-c values of the plate current I_P and the plate-to-cathode voltage V_{PK}.

Since the current i_p in the thermionic diode is limited by the negative space charge between the electrodes, it is reasonable to expect that *external* control of i_p could be achieved by placing a negative potential on a third electrode located between P and K. If a mesh or grid of fine wire is placed between the anode and cathode, electrons can pass through the grid in transit from K to P, but the electrostatic potential in the neighborhood of the grid can be controlled externally by an applied bias. As a result, the current i_p at a particular value of v_{PK} will be increased or decreased, depending on the

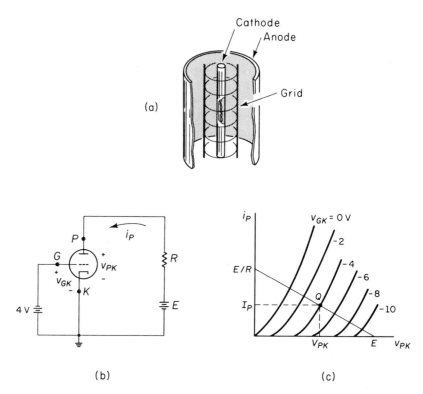

FIGURE 8-2. A thermionic triode: (a) typical construction; (b) schematic circuit representation; (c) plate characteristics and load line.

voltage applied to the grid. This third electrode, called the *control grid* (or simply grid), carries the input signal for amplification; a small change in grid voltage can cause a rather large change in plate current and voltage. The resulting three-terminal device is the *thermionic triode* (Fig. 8-2a). The *plate characteristics* $i_P = f(v_{PK}, v_{GK})$ of Fig. 8-2c take the form of a family of diode-like curves, each depending on the grid-to-cathode voltage v_{GK}. The plate characteristics illustrate the fact that the current flow through the triode for a given v_{PK} decreases as v_{GK} becomes more negative. The quiescent operating point Q is located along the load line at the particular curve corresponding to the applied d-c grid voltage V_{GK}. For convenience of notation, we shall use the symbols of Fig. 8-3 to distinguish between total quantities and a-c or d-c components.

When the triode is used in a simple amplifier circuit, the grid is biased negative with respect to the cathode. Once the Q-point is established, a-c variations can be made in the grid voltage v_{gk} to induce large changes in

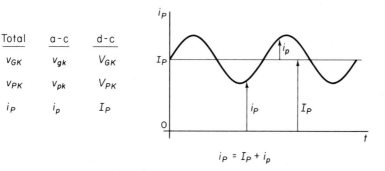

Total	a-c	d-c
v_{GK}	v_{gk}	V_{GK}
v_{PK}	v_{pk}	V_{PK}
i_P	i_p	I_P

$$i_P = I_P + i_p$$

FIGURE 8-3. Notation for total quantities and a-c and d-c components of voltages and current in a triode.

i_p and v_{pk}. In most triode circuits, the d-c bias V_{GK} is obtained by placing a cathode resistor from K to ground (Prob. 8.2), thus eliminating the battery in the grid circuit.

8.1.2 Amplification. If an a-c signal is applied to the grid of a triode, the grid-to-cathode voltage v_{GK} varies about its quiescent value (Fig. 8-4). The plate-to-cathode voltage v_{PK} varies in response, and the instantaneous value v_{PK} corresponds to the instantaneous v_{GK} along the load line [note that Eq. (8-1) is still valid]. Since a very small a-c voltage v_s applied to the grid results in large variations in v_{pk}, there is an a-c *voltage gain A_v* between the input and output of the amplifier. The value of A_v can be found graphically as the ratio of the change in the plate-to-cathode voltage Δv_{PK} corresponding to a given change in Δv_{GK}

(8-2)
$$A_v = \frac{v_o}{v_s} = -\frac{\Delta v_{PK}}{\Delta v_{GK}}$$

The minus sign in Eq. (8-2) indicates that the output voltage is 180° out of phase with the input. That is, as v_{GK} increases (becomes less negative) v_{PK} decreases, and vice versa. If the plate characteristic curves are equally spaced along the load line, the a-c output voltage v_{pk} will vary linearly with the input signal v_s. The output of the amplifier will then be a perfect replica of the input except that it will be larger and 180° out of phase. In real triodes, however, the spacing of the curves in the plate characteristic varies somewhat over the (i_P, v_{PK}) plane, and some distortion is introduced to the output signal.

For purposes of small-signal a-c analysis we can often approximate the triode plate characteristics by a series of straight lines separated by a constant spacing and with a constant slope (Fig. 8-5a). For best accuracy in a-c cal-

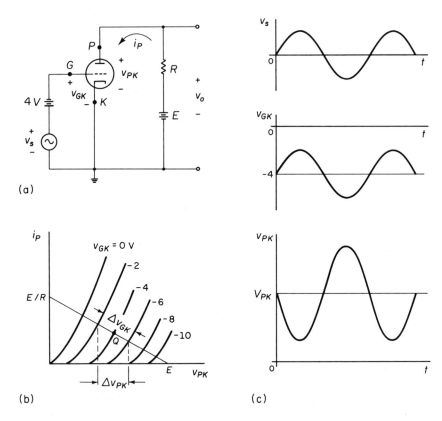

(a)

(b)

(c)

FIGURE 8-4. Voltage gain in a simple triode circuit: (a) triode amplifier with a-c signal input v_s; (b) plate characteristics with variations of grid voltage Δv_{GK} and plate voltage Δv_{PK} along the load line; (c) variations in the output voltage of the amplifier with a sinusoidal input.

culations, we choose the slope and spacing corresponding to the particular region of the characteristics neighboring the Q-point for the circuit under consideration. We can write the equation for each straight line in the approximated characteristic by relating v_{pk} to i_p linearly via the slope of the line, keeping in mind that the abscissa crossing of each line is displaced from $v_{pk} = 0$ by a constant factor μ times the grid voltage

$$(8\text{-}3) \qquad\qquad v_{pk} = i_p r_p - \mu v_{gk}$$

Since Eq. (8-3) relates the pertinent parameters v_{pk}, i_p, and v_{gk}, we can conveniently draw a *small-signal circuit model* for the triode to represent this

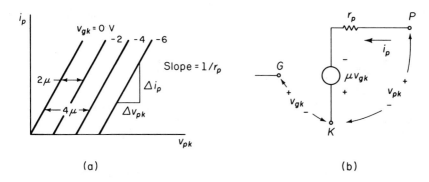

Figure 8-5. Small-signal a-c model for the triode: (a) approx-imated plate characteristics; (b) equivalent circuit model for the triode.

equation (Fig. 8-5b). In this circuit model a voltage summation from plate to cathode gives Eq. (8-3). Thus in many calculations requiring analysis of a-c signals, we can use the model instead of performing a graphical analysis of the actual triode charateristics. Of course, if relatively large signals are superimposed on the quiescent quantities, no single set of values for r_p and μ correctly describes the characteristics along a given load line. Therefore, for analysis of distortion in large-signal a-c amplification, we resort to the graphical analysis described above.

In Fig. 8-5b the mutual relations among v_{pk}, i_p, and v_{gk} for the triode are synthesized by the *controlled voltage source* μv_{gk} and the resistor r_p. The voltage generator in the model is called a controlled source because its value is dictated by v_{gk}. The proportionality factor μ is the *voltage amplification factor* for the triode, and r_p is called the *plate resistance*

$$(8\text{-}4) \qquad \mu = \frac{\Delta v_{PK}}{\Delta v_{GK}}\bigg|_{i_P \text{ (const.)}}, \qquad r_p = \frac{\Delta v_{PK}}{\Delta i_P}\bigg|_{v_{GK} \text{ (const.)}}$$

We have shown no electrical connection between the grid and the plate–cathode circuit in Fig. 8-5b, since no current flows in the grid when its bias is negative with respect to the cathode.† Thus the triode with negative grid bias presents a relatively large input impedance to the grid circuit, which is a desirable feature in most amplifiers.

In an a-c amplifier we can replace the triode with its equivalent circuit model to calculate the gain (Fig. 8-6). In the example shown, the a-c voltage

---

†This assumption is invalid at high frequencies, when interelectrode capacitance becomes important.

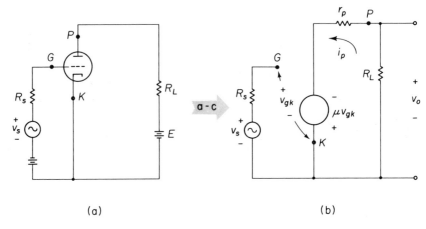

(a)   (b)

FIGURE 8-6. Application of the small-signal a-c model for the triode in a simple amplifier circuit: (a) schematic diagram of the amplifier circuit; (b) equivalent circuit for a-c signals.

v_{gk} is simply v_s if the grid current is zero, and the voltage gain A_v is found by writing a loop equation around the plate circuit

$$\mu v_{gk} = i_p(R_L + r_p)$$

$$i_p = \frac{\mu v_{gk}}{R_L + r_p} = \frac{\mu v_s}{R_L + r_p}$$

$$v_o = -i_p R_L = -\frac{\mu v_s R_L}{R_L + r_p}$$

(8-5)
$$A_v = \frac{v_o}{v_s} = -\frac{\mu R_L}{R_L + r_p}$$

The voltage gain calculated by this method is negative, as in the graphical analysis, and it is larger than unity if μ is large (Prob. 8.1).

We have described voltage amplification in the thermionic triode in terms of a graphical analysis of the plate characteristics and by use of a simple a-c equivalent circuit model. We have touched only the high points of amplifier analysis and the techniques of obtaining and using circuit models. These topics can be studied in detail in other texts. It should be clear, however, that amplification can indeed be obtained using a device in which small changes in one voltage cause large changes in another voltage. However, there are other variations in amplification. For example, the idealized characteristics of Fig. 8-7 suggest two types of amplifying devices. In Fig. 8-7a the current in a device (i_d) is controlled by a voltage v_c; in Fig. 8-7b the device current is controlled by a current i_c. As we shall see in the following sections, the two basic types of transistors have characteristics approximating those

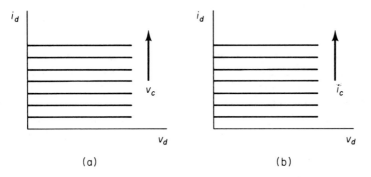

FIGURE 8-7. Idealized device characteristics for (a) a voltage-controlled amplifier, and (b) a current-controlled amplifier.

of Fig. 8-7. The field-effect transistor is a voltage-controlled device (as is the triode), and the bipolar junction transistor is a current-controlled device. In each case small changes in the control parameter result in large changes in the output signal.

8.2 Field-Effect Transistors

The first solid state amplifying device we shall study is the *field-effect transistor (FET)*. As the name implies, amplification in this device occurs when the current through two terminals is varied by an electric field arising from voltage applied to a third terminal. Thus, the FET is a voltage-controlled device (Fig. 8-7a). Basically, there are two types of FET's, depending on whether the controlling field appears at a reverse-biased junction or at an insulating layer. In either case, variations in the field cause corresponding variations in the current through the device. Because the input or control voltage is applied across a reverse-biased junction or an insulator, the FET is characterized by a high input impedance.

In this section we shall discuss the two types of FET's from a phenomenological viewpoint, and then we shall include sufficient analytical treatment to establish the basis of operation. There is considerable variation in the terminology used in connection with FET's, and the proliferation of terms and abbreviations may seem confusing at first. In cases where several commonly accepted terms are used interchangeably, we shall adopt one term for our discussions.

8.2.1 The Junction FET. In a *junction FET (JFET)* the voltage-variable depletion region width of a junction is used to control the effective cross-sectional area of a conducting *channel*. In the device of Fig. 8-8, the current I_{DS} flows through an n-type channel between two p$^+$ regions. A reverse bias between these p$^+$ regions and the channel causes the depletion regions to intrude into the n material, and therefore the effective width of the channel can be restricted. Since the resistivity of the channel region is fixed by its doping, the channel resistance varies with changes in the effective cross-sectional area. By analogy, the variable depletion regions serve as the two doors of a gate which open and close on the conducting channel.

In Fig. 8-8 electrons in the n-type channel drift from right to left, opposite to the current flow. The end of the channel from which electrons flow is called the *source*, and the end toward which they flow is called the *drain*. The p$^+$ regions are called *gates*. If the channel were p-type, holes would flow from the source to the drain, in the same direction as the current flow, and the gate regions would be n$^+$. It is common practice to connect the two gate regions electrically; therefore, the voltage V_{GS} refers to the potential from each gate region G to the source S. Since the conductivity of the heavily doped regions is high, we can assume that the potential is uniform throughout each gate. In the lightly doped channel material, however, the potential varies with position (Fig. 8-8b). If the channel of Fig. 8-8 is considered as a distri-

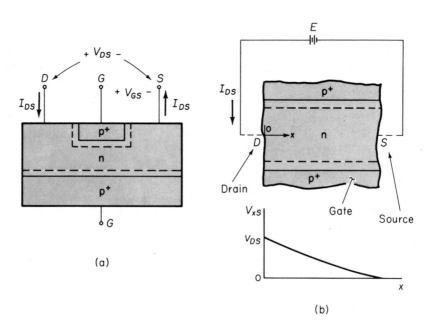

FIGURE 8-8. Simplified cross-sectional view of a junction FET: (a) transistor geometry; (b) detail of the channel and voltage variation along the channel with $V_{GS} = 0$ and small I_{DS}.

buted resistor carrying a current I_{DS}, it is clear that the voltage from the drain end of the channel D to the source electrode S must be greater than the voltage from a point near the source end to S. For low values of current we can assume a linear variation of voltage V_{xS} in the channel, varying from V_{DS} at the drain end to zero at the source end (Fig. 8-8b).

In Fig. 8-9 we consider the channel in a simplified way by neglecting voltage drops between the source and drain electrodes and the respective ends of the channel. For example, we assume that the potential at the drain end of the channel is the same as the potential at the electrode D. This is a good approximation if the source and drain regions are relatively large, so that there is little resistance between the ends of the channel and the electrodes. In Fig. 8-9 the gates are short circuited to the source ($V_{GS} = 0$), such that the potential at $x = L$ is the same as the potential everywhere in the gate regions. For very small currents, the widths of the depletion regions are close to the equilibrium values (Fig. 8-9a). As the current I_{DS} is increased,

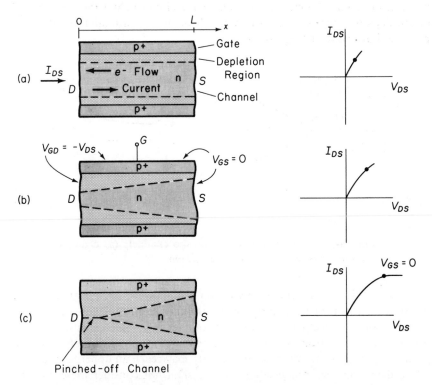

FIGURE 8-9. Depletion regions in the channel of a JFET with zero gate bias for several values of V_{DS}: (a) linear range; (b) near pinch-off; (c) beyond pinch-off.

however, it becomes important that V_{xS} is large near the drain end and small near the source end of the channel. Since the reverse bias across each point in the gate-to-channel junction is simply V_{xS} when V_{GS} is zero, we can estimate the shape of the depletion regions as in Fig. 8-9b. The reverse bias is relatively large near the drain ($V_{GD} = -V_{DS}$) and decreases toward zero near the source. As a result, the depletion region intrudes into the channel near the drain, and the effective channel area is constricted.

Since the resistance of the constricted channel is higher, the *I–V* plot for the channel begins to depart from the straight line which was valid at low current levels. As the voltage V_{DS} and current I_{DS} are increased still further, the channel region near the drain becomes more constricted by the depletion regions and the channel resistance continues to increase. As V_{DS} is increased, there must be some bias voltage at which the depletion regions meet near the drain and essentially *pinch off* the channel (Fig. 8-9c). When this happens, the current I_{DS} cannot increase significantly with further increase in V_{DS}. Beyond pinch-off the current is *saturated* approximately at its value at pinch-off. Once electrons from the channel enter the electric field of the depletion region, they are swept through and ultimately flow to the positive drain contact (Fig. 8-10).

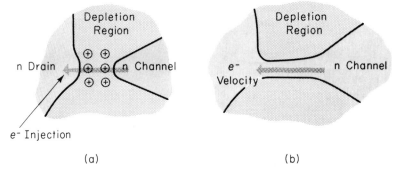

(a)   (b)

FIGURE 8-10. Pinched-off region of the channel: (a) channel depletion; (b) electron velocity saturation.

The nature of the pinched-off region is somewhat complicated in that the limiting factor for the saturation current may be electron velocity saturation (Section 3.4.3) rather than the effect of injection into a depletion region. For a given saturation current, the current density is very high near the pinched-off region, and the electron velocity may reach an upper limit before a true depletion region is established. We shall not consider these complications in detail here; the important fact is that the current does saturate beyond pinch-off, and the differential channel resistance dV_{DS}/dI_{DS} becomes very high. To a good approximation, we can calculate the current at the

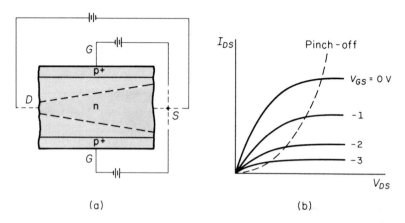

FIGURE 8-11. Effects of a negative gate bias: (a) increase of depletion region widths with V_{GS} negative; (b) family of current-voltage curves for the channel as V_{GS} is varied.

critical pinch-off voltage and assume there is no further increase in current as V_{DS} is increased.

The effect of a negative gate bias $-V_{GS}$ is to increase the resistance of the channel and induce pinch-off at a lower value of current (Fig. 8-11). Since the depletion regions are larger with V_{GS} negative, the effective channel width is smaller and its resistance is higher in the low-current range of the characteristic. Therefore, the slopes of the I_{DS} vs. V_{DS} curves below pinch-off become smaller as the gate voltage is made more negative (Fig. 8-11b). As the current I_{DS} increases, the pinch-off condition is reached at a lower drain-to-source voltage, and the saturation current is lower than for the case of zero gate bias. As V_{GS} is varied, a family of curves is obtained for the I–V characteristic of the channel, as in Fig. 8-11b.

Beyond the pinch-off voltage the channel characteristics approximate those of the ideal voltage-controlled amplifying device of Fig. 8-7a. Clearly, by varying the gate bias we can obtain amplification of an a-c signal. Since the input control voltage V_{GS} appears across the reverse-biased gate junctions, the input impedance of the device is high.

We can calculate the pinch-off voltage rather simply by representing the channel in the approximate form of Fig. 8-12. If the channel is symmetrical and the effects of the gates are the same in each half of the channel region, we can restrict our attention to the channel half-width $h(x)$, measured from the center line ($y = 0$). The metallurgical half-width of the channel (i.e., neglecting the depletion region) is a. We can find the pinch-off voltage by calculating the reverse bias between the n channel and the p$^+$ gate at the drain end of the channel ($x = 0$). For simplicity we shall assume that the

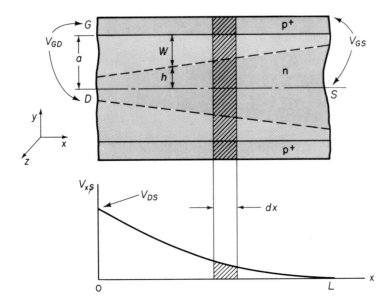

FIGURE 8-12. Simplified diagram of the channel with definitions of dimensions and differential volume for calculations.

channel width at the drain decreases uniformly as the reverse bias increases to pinch-off. If the reverse bias between the gate and the drain is $-V_{GD}$, the width of the depletion region at $x = 0$ can be found from Eq. (5-62)

$$(8\text{-}6) \qquad W(x = 0) = \left[\frac{2\epsilon(-V_{GD})}{qN_d}\right]^{1/2}, \qquad (V_{GD} \text{ negative})$$

In this expression we assume the equilibrium contact potential V_0 is negligible compared with V_{GD} and the depletion region extends primarily into the channel for the p$^+$-n junction.

Pinch-off occurs at the drain end of the channel when

$$(8\text{-}7) \qquad\qquad h(x = 0) = a - W(x = 0) = 0$$

that is, when $W(x = 0) = a$. If we define the value of $-V_{GD}$ at pinch-off as V_P, we have

$$(8\text{-}8) \qquad\qquad \left[\frac{2\epsilon V_P}{qN_d}\right]^{1/2} = a$$

$$(8\text{-}9) \qquad\qquad V_P = \frac{qa^2 N_d}{2\epsilon}$$

The pinch-off voltage V_P is a positive number; its relation to V_{DS} and

V_{GS} is

(8-10) $V_P = -V_{GD}$ (pinch-off) $= -V_{GS} + V_{DS}$

where V_{GS} is zero or negative for proper device operation. A forward bias on the gate would cause hole injection from the p$^+$ regions into the channel, eliminating the field-effect control of the device. From Eq. (8-10) it is clear that pinch-off results from a combination of gate-to-source voltage and drain-to-source voltage. Pinch-off is reached at a lower value of V_{DS} (and therefore a lower I_{DS}) when a negative gate bias is applied, in agreement with Fig. 8-11b.

Calculation of the channel current is slightly more complicated, although the mathematics is relatively straightforward below pinch-off. The approach we shall take is to find the expression for I_{DS} just at pinch-off, and then assume the saturation current beyond pinch-off remains fairly constant at this value.

In the coordinate system defined in Fig. 8-12, the center of the channel at the drain end is taken as the origin. The length of the channel in the x-direction is L, and the depth of the channel in the z-direction is Z. We shall call the resistivity of the n-type channel material ρ (valid only in the neutral n material, outside the depletion regions). If we consider the differential volume of neutral channel material $Z2h(x)\,dx$, the resistance of the volume element is $\rho dx/Z2h(x)$ [see Eq. (3-44)]. Since the current does not change with distance along the channel, I_{DS} is related to the differential voltage drop in the element $-dV_{xS}$ by the conductance of the element

(8-11) $$I_{DS} = -\frac{Z2h(x)}{\rho}\frac{dV_{xS}}{dx}$$

The minus sign associated with dV_{xS} simply indicates that V_{xS} decreases as x increases along the channel. The term $2h(x)$ is the channel width at x.

The half-width of the channel at point x depends on the local reverse bias between gate and channel $-V_{Gx}$

(8-12) $h(x) = a - W(x) = a - \left[\frac{2\epsilon(-V_{Gx})}{qN_d}\right]^{1/2} = a\left[1 - \left(\frac{V_{xS} - V_{GS}}{V_P}\right)^{1/2}\right]$

since $V_{Gx} = V_{GS} - V_{xS}$ and $V_P = qa^2N_d/2\epsilon$. Implicit in Eq. (8-12) is the assumption that the expression for $W(x)$ can be obtained by a simple extension of Eq. (8-6) to point x in the channel. This is called the *gradual approximation*; it is valid if $h(x)$ does not vary abruptly in any element dx.

The voltage V_{Gx} will be negative since the gate voltage V_{GS} is chosen zero or negative for proper operation. Substituting Eq. (8-12) into Eq. (8-11) we have

(8-13) $$\frac{2Za}{\rho}\left[1 - \left(\frac{V_{xS} - V_{GS}}{V_P}\right)^{1/2}\right]dV_{xS} = -I_{DS}\,dx$$

Chap. 8

We can solve this equation (Prob. 8.4) to obtain

(8-14) $I_{DS} = G_0 V_P \left[\dfrac{V_{DS}}{V_P} + \dfrac{2}{3} \left(-\dfrac{V_{GS}}{V_P} \right)^{3/2} - \dfrac{2}{3} \left(\dfrac{V_{DS} - V_{GS}}{V_P} \right)^{3/2} \right]$

where V_{GS} is negative and $G_0 \equiv 2aZ/\rho L$ is the conductance of the channel for negligible $W(x)$, i.e., with no gate voltage and low values of I_{DS}. This equation is valid only up to pinch-off, where $V_{DS} - V_{GS} = V_P$. If we assume the saturation current remains essentially constant at its value at pinch-off, we have

(8-15) $I_{DS} \text{ (sat.)} = G_0 V_P \left[\dfrac{V_{DS}}{V_P} + \dfrac{2}{3} \left(-\dfrac{V_{GS}}{V_P} \right)^{3/2} - \dfrac{2}{3} \right]$

$= G_0 V_P \left[\dfrac{V_{GS}}{V_P} + \dfrac{2}{3} \left(-\dfrac{V_{GS}}{V_P} \right)^{3/2} + \dfrac{1}{3} \right]$

where

$$\dfrac{V_{DS}}{V_P} = 1 + \dfrac{V_{GS}}{V_P}$$

The resulting family of I–V curves for the channel (Fig. 8-13a) agrees with the results we predicted qualitatively. The saturation current is greatest when V_{GS} is zero and becomes smaller as V_{GS} is made negative.

We can represent the device biased in the saturation region by the equivalent circuit of Fig. 8-13b, where

(8-16) $g_m \text{ (sat.)} = \dfrac{\partial I_{DS} \text{ (sat.)}}{\partial V_{GS}} = G_0 \left[1 - \left(-\dfrac{V_{GS}}{V_P} \right)^{1/2} \right]$

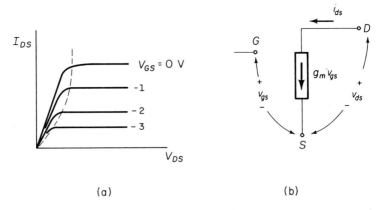

(a)   (b)

FIGURE 8-13. The JFET in the saturation region: (a) characteristic curves; (b) simple circuit model for a-c.

The quantity g_m relating changes in i_{DS} to changes in v_{GS} is called the *mutual transconductance*. When evaluated in the saturation range, g_m varies with the square root of the gate voltage as in Eq. (8-16). In the equivalent circuit of Fig. 8-13b, an open circuit is shown from gate to source, thereby neglecting reverse saturation current in the reverse-biased p$^+$-n gate junctions. The drain current i_{ds} is shown to be dictated by the current generator $g_m v_{gs}$, and thus to be independent of v_{ds}. Such a simple equivalent circuit is valuable for many applications, although refinements can be made. For example, in an a-c model, capacitances should be added between the gate electrode and the drain and source electrodes to account for junction capacitance at high frequencies.

It is important to note that transconductance varies with bias, since g_m (sat.) depends on $V_{GS}^{1/2}$. This nonlinearity should be considered in amplifier design.

8.2.2 The Insulated-Gate FET.

In the second class of FET's the channel current is controlled by a voltage at a gate electrode which is isolated from the channel by an insulator. The resulting device is called an *insulated-gate field-effect transistor (IGFET)*. In the most common configuration, an oxide layer is grown or deposited on the semiconductor surface, and the metal gate electrode is evaporated onto this oxide layer. This structure is commonly called a *metal-oxide-semiconductor (MOS)* structure. If the device includes a source and drain, it represents an *MOS* transistor, or *MOST*. The general term IGFET is retained to include devices in which the insulator may be some material other than an oxide layer. For convenience, we shall discuss the MOS transistor, which is the most common configuration.

The n$^+$ source and drain regions of Fig. 8-14 are diffused into a high-resistivity p substrate. The channel region may be a thin diffused n layer (Fig. 8-14a) or an *induced inversion region* (Fig. 8-14b). If an n-type diffused channel is included, the effect of the field is to raise or lower the conductance of the channel by either depleting or enhancing the electron density in the channel. Let us examine what happens at the oxide–semiconductor interface when a positive voltage is applied to the gate (Fig. 8-14c). The electric field in the oxide layer exists between positive charge on the metal gate electrode and negative charge in the semiconductor. The negative charge is composed of an accumulation of mobile electrons into the channel and fixed ionized acceptor atoms in the depleted p material. If the gate-to-source voltage is positive, the conductivity of the channel is enhanced, while a negative gate voltage tends to deplete the channel of electrons. Thus a diffused-channel MOST can be operated in either the *depletion* or *enhancement modes* (Fig. 8-15a).

We shall concentrate our analysis on the interesting case of the *induced-channel MOST*, in which no diffused n-type region exists between source and

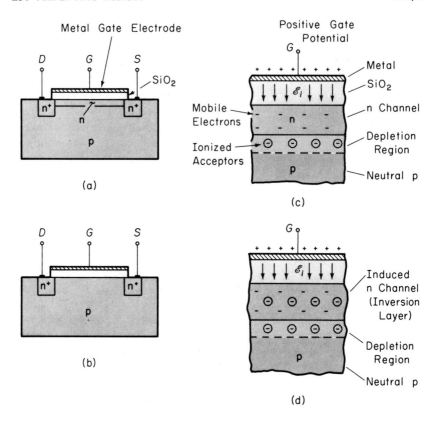

FIGURE 8-14. Cross-sectional views of MOS transistor structures with expanded diagrams of the metal-oxide-semiconductor boundaries: (a) diffused-channel device; (b) induced-channel device; (c) expanded view of the diffused-channel device with positive gate bias; (d) expanded view of the induced-channel device with positive gate bias. The relative thickness of the oxide layer is exaggerated for illustrative purposes.

drain at equilibrium (Fig. 8-14b). As with the JFET of the previous section, we shall consider an n-channel device; the corresponding p-channel case can be obtained easily from this analysis. When a positive gate voltage is applied to this structure, a depletion region is formed in the p material, and a thin layer of mobile electrons is drawn from the source and drain into the channel. Where the mobile electrons dominate, the material is effectively n-type. This is called an *inversion layer*, since the material was originally p-type.† Once an inversion layer is formed near the semiconductor surface,

† The term "inversion layer" has a different implication here compared with the case of a laser junction (Section 7.4), in which a population inversion of electrons and holes is indicated.

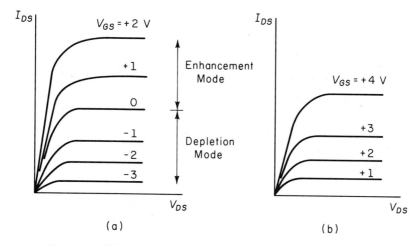

FIGURE 8-15. Characteristic curves for (a) diffused-channel device and (b) induced-channel device, showing enhancement and depletion modes. Surface charge in the oxide is neglected in (b).

a conducting channel exists from source to drain. The operation of the device is then quite similar to the JFET. The channel conductance is controlled by the field $\mathscr{E}_i$ in the insulator, but the magnitude of this field varies along the channel as V_{Gx} varies from $V_{GS} - V_{DS}$ at the drain to V_{GS} at the source (Fig. 8-16c); this corresponds qualitatively to the variation in junction reverse bias along the channel for the JFET. Since a positive voltage is required between the gate and each point x in the channel to maintain inversion, a large enough value of V_{DS} can cause $\mathscr{E}_i$ to go to zero at the drain (Fig. 8-16d). As a result there is a small depleted region at the drain end of the channel through which electrons are injected in the saturation current. Once pinch-off is reached, the saturation current remains essentially constant, as in the JFET.

Before discussing the induced-channel MOST analytically, we must consider the important effects of fixed *surface charge* at the semiconductor–oxide interface and in the oxide layer. It is found experimentally that a layer of net positive charge generally exists at this boundary for oxidized Si. Much of this surface charge is unavoidable, although it can be reduced or enhanced by proper surface treatment in the fabrication process. The surface charge can be exploited to allow both depletion and enhancement operation of the induced n-channel device, since the fixed positive space charge can induce an inversion layer in the semiconductor at equilibrium (Fig. 8-17). In the presence of positive surface charge, no threshold value of V_{GS} is needed to establish a conducting channel; the device operates not only in the enhance-

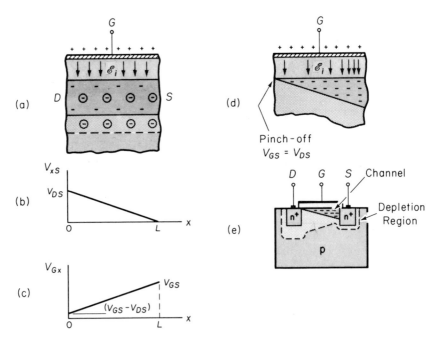

FIGURE 8-16. Properties of an induced-channel MOST: (a) detail of channel with positive gate bias and low current; (b) voltage variation along the channel for low I_{DS}; (c) voltage from gate to channel as a function of x near pinch-off; (d) view of the channel at pinch-off; (e) shape of the channel and depletion region at pinch-off.

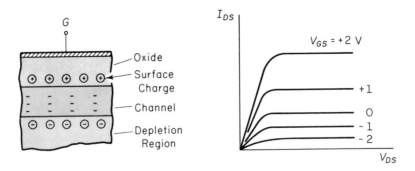

FIGURE 8-17. The effect of positive surface charge at the oxide–semiconductor interface.

ment mode with positive V_{GS}, but also in the depletion mode over a narrow range of negative V_{GS}.[†]

We can calculate the current I_{DS} in the induced-channel MOST by an approach analogous to the JFET analysis. One important difference is that in this case the conductance of the channel involves not only the channel width but also the electron density. For a differential volume of the channel (Fig. 8-18), the conductance (conductivity times area divided by length) can be written

(8-17) $$\text{Conductance} = \frac{q\bar{\mu}_n Z \int n\, dy}{dx}$$

where the integral is carried out over the channel width in the y-direction. As before, Z is the depth of the channel in the z-direction. The integral takes into account the width of the channel at x and the fact that the electron

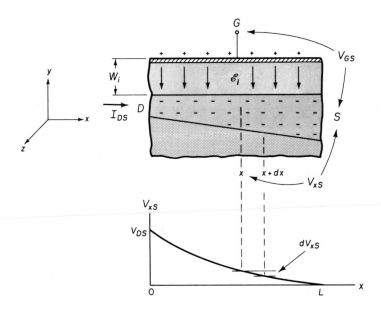

FIGURE 8-18. Differential volume of the channel for calculations.

[†]MOS transistors with characteristics such as Fig. 8-15a and Fig. 8-17 are commonly called *depletion* devices. Since a channel exists at zero gate voltage, these are also referred to as *normally on* devices. On the other hand, devices which require a gate voltage to induce a channel (Fig. 8-15b) are called *enhancement* or *normally off* transistors.

density n varies in the y-direction. In this equation we use an average *surface mobility* $\bar{\mu}_n$, since the scattering mechanisms in a surface layer are more complicated than in bulk material. In a very careful analysis the mobility should be kept inside the integral, since μ_n varies with distance from the surface into the bulk material. We shall neglect this effect to simplify the calculations.

If we assume that dV_{xS}/dx is fairly uniform over the channel width, the current I_{DS} can be written

$$(8\text{-}18) \qquad I_{DS} = -q\bar{\mu}_n Z \frac{dV_{xS}}{dx} \int n \, dy$$

This is analogous to Eq. (8-11) in the JFET analysis.

We can evaluate $q \int n \, dy$ from Gauss's law. This term describes the negative charge in the semiconductor due to mobile electrons and represents one set of negative charges on which the electric field lines terminate. Other charges include the bulk negative charge Q_b due to the ionized acceptor atoms and the positive surface charge Q_s discussed above. With the electric field defined in the $-y$-direction, Gauss's law gives

$$(8\text{-}19) \qquad \epsilon_i \mathscr{E}_i = \epsilon \mathscr{E}_s = -\int (\text{charge density}) \, dy = q \int n \, dy + \frac{Q_b - Q_s}{LZ}$$

In this equation ϵ_i and $\mathscr{E}_i$ refer to the dielectric constant and electric field in the insulator, while ϵ and $\mathscr{E}_s$ are the semiconductor dielectric constant and the field at the semiconductor surface.

For simplicity we shall neglect the fixed charges compared with the mobile electron density. The effects of Q_b and Q_s can be included in the analysis with some additional mathematical effort. With this simplification Eq. (8-19) can be related to the voltage across the insulator and its thickness W_i

$$(8\text{-}20) \qquad q \int n \, dy = \epsilon_i \mathscr{E}_i = \epsilon_i \frac{V_{GS} - V_{xS}}{W_i}$$

Substituting this expression into Eq. (8-18) and integrating along the channel from drain to source, we obtain

$$(8\text{-}21\text{a}) \qquad \int_0^L I_{DS} \, dx = -\bar{\mu}_n Z \epsilon_i \int_{V_{DS}}^0 \frac{V_{GS} - V_{xS}}{W_i} \, dV_{xS}$$

$$(8\text{-}21\text{b}) \qquad I_{DS} = \frac{\bar{\mu}_n Z \epsilon_i}{L W_i} \left[V_{GS} V_{DS} - \frac{1}{2} V_{DS}^2 \right]$$

Pinch-off occurs when the inversion layer is lost at the drain end of the channel, that is, when $V_{DG} = V_{DS} - V_{GS} = 0$. Assuming the saturation current remains constant at the pinch-off value, we can put $V_{DS} = V_{GS}$ in Eq. (8-21) to obtain

$$(8\text{-}22) \qquad I_{DS}\,(\text{sat.}) = \frac{\bar{\mu}_n Z \epsilon_i}{2LW_i} V_{GS}{}^2, \qquad V_{GS} \text{ positive}$$

Thus the channel current I_{DS} varies as the square of the gate voltage at saturation. The transconductance in the saturation range is simply

$$(8\text{-}23) \qquad g_m\,(\text{sat.}) = \frac{\partial I_{DS}}{\partial V_{GS}} = \frac{\bar{\mu}_n Z \epsilon_i}{LW_i} V_{GS}$$

For this example there is no current unless a positive gate voltage is present to induce the inversion layer. Pinch-off occurs when V_{GS} becomes equal to V_{DS}. If a surface charge is present (Prob. 8.9),† there will be a threshold voltage V_T related to the surface charge Q_s by

$$(8\text{-}24) \qquad V_T = -\frac{W_i Q_s}{\epsilon_i Z L}$$

The expression for the channel current in this case is modified to

$$(8\text{-}25) \qquad I_{DS} = \frac{\bar{\mu}_n Z \epsilon_i}{LW_i}\left[(V_{GS} - V_T)V_{DS} - \frac{1}{2}V_{DS}{}^2\right]$$

and pinch-off occurs when $V_{GS} - V_T = V_{DS}$. The saturation current is

$$(8\text{-}26) \qquad I_{DS}\,(\text{sat.}) = \frac{\bar{\mu}_n Z \epsilon_i}{2LW_i}(V_{GS} - V_T)^2$$

when surface charges are present.

The MOS transistor has the advantage of extremely high input impedance between the gate and source electrodes, since these terminals are separated by an oxide layer. The input impedance of a MOST can be on the order of $10^{14}\ \Omega$, making this device very useful in amplifying signals delivered by circuits which are sensitive to loading. As an example, if the voltage across a 1-MΩ resistor is measured with a voltmeter having a 1-MΩ input impedance, current flows through the voltmeter and a false reading is displayed. We say that the meter has *loaded* the circuit. On the other hand, if a MOST is used in the input of the voltmeter, negligible current is drawn by the meter.

---

†In a proper calculation, it is important to include the bulk charge Q_b due to ionized acceptors in the channel. The charge Q_b varies with x, however, and involves surface effects which are beyond the scope of this book. A good treatment including these effects is given in Chapters 10 and 11 of the book by Grove in the reading list.

There are many such applications for amplifying devices with high input impedance.

The MOST is useful in integrated circuits utilizing silicon planar technology (Chapter 10). For many years it was difficult to control the surface charge and other effects in MOS structures, but now these devices can be manufactured in large numbers with good reproducibility.

Transistors of the IGFET type can be fabricated by evaporation or sputtering† of thin films of metal, insulator, and semiconductor materials onto an insulating substrate. One possible configuration of such a *thin-film transistor (TFT)* is shown in Fig. 8-19. By evaporation in a vacuum through

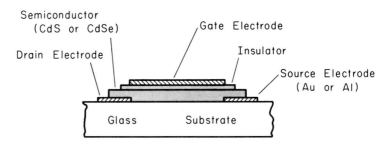

FIGURE 8-19. Typical configuration of a thin-film MOS transistor.

metal masks, the drain and source electrodes are deposited on a glass substrate. Commonly used metals for evaporated electrodes are Au and Al. The semiconductor layer (typically CdS or CdSe) is deposited, and then the insulator (an oxide of Si or Al) is deposited on the semiconductor. Finally, a gate electrode is evaporated onto the insulator to complete the structure. There are other versions of the TFT; for example, the entire structure of Fig. 8-19 can be inverted on the substrate by depositing the gate electrode first.

Another important MOS structure is the *charge-coupled device (CCD)*. This device is composed of a closely-spaced array of metal electrodes on the oxide layer (or some other insulator). By applying appropriate voltages to the electrodes, depletion regions are created in the underlying semiconductor. The result is a series of potential wells under the electrodes. Minority carriers can then be stored in selected potential wells, and this charge can also be shifted from one well to another by properly varying the voltage on adjacent electrodes. This type of device is very useful in computer memory

---

†Sputtering is a process in which an electrode of the material to be deposited is bombarded by ions of argon or other gases in a low-pressure electric discharge.

and also as an optical image sensor. Since our primary interest in this chapter is in transistor structures, we shall leave study of this interesting device to outside reading.†

8.3 Bipolar Junction Transistors

The most widely used transistor structure is the bipolar type, in which both electrons and holes play important roles. This device is the "workhorse" of modern electronics, performing functions in amplification, switching, logic, and other applications. When the word "transistor" is used alone, it is usually understood that the *bipolar junction transistor* (*BJT*) is intended. We shall follow this convention unless the more specific reference BJT is required to avoid confusion with the FET.

The BJT is a current-controlled device (Fig. 8-7b), in which the current through two terminals is controlled by a relatively small current in a third terminal. In this section we shall discuss the general operation of the BJT in qualitative terms and point out the manner in which it provides amplification in simple circuits. Most of the detailed analysis of the device operation will be left for Chapter 9, but in this section we shall lay the groundwork for that discussion.

8.3.1 Fundamentals of BJT Operation. Let us begin the discussion of bipolar transistors by considering the reverse-biased p-n junction diode of Fig. 8-20. According to the theory of Chapter 5, the reverse saturation current through this diode depends on the rate at which minority carriers are generated in the neighborhood of the junction. We found, for example, that the reverse current due to holes being swept from n to p is essentially independent of the size of the junction $\mathscr{E}$ field and hence is independent of the reverse bias. The reason given was that the hole current depends on how often minority holes are generated by EHP creation within a diffusion length of the junction—not upon how fast a particular hole is swept across the depletion layer by the field. As a result, it is possible to increase the reverse current through the diode by increasing the rate of EHP generation (Fig. 8-20b). One convenient method for accomplishing this is optical excitation of EHP's with light ($h\nu > E_g$), as in the photodetector of Section 6.3. With steady photoexcitation the reverse current will still be essentially independent of bias voltage, and if the dark saturation current is negligible, the reverse

---

†For an introduction to charge-coupled devices see articles in *Electronics*, 21 June 1971, pp. 50-59, and *IEEE Spectrum*, July 1971, pp. 18-27.

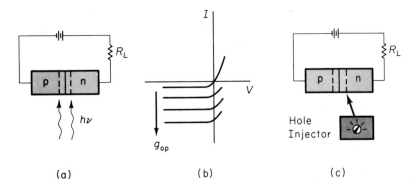

FIGURE 8-20. External control of the current in a reverse-biased p-n junction: (a) optical generation; (b) junction *I-V* characteristics as a function of EHP generation; (c) minority carrier injection by a hypothetical device.

current is directly proportional to the optical generation rate g_{op} [see Eq. (6-9)].

The example of external control of current through a junction by optical generation raises an interesting question: Is it possible to inject minority carriers into the neighborhood of the junction *electrically* instead of optically? If so, we could control the junction reverse current simply by varying the rate of minority carrier injection. For example, let us consider a hypothetical *hole injection device* as in Fig. 8-20c. If we can inject holes at a predetermined rate into the n side of the junction, the effect on the junction current will resemble the effects of optical generation. The current from n to p will depend on the hole injection rate and will be essentially independent of the bias voltage. There are several obvious advantages to such external control of a current; for example, the current through the reverse-biased junction would vary very little if the load resistor R_L were changed, since the magnitude of the junction voltage is relatively unimportant. Therefore, such an arrangement should be a good approximation to a controllable constant current source.

A convenient hole injection device is a forward-biased p^+-n junction. According to Section 5.3.2, the current in such a junction is due primarily to holes injected from the p^+ region into the n material. If we make the n side of the forward-biased junction the same as the n side of the reverse-biased junction, the p^+-n-p structure of Fig. 8-21 results. With this configuration, injection of holes from the p^+-n junction into the center n region supplies the minority carrier holes to participate in the reverse current through the n-p junction. Of course, it is important that the injected holes do not recombine in the n region before they can diffuse to the depletion layer of the

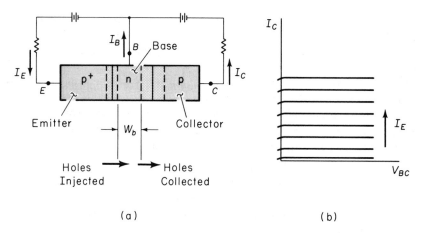

$$(a) \qquad\qquad\qquad\qquad (b)$$

FIGURE 8-21. A p-n-p transistor: (a) schematic representation of
a p-n-p device with a forward-biased emitter junction and a
reverse-biased collector junction; (b) *I-V* characteristics of the
reverse-biased n-p junction as a function of emitter current.

reverse-biased junction. Thus we must make the n region narrow compared
with a hole diffusion length.

The structure we have described is a p-n-p bipolar junction transistor.
The forward-biased junction which injects holes into the center n region is
called the *emitter junction,* and the reverse-biased junction which collects
the injected holes is called the *collector junction.* The p$^+$ region which serves
as the source of injected holes is called the *emitter,* and the p region into which
the holes are swept by the reverse-biased junction is called the *collector.* The
center n region is called the *base,* for reasons which will become clear in Section 8.3.3, when we discuss the historical development of transistor fabrication. The biasing arrangement in Fig. 8-21 is called the *common base* configuration, since the base electrode *B* is common to the emitter and collector
circuits.

To have a good p-n-p transistor, we would prefer that almost all the
holes injected by the emitter into the base be collected. Thus the n-type base
region should be narrow, and the hole lifetime τ_p should be long. This requirement is summed up by specifying $W_b \ll L_p$, where W_b is the length of the
neutral n material of the base (measured between the depletion regions of
the emitter and collector junctions), and L_p is the diffusion length for holes
in the base $(D_p\tau_p)^{1/2}$. With this requirement satisfied, an average hole injected
at the emitter junction will diffuse to the depletion region of the collector
junction without recombination in the base. A second requirement is that the
current I_E crossing the emitter junction should be composed almost entirely
of holes injected into the base, rather than electrons crossing from base to

emitter. This requirement is satisfied by doping the base region lightly compared with the emitter, so that the p^+-n emitter junction of Fig. 8-21 results.

It is clear that current I_E flows into the emitter of a properly biased p-n-p transistor and that I_C flows out at the collector, since the direction of hole flow is from emitter to collector. However, the base current I_B requires a bit more thought. In a good transistor the base current will be very small since I_E is essentially hole current, and the collected hole current I_C is almost equal to I_E. There must be some base current, however, due to requirements of electron flow into the n-type base region (Fig. 8-22). We can account for I_B physically by three dominant mechanisms:

(a) There must be some recombination of injected holes with electrons in the base, even with $W_b \ll L_p$. The electrons lost to recombination must be resupplied through the base contact.

(b) Some electrons will be injected from n to p in the forward-biased emitter junction, even if the emitter is heavily doped

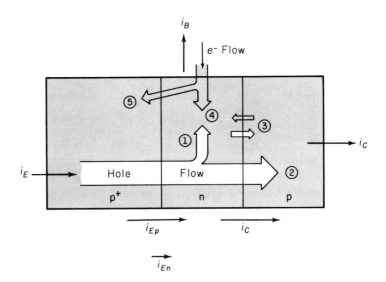

FIGURE 8-22. Summary of hole and electron flow in a p-n-p transistor with proper biasing: (1) injected holes lost to recombination in the base; (2) holes reaching the reverse-biased collector junction; (3) thermally generated electrons and holes making up the reverse saturation current of the collector junction; (4) electrons supplied by the base contact for recombination with holes; (5) electrons injected across the forward-biased emitter junction.

compared to the base. These electrons must also be supplied by I_B.

(c) Some electrons are swept into the base at the reverse-biased collector junction due to thermal generation in the collector. This small current reduces I_B by supplying electrons to the base.

The dominant mechanism in the base current is usually recombination, and we can often approximate the base current by calculating the recombination rate in the base. In a well-designed transistor, I_B will be a very small fraction (perhaps one-hundredth) of I_E.

In an n-p-n transistor the three current directions are reversed, since electrons flow from emitter to collector and holes must be supplied to the base. The physical mechanisms for operation of the n-p-n can be understood simply by reversing the roles of electrons and holes in the p-n-p discussion.

8.3.2 Amplification with BJT's. In this section we shall discuss rather simply the various factors involved in transistor amplification. Basically, the transistor is useful in amplifiers because the currents at the emitter and collector are controllable by the relatively small base current. The essential mechanisms are easy to understand if various secondary effects are neglected. We shall use total current (d-c plus a-c) in this discussion, with the understanding that the simple analysis applies only to d-c and to small-signal a-c at low frequencies. We can relate the terminal currents of the transistor i_E, i_B, and i_C by several important factors. In this introduction we shall neglect the saturation current at the collector (Fig. 8-22, component 3) and such effects as recombination in the transition regions. Under these assumptions, the collector current is made up entirely of those holes injected at the emitter which are not lost to recombination in the base. Thus i_C is proportional to the hole component of the emitter current i_{Ep}

(8-27) $$i_C = Bi_{Ep}$$

The proportionality factor B is simply the fraction of injected holes which make it across the base to the collector; B is called the *base transport factor*. The total emitter current i_E is made up of the hole component i_{Ep} and an electron component i_{En}, due to electrons injected from base to emitter (component 5 in Fig. 8-22). The *emitter injection efficiency* γ is

(8-28) $$\gamma = \frac{i_{Ep}}{i_{En} + i_{Ep}}$$

For an efficient transistor we would like B and γ to be very near unity;

that is, the emitter current should be due mostly to holes ($\gamma \simeq 1$), and most of the injected holes should eventually participate in the collector current ($B \simeq 1$). The relation between the collector and emitter currents is

$$(8\text{-}29) \qquad \frac{i_C}{i_E} = \frac{Bi_{Ep}}{i_{En} + i_{Ep}} = B\gamma \equiv \alpha$$

The product $B\gamma$ is defined as the factor α, which represents the emitter-to-collector current amplification. There is no real amplification between these currents, since α is smaller than unity. On the other hand, the relation between i_C and i_B is more promising for amplification.

In accounting for the base current, we must include the rates at which electrons are lost from the base by injection across the emitter junction (i_{En}) and the rate of electron recombination with holes in the base. In each case, the lost electrons must be resupplied through the base current i_B. If the fraction of injected holes making it across the base *without* recombination is B, then it follows that $(1 - B)$ is the fraction *recombining* in the base. Thus the base current is

$$(8\text{-}30) \qquad i_B = i_{En} + (1 - B)i_{Ep}$$

neglecting the collector saturation current. The relation between the collector and base currents is found from Eq. (8-27) and Eq. (8-30)

$$\frac{i_C}{i_B} = \frac{Bi_{Ep}}{i_{En} + (1 - B)i_{Ep}} = \frac{B[i_{Ep}/(i_{En} + i_{Ep})]}{1 - B[i_{Ep}/(i_{En} + i_{Ep})]}$$

$$(8\text{-}31) \qquad \frac{i_C}{i_B} = \frac{B\gamma}{1 - B\gamma} = \frac{\alpha}{1 - \alpha} \equiv \beta$$

The factor β relating the collector current to the base current is the *base-to-collector current amplification factor*. Since α is near unity, it is clear that β can be large for a good transistor, and the collector current is large compared with the base current.

It remains to be shown that the collector current i_C can be controlled by variations in the small current i_B. In the discussion up to this point, we have indicated the control of i_C by the emitter current i_E, with the base current characterized as a small side effect. In fact, we can show from space charge neutrality arguments that i_B can indeed be used to determine the magnitude of i_C. Let us consider the transistor of Fig. 8-23, in which i_B is determined by a biasing circuit. For simplicity, we shall assume unity emitter injection efficiency and negligible collector saturation current. Since the n-type base region is electrostatically neutral between the two transition regions, the presence of excess holes in transit from emitter to collector calls for com-

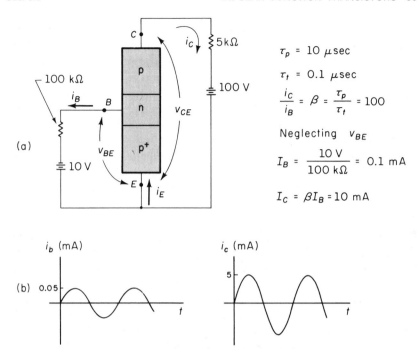

$\tau_p = 10 \ \mu\text{sec}$

$\tau_t = 0.1 \ \mu\text{sec}$

$\dfrac{i_C}{i_B} = \beta = \dfrac{\tau_p}{\tau_t} = 100$

Neglecting v_{BE}

$I_B = \dfrac{10 \text{ V}}{100 \text{ k}\Omega} = 0.1 \text{ mA}$

$I_C = \beta I_B = 10 \text{ mA}$

FIGURE 8-23. Example of amplification in a common-emitter transistor circuit: (a) biasing circuit; (b) addition of an a-c variation of base current i_b to the d-c value of I_B, resulting in an a-c component i_c.

pensating excess electrons from the base contact. However, there is an important difference in the times which electrons and holes spend in the base. The average hole spends a time τ_t, defined as the *transit time* from emitter to collector. Since the base width W_b is made small compared with L_p, this transit time is much less than the average hole lifetime τ_p in the base. On the other hand, an average excess electron supplied from the base contact spends τ_p sec in the base (for simple recombination and equal excess carrier densities, τ_n and τ_p are equal). While the average electron waits τ_p sec for recombination, many holes can enter and leave the base region, each with an average transit time τ_t. In particular, for each electron entering from the base contact, τ_p/τ_t holes can pass from emitter to collector while maintaining space charge neutrality. Thus the ratio of collector current to base current is simply

(8-32) $$\frac{i_C}{i_B} = \beta = \frac{\tau_p}{\tau_t}$$

for $\gamma = 1$ and negligible collector saturation current.

If the electron supply to the base (i_B) is restricted, the traffic of holes from emitter to base is correspondingly reduced. This can be argued simply by supposing that the hole injection does continue despite the restriction on electrons from the base contact. The result would be a net buildup of positive charge in the base and a loss of forward bias (and therefore a loss of hole injection) at the emitter junction. Clearly, the supply of electrons through i_B can be used to raise or lower the hole flow from emitter to collector.

The base current is controlled independently in Fig. 8-23. This is called a *common-emitter* circuit, since the emitter electrode is common to the base and collector circuits. The emitter junction is clearly forward biased by the battery in the base circuit. The voltage drop in the forward-biased emitter junction is small, however, so that almost all of the voltage from collector to emitter appears across the reverse-biased collector junction. Since v_{BE} is small for the forward-biased junction, we can neglect it and approximate the base current as 10 V/100 kΩ = 0.1 mA. If $\tau_p = 10$ μsec and $\tau_t = 0.1$ μsec, β for the transistor is 100 and the collector current I_C is 10 mA. It is important to note that i_C is determined by β and the base current, rather than by the battery and resistor in the collector circuit (as long as these are of reasonable values to maintain a reverse-biased collector junction). In this example 50 V of the collector circuit battery voltage appears across the 5 kΩ resistor, and 50 V serves to reverse bias the collector junction.

If a small a-c current i_b is superimposed on the steady state base current of Fig. 8-23a, a corresponding a-c current i_c appears in the collector circuit. The time-varying portion of the collector current will be i_b multiplied by the factor β, and current gain results.

We have neglected a number of important properties of the transistor in this introductory discussion, and many of these properties will be treated in detail in Chapter 9. We have established, however, the fundamental basis of operation for the bipolar transistor and have indicated in a simplified way how it can be used to produce current gain in an electronic circuit.

8.3.3 Fabrication.

The first junction transistors were fabricated by alloying techniques. For example, alloyed p-type regions on opposite sides of an n-type Ge sample result in a p^+-n-p^+ structure (Fig. 8-24). Whether such a structure can serve as a useful transistor depends largely on the accuracy with which the depth of the alloyed regions can be controlled. Since the n-type base region must be thin for proper transistor action, the p^+ collector and emitter regrown regions must be very close together. Obviously, the temperature and duration of the alloying process must be controlled very carefully if reproducible transistor units are to be manufactured. The emitter and collector regrown regions can be maintained fairly flat and parallel by orienting the Ge crystal such that alloying takes place along crystal planes. The importance of crystal planes in defining the boundaries of alloyed regions is demonstrated by the fact that flat junctions can be obtained (Figs. 8-24b and 1-10). In one method of transistor fabrication, the thin n-type Ge slab is placed in a carbon holder with alloy preforms in contact with both sides. The preforms contain In (perhaps mixed with Ga) to provide acceptor impurities for the collector and emitter regions. A large number of such arrangements of Ge and alloy preforms can be placed in a carefully controlled furnace for a specified time. This results in transistors such as the one shown in Fig. 8-24. In this device the active part of the base is the thin region between the alloyed emitter and collector contacts.

The alloyed Ge power transistor illustrated in Fig. 8-24 has two base contacts—a dot in the center of the wafer and a ring near the periphery. The p^+ emitter is an alloyed In ring between the base contacts, and the collector junction covers most of the bottom surface of the wafer. The advantages of base contacts on both sides of the emitter are discussed in Section 9.3.5. The packaging of this transistor in a usable structure is illustrated in Fig. 8-25. The two base contacts are contacted by a ring tab (*d*) and a ribbon (*e*); these two contacts are then bonded to one post on the header (*i*). The emitter ring (*f*) is connected to a second post by a metal strip (*g*). The collector contact (*h*) is alloyed to a pedestal on the header body. After incapsulation (*j*) the active part of the transistor is protected from the atmosphere by a hermetically sealed package. Figure 8-25 illustrates that the active transistor is much smaller than the incapsulated device. Since this device is designed for use at relatively high power levels, the bulky incapsulation is necessary for efficient heat transfer away from the junctions.

In the early versions of alloyed transistors, it was common practice to mount the semiconductor sample on a header and connect the emitter and collector alloyed regions by wires to the appropriate header posts. In this configuration the material common to the base region of the transistor provides mechanical support for the structure. In the point contact transistor, which preceded alloyed structures, the emitter and collector were sharp wires pressed upon the surface of a semiconductor which served as the base of the device. This is the origin of the term "base" as applied to the central region of a transistor. Although the base region is not used to provide mechanical support in modern transistors, the terminology is retained.

Although alloyed junction transistors served a useful purpose for many years, they have been largely replaced with transistors fabricated by the more precise and convenient diffusion process. Most transistors are now made in Si, utilizing selective diffusion and oxide masking. A simple example of an n-p-n diffused transistor is shown in Fig. 8-26. Beginning with an n-type Si substrate, an oxide layer is grown on the surface and a window is opened

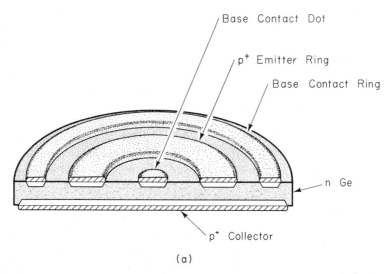

Base Contact Dot

p⁺ Emitter Ring

Base Contact Ring

n Ge

p⁺ Collector

(a)

(b)

FIGURE **8-24.** An alloyed junction transistor: (a) cross section of Ge wafer with alloyed collector and emitter regions and base contacts; this transistor has two base contacts (central dot and outer ring) separated by the emitter ring; (b) microscope view of the wafer cross section, showing the emitter region and part of the collector. Photograph courtesy of Delco Electronics Division, General Motors Corporation.

by the photolithographic techniques described in Section 5.1.3. The sample is placed in a furnace for a boron diffusion, forming the p-type base region. After reoxidation a new window is opened in the oxide for a phosphorus diffusion, forming the n^+ emitter region. The time and temperature of each

FIGURE 8-25.   Assembly of the alloyed junction transistor of Fig.
8-24: (a) indium alloy preform for collector junction; (b) n-type
Ge wafer; (c) alloy preform for emitter ring; (d) contact tab for
base ring; (e) contact ribbon for central base dot; (f) emitter ring;
(g) emitter contact strip; (h) alloyed collector contact; (i) assem-
bled device without top; (j) hermetically sealed device. Photograph courtesy of Delco Electronics Division, General Motors
Corporation.

diffusion can be controlled quite accurately to ensure reproducible junction
depths and base region thickness. After opening windows to the top surface
of the p and n^+ regions, aluminum is evaporated onto the wafer. The final
metallization pattern is defined by photolithography, and the unwanted
Al is etched away. Several hundred such transistors can be made on a single
wafer of Si and then separated by scribing and breaking into individual
devices. In this geometry the n-type substrate is alloyed to a header to provide
contact to the collector region, and Au or Al wires are bonded to the metallized regions to provide leads to the emitter and base. The final encapsulation
can be a hermetically sealed header or a molded plastic case, depending on
the power rating and environmental specifications.

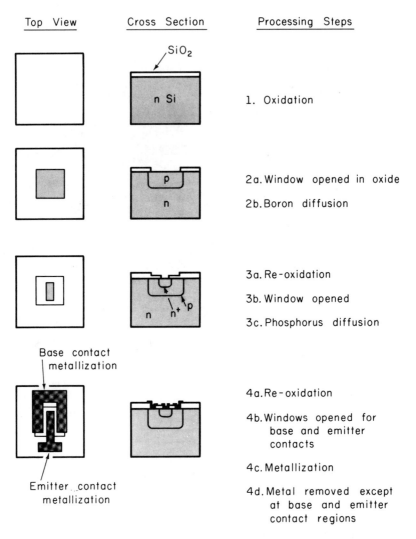

Top View   Cross Section   Processing Steps

SiO$_2$

n Si   1. Oxidation

2a. Window opened in oxide

2b. Boron diffusion

3a. Re-oxidation

3b. Window opened

3c. Phosphorus diffusion

Base contact
metallization

4a. Re-oxidation

4b. Windows opened for
    base and emitter
    contacts

4c. Metallization

Emitter contact
metallization

4d. Metal removed except
    at base and emitter
    contact regions

FIGURE 8-26. Steps in fabricating an n-p-n transistor by double-diffusion of boron and phosphorus in an n-type Si substrate.

Many alterations can be made in this basic process to improve the transistor characteristics. For example, the n-type collector region is often an epitaxial layer grown on an n$^+$ substrate. This improves the properties of the collector as discussed in Chapter 9. Special requirements exist for high-power and high-frequency devices (Sections 9.3 and 9.4). Further changes are required in fabricating transistors in integrated circuits. In fact, many discrete (single-unit) transistors are made using the techniques of IC technology because of the requirements of mass production. We shall discuss these processes in Chapter 10, where IC fabrication will be treated in detail.

READING LIST

Special report, "The Field-Effect Transistor," *Electronics*, vol. 37, no. 30, pp. 45–68, 30 November 1964.

C. FELDMAN, "Thin-Film Active Devices," *Electronics*, vol. 37, no. 4, pp. 23–26, 24 January 1964.

H. BORKAN, "Depositing Active and Passive Thin-Film Elements on One Chip," *Electronics*, vol. 37, no. 14, pp. 53–60, 20 April 1964.

Special report, "The Transistor: Two Decades of Progress," *Electronics*, vol. 41, no. 4, pp. 77–130, 19 February 1968.

J. D. RYDER, *Electronic Fundamentals and Applications* (4th ed.). Englewood Cliffs, N.J.: Prentice-Hall, Inc., 1970.

E. J. ANGELO, JR., *Electronics: BJTs, FETs, and Microcircuits.* New York: McGraw-Hill Book Company, 1969, pp. 9–23, 97–108, 213–320.

P. E. GRAY and C. L. SEARLE, *Electronic Principles: Physics, Models, and Circuits.* New York: John Wiley & Sons, Inc., 1969, pp. 311–362.

J. T. WALLMARK and H. JOHNSON, *Field-Effect Transistors: Physics, Technology and Applications.* Englewood Cliffs, N.J.: Prentice-Hall, Inc., 1966.

J. R. HAUSER, "Unipolar Transistors," Sec. II in *Fundamentals of Silicon Integrated Device Technology, Vol. II: Bipolar and Unipolar Transistors*, eds. R. M. BURGER and R. P. DONOVAN. Englewood Cliffs, N.J.: Prentice-Hall, Inc., 1968, pp. 265–452.

A. S. GROVE, *Physics and Technology of Semiconductor Devices.* New York: John Wiley & Sons, Inc., 1967, pp. 208–259, 317–333.

S. M. SZE, *Physics of Semiconductor Devices.* New York: John Wiley & Sons, Inc., 1969, Ch. 6, "Junction Transistors," pp. 261–318; Part III, "Interface and Thin-Film Devices," pp. 363–624.

PROBLEMS

8.1 The triode in Fig. 8-4a has the characteristics given in Fig. P8-1. The negative grid bias battery is 8 V, E is 300 V, and $R = 35$ kΩ.

(a) Find the a-c voltage gain A_v graphically. *(Continued)*

(Continued 8.1)

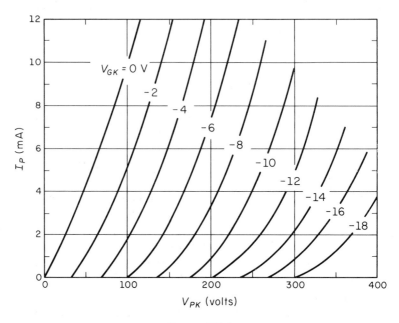

V_{GK} = 0 V

FIGURE P8-1

(b) Given $\mu = 20$ and $r_p = 13.5$ kΩ, draw the small-signal a-c equivalent circuit and calculate A_v analytically.

8.2 In this problem we demonstrate that the cathode resistor R_k in Fig. P8-2 can be used to provide negative bias to the grid. The triode characteristics are given in Fig. P8-1. Assuming zero grid current through R_g, find the Q-point (V_{PK}, V_{GK}, I_p) for this circuit. *Hint*: three graphical equations are needed—the triode characteristics, and loop equations around the plate and grid circuits.

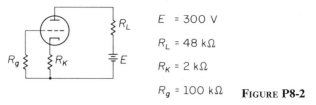

E = 300 V

R_L = 48 kΩ

R_K = 2 kΩ

R_g = 100 kΩ FIGURE P8-2

8.3 Given $\mu = 20$ and $r_p = 13.5$ kΩ for the triode in Fig. P8-3, draw the small-signal a-c equivalent circuit and calculate the voltage gain A_v. Assume all capacitors have negligible a-c impedance at the operating frequency. What is the advantage of bypassing R_k with a capacitor?

(Continued 8.3)

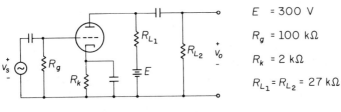

E = 300 V

R_g = 100 kΩ

R_k = 2 kΩ

$R_{L_1} = R_{L_2}$ = 27 kΩ

FIGURE P8-3

8.4 Show that Eq. (8-14) results from Eqs. (8-11) and (8-12).

8.5 The current I_{DS} varies almost linearly with V_{DS} in a JFET for low values of V_{DS}.
   (a) Use the binomial expansion with $V_{DS}/(-V_{GS}) < 1$ to rewrite Eq. (8-14) as an approximation to this case.
   (b) Show that the expression for the channel conductance I_{DS}/V_{DS} in the linear range is the same as g_m (sat.) given by Eq. (8-16).
   (c) What value of gate voltage V_{GS} turns the device off such that the channel conductance goes to zero?

8.6 Modify Eqs. (8-14) and (8-15) to include effects of the contact potential V_0 at the p$^+$-n junction on either side of the channel.

8.7 Calculate the incremental channel resistance $\partial V_{DS}/\partial I_{DS}$ for a JFET below saturation. What does the simple theory imply about this resistance beyond pinch-off?

8.8 Show that the voltage with respect to the source V_{xS} at any point x in the channel of a MOSFET at pinch-off is $V_{xS} = V_{DS}(1 - \sqrt{x/L})$. *Hint:* integrate Eq. (8-21a) from point x to the source, and solve the resulting quadratic equation for V_{xS}. You may assume $V_{DS} = V_{GS}$ at pinch-off and use Eq. (8-22).

8.9 Modify Eq. (8-21) to include effects of the surface charge Q_s described in Eq. (8-19). The result is Eq. (8-25). Relate the threshold voltage V_T to the capacitance of the oxide layer. Notice from the conductance in the linear range that a sufficient positive charge Q_s can result in a conducting channel at zero gate voltage, and a negative Q_s requires $V_{GS} > V_T$ to induce a channel.

8.10 (a) Show that Eq. (8-32) is valid from arguments of the steady state replacement of stored charge.
   (b) What is the steady state charge $Q_n = Q_p$ due to excess electrons and holes in the neutral base region for the transistor of Fig. 8-23?

8.11 (a) Sketch the energy band diagrams for a p-n-p and an n-p-n transistor at equilibrium and under normal bias conditions.

(b) Use the n-p-n band diagrams and a figure similar to Fig. 8-22 to discuss the injection and collection of electrons in this structure; what are the important components of base current in the n-p-n?

8.12 Sketch the ideal collector characteristics (i_C, $-v_{CE}$) for the transistor of Fig. 8-23; let i_B vary from zero to 0.2 mA in increments of 0.02 mA, and let $-v_{CE}$ vary from 0 to 100 V. Draw a load line on the resulting characteristics for the circuit of Fig. 8-23, and find the steady state value of $-V_{CE}$ graphically for $I_B = 0.1$ mA.

8.13 Sketch the contact masks required for fabrication of the double-diffused transistor of Fig. 8-26. Refer to Fig. 5-6 as an aid. For simplicity, sketch the masks for a single transistor rather than for an array.

8.14 Given the data of Prob. 5.22, plot the doping profiles $N_a(x)$ and $N_d(x)$ for the following double-diffused transistor: The starting wafer is n-type Si with $N_d = 5 \times 10^{16}$ cm^{-3}; $N_s = 5.82 \times 10^{13}$ cm^{-2} boron atoms are deposited on the surface, and these atoms are diffused into the wafer at 1100°C for 1 hr ($D = 3 \times 10^{-13}$ cm^2/sec for B in Si at 1100°C); then the wafer is placed in a phosphorus diffusion furnace at 1000°C for 15 min ($D = 3 \times 10^{-14}$ cm^2/sec for P in Si at 1000°C); during the emitter diffusion the surface concentration is held constant at 5×10^{20} cm^{-3}. You may assume the base doping profile does not change appreciably during the emitter diffusion, which takes place at a lower temperature and for a shorter time. Find the width of the base region from plots of $N_a(x)$ and $N_d(x)$. *Hint*: use five-cycle semilog paper and let x vary from zero to about 1.5 μm in steps which are chosen to be simple multiples of $2\sqrt{Dt}$.

ANALYSIS OF THE BIPOLAR JUNCTION TRANSISTOR 9

Since the bipolar transistor is basic to modern electronics, it deserves a more detailed analysis than the introductory treatment of Chapter 8. In this chapter we shall investigate carefully the charge distributions in the transistor and relate the three terminal currents to the physical characteristics of the device. Our aim is to gain a solid understanding of the current flow and control of the transistor and to discover the most important secondary effects which influence its operation. We shall discuss the properties of the transistor with proper biasing for amplification and then consider the effects of more general biasing, as encountered in switching circuits. We shall relate the properties of the device to certain circuit parameters and consider variations such as thermal effects and high-frequency operation. The analysis will be devoted primarily to the abrupt-junction transistor, in which the base is a uniformly doped n region terminated by sharp n-p junctions. The more general case including graded doping of the base will then be considered as an alteration of the basic theory. This approach allows us to establish firmly the fundamentals of transistor operation before including the complications of carrier drift in the base and other effects.

In this chapter we shall concentrate primarily on the p-n-p transistor. This choice is completely arbitrary. The main advantage of the p-n-p for analysis is that hole flow and current are in the same direction. This makes the various mechanisms of charge transport somewhat easier to visualize in a preliminary explanation. Once these basic ideas are established for the p-n-p device, it is simple to relate them to the n-p-n.

9.1 Minority Carrier Distributions and Terminal Currents

We begin our analysis of the transistor by applying the techniques of previous chapters to the problem of hole injection into a narrow n-type base region. The mathematics is very similar to that used in the problem of the narrow-base diode (Prob. 6.5). Basically, we assume holes are injected into the base at the forward-biased emitter, and these holes diffuse to the collector junction. The first step is to solve for the excess hole distribution in the base, and the second step is to evaluate the emitter and collector currents (I_E, I_C) from the gradient of the hole distribution on each side of the base. Then the base current (I_B) can be found from a current summation or from a charge control analysis of recombination in the base.

In this section we shall simplify the calculations by making several assumptions:

1. Holes diffuse from emitter to collector; drift is negligible in the base region.
2. The emitter current is made up entirely of holes; the emitter injection efficiency is $\gamma = 1$.
3. The collector saturation current is negligible.
4. The active part of the base and the two junctions are of uniform cross-sectional area A; current flow in the base is essentially one-dimensional from emitter to collector.
5. All currents and voltages are steady state.

In later sections we shall consider the implications of drift due to nonuniform doping in the base, the addition of I_{En} and collector saturation current, structural effects such as different areas for the emitter and collector junctions, and capacitance and transit time effects in a-c operation.

9.1.1 Solution of the Diffusion Equation in the Base Region.
Since the injected holes are assumed to flow from emitter to collector by diffusion, we can evaluate the currents crossing the two junctions by techniques used in Chapter 5. Neglecting recombination in the two depletion regions, the hole current entering the base at the emitter junction is the current I_E, and the hole current leaving the base at the collector is I_C. If we can solve for the distribution of excess holes in the base region, it is simple to evaluate the gradient of the distribution at the two ends of the base to find the currents. We shall consider the simplified geometry of Fig. 9-1, in which

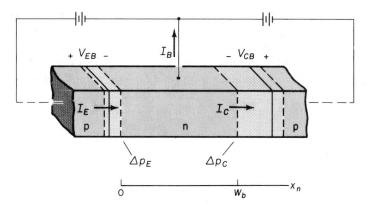

FIGURE 9-1. Simplified p-n-p transistor geometry used in the calculations.

the base width is W_b between the two depletion regions, and the uniform cross-sectional area is A. The excess hole density at the edge of the emitter depletion region Δp_E and the corresponding density on the collector side of the base Δp_C are found from Eq. (5-29)

(9-1a) $$\Delta p_E = p_n(e^{qV_{EB}/kT} - 1)$$

(9-1b) $$\Delta p_C = p_n(e^{qV_{CB}/kT} - 1)$$

If the emitter junction is strongly forward biased ($V_{EB} \gg kT/q$) and the collector junction is strongly reverse biased ($V_{CB} \ll 0$), these excess densities simplify to

(9-2a) $$\Delta p_E \simeq p_n e^{qV_{EB}/kT}$$

(9-2b) $$\Delta p_C \simeq -p_n$$

We can solve for the excess hole density as a function of distance in the base $\delta p(x_n)$ by using the proper boundary conditions in the diffusion equation, Eq. (4-37b)

(9-3) $$\frac{d^2\, \delta p(x_n)}{dx_n{}^2} = \frac{\delta p(x_n)}{L_p{}^2}$$

The solution of this equation is

(9-4) $$\delta p(x_n) = C_1 e^{x_n/L_p} + C_2 e^{-x_n/L_p}$$

where L_p is the diffusion length of holes in the base region. Unlike the simple problem of injection into a long n region, we cannot eliminate one of the constants by assuming the excess holes disappear for large x_n. In fact, since $W_b \ll L_p$ in a properly designed transistor, most of the injected holes reach the collector at W_b. The solution is very similar to that of the narrow base diode problem. In this case the appropriate boundary conditions are

(9-5a) $$\delta p(x_n = 0) = C_1 + C_2 = \Delta p_E$$

(9-5b) $$\delta p(x_n = W_b) = C_1 e^{W_b/L_p} + C_2 e^{-W_b/L_p} = \Delta p_C$$

Solving for the parameters C_1 and C_2 we obtain

(9-6a) $$C_1 = \frac{\Delta p_C - \Delta p_E e^{-W_b/L_p}}{e^{W_b/L_p} - e^{-W_b/L_p}}$$

(9-6b) $$C_2 = \frac{\Delta p_E e^{W_b/L_p} - \Delta p_C}{e^{W_b/L_p} - e^{-W_b/L_p}}$$

These parameters applied to Eq. (9-4) give the full expression for the excess hole density in the base region. For example, if we assume the collector junction is strongly reverse biased [Eq. (9-2b)] and the equilibrium hole density p_n is negligible compared with the injected density Δp_E, the excess hole distribution simplifies to

(9-7) $$\delta p(x_n) = \Delta p_E \frac{e^{W_b/L_p} e^{-x_n/L_p} - e^{-W_b/L_p} e^{x_n/L_p}}{e^{W_b/L_p} - e^{-W_b/L_p}}$$

The various terms in Eq. (9-7) are sketched in Fig. 9-2, and the corresponding excess hole density in the base region is demonstrated for a moderate value of W_b/L_p. Note that $\delta p(x_n)$ varies almost linearly between the emitter and collector junction depletion regions (Prob. 9.1). As we shall see below, the slight deviation from linearity of the distribution indicates the small value of I_B caused by recombination in the base region.

9.1.2 Evaluation of the Terminal Currents.
Having solved for the excess hole distribution in the base region, we can evaluate the emitter and collector currents from the gradient of the hole density at each depletion region edge. From Eq. (4-22b) we have

(9-8) $$I_p(x_n) = -qAD_p \frac{d\delta p(x_n)}{dx_n}$$

For unity emitter injection efficiency, this expression evaluated at $x_n = 0$ gives the emitter current

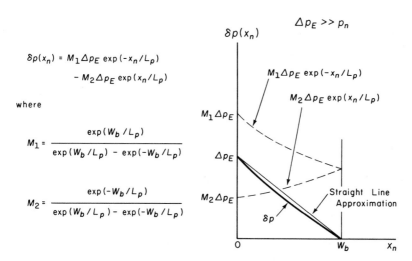

$$\delta p(x_n)$$

$$\Delta p_E \gg p_n$$

$$\delta p(x_n) = M_1 \Delta p_E \exp(-x_n/L_p)$$
$$- M_2 \Delta p_E \exp(x_n/L_p)$$

where

$$M_1 = \frac{\exp(W_b/L_p)}{\exp(W_b/L_p) - \exp(-W_b/L_p)}$$

$$M_2 = \frac{\exp(-W_b/L_p)}{\exp(W_b/L_p) - \exp(-W_b/L_p)}$$

$$M_1 \Delta p_E \exp(-x_n/L_p)$$
$$M_2 \Delta p_E \exp(x_n/L_p)$$

$M_1 \Delta p_E$

Δp_E

$M_2 \Delta p_E$ Straight Line Approximation

δp

0 W_b x_n

FIGURE 9-2. Sketch of the terms in Eq. (9-7), illustrating the linearity of the hole distribution in the base region. In this example, $W_b/L_p = 1/2$.

$$(9\text{-}9) \qquad I_E = I_p(x_n = 0) = qA\frac{D_p}{L_p}(C_2 - C_1)$$

Similarly, if we neglect the electrons crossing from collector to base in the collector reverse saturation current, I_C is made up entirely of holes entering the collector depletion region from the base. Evaluating Eq. (9-8) at $x_n = W_b$ we have the collector current

$$(9\text{-}10) \qquad I_C = I_p(x_n = W_b) = qA\frac{D_p}{L_p}(C_2 e^{-W_b/L_p} - C_1 e^{W_b/L_p})$$

When the parameters C_1 and C_2 are substituted from Eqs. (9-6), the emitter and collector currents take a form which is most easily written in terms of hyperbolic functions

$$(9\text{-}11) \qquad I_E = qA\frac{D_p}{L_p}\left[\frac{\Delta p_E(e^{W_b/L_p} + e^{-W_b/L_p}) - 2\Delta p_C}{e^{W_b/L_p} - e^{-W_b/L_p}}\right]$$

$$= qA\frac{D_p}{L_p}\left(\Delta p_E \operatorname{ctnh}\frac{W_b}{L_p} - \Delta p_C \operatorname{csch}\frac{W_b}{L_p}\right)$$

$$(9\text{-}12) \qquad I_C = qA\frac{D_p}{L_p}\left(\Delta p_E \operatorname{csch}\frac{W_b}{L_p} - \Delta p_C \operatorname{ctnh}\frac{W_b}{L_p}\right)$$

Now we can obtain the value of I_B by a current summation, noting that the sum of the base and collector currents leaving the device must equal the emitter current entering

$$(9\text{-}13) \quad I_B = I_E - I_C = qA\frac{D_p}{L_p}\left[(\Delta p_E + \Delta p_C)\left(\text{ctnh}\,\frac{W_b}{L_p} - \text{csch}\,\frac{W_b}{L_p}\right)\right]$$

$$= qA\frac{D_p}{L_p}\left[(\Delta p_E + \Delta p_C)\tanh\frac{W_b}{2L_p}\right]$$

By using the techniques of Chapter 5 we have evaluated the three terminal currents of the transistor in terms of the material parameters, the base width, and the excess densities Δp_E and Δp_C. Furthermore, since these excess densities are related in a straightforward way to the emitter and collector junction bias voltages by Eq. (9-1), it should be simple to evaluate the transistor performance under various biasing conditions. It is important to note here that Eqs. (9-11) through (9-13) are not restricted to the case of the usual transistor biasing. For example, Δp_C may be $-p_n$ for a strongly reverse-biased collector, or it may be a significant positive number if the collector is positively biased. The generality of these equations will be used in Section 9.2 in considering the application of transistors to switching circuits.

9.1.3 Approximations of the Terminal Currents. The general equations of the previous section can be simplified for the case of normal transistor biasing, and such simplification allows us to gain insight into the current flow. For example, if the collector is reverse biased, $\Delta p_C = -p_n$ from Eq. (9-2b). Furthermore, if the equilibrium hole density p_n is small, we can neglect the terms involving Δp_C. For this case the terminal currents reduce to

$$(9\text{-}14a) \qquad\qquad I_E \simeq qA\frac{D_p}{L_p}\,\Delta p_E\,\text{ctnh}\,\frac{W_b}{L_p}$$

$$(9\text{-}14b) \qquad\qquad I_C \simeq qA\frac{D_p}{L_p}\,\Delta p_E\,\text{csch}\,\frac{W_b}{L_p}$$

$$(9\text{-}14c) \qquad\qquad I_B \simeq qA\frac{D_p}{L_p}\,\Delta p_E\,\tanh\frac{W_b}{2L_p}$$

Series expansions of the hyperbolic functions are given in Table 9-1. For small values of W_b/L_p, we can neglect terms above the first order of the argument. It is clear from this table that I_C is only slightly smaller than I_E, as expected. The first-order approximation of tanh y is simply y, so that the base current is

$$(9\text{-}15) \qquad\qquad I_B \simeq qA\frac{D_p}{L_p}\,\Delta p_E\frac{W_b}{2L_p} = \frac{qAW_b\,\Delta p_E}{2\tau_p}$$

TABLE 9-1. Expansions of several pertinent hyperbolic functions.

$$\operatorname{sech} y = \quad 1 - \frac{y^2}{2} + \frac{5y^4}{24} - \cdots$$

$$\operatorname{ctnh} y = \quad \frac{1}{y} + \frac{y}{3} - \frac{y^3}{45} + \cdots$$

$$\operatorname{csch} y = \quad \frac{1}{y} - \frac{y}{6} + \frac{7y^3}{360} - \cdots$$

$$\tanh y = \quad y - \frac{y^3}{3} + \cdots$$

The same approximate expression for the base current is found from the difference in the first-order approximations to I_E and I_C

$$
\begin{aligned}
(9\text{-}16) \quad I_B &= I_E - I_C \\
&\simeq qA\frac{D_p}{L_p}\Delta p_E\left[\left(\frac{1}{W_b/L_p} + \frac{W_b/L_p}{3}\right) - \left(\frac{1}{W_b/L_p} - \frac{W_b/L_p}{6}\right)\right] \\
&\simeq \frac{qAD_pW_b\,\Delta p_E}{2L_p^{\,2}} = \frac{qAW_b\,\Delta p_E}{2\tau_p}
\end{aligned}
$$

It is interesting to compare this expression for the base current with that obtained from the charge control model, assuming an essentially straight-line hole distribution in the base (Fig. 9-3b). Since the hole distribution diagram appears as a triangle in this approximation, we have

$$(9\text{-}17) \qquad\qquad Q_p \simeq \tfrac{1}{2}qA\,\Delta p_E W_b$$

If we consider that this stored charge must be replaced every τ_p sec and relate the recombination rate to the rate at which electrons are supplied by the base current, I_B becomes

$$(9\text{-}18) \qquad\qquad I_B \simeq \frac{Q_p}{\tau_p} = \frac{qAW_b\,\Delta p_E}{2\tau_p}$$

which is the same as that found in Eqs. (9-15) and (9-16).

Since we have neglected the collector saturation current and have assumed $\gamma = 1$ in these approximations, the difference between I_E and I_C is accounted for by the requirements of recombination in the base. In Eq. (9-18) we have a clear demonstration that the base current is reduced for

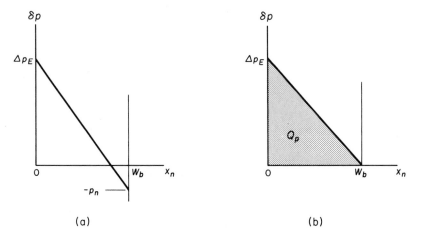

FIGURE 9-3. Approximate excess hole distributions in the base: (a) forward-biased emitter, reverse-biased collector; (b) triangular distribution for $V_{CB} = 0$ or for negligible p_n.

small W_b and large τ_p. We can increase τ_p by using light doping in the base region, which of course also improves the emitter injection efficiency.

The approximations of Table 9-1 can be used to calculate the current transfer ratios α and β, discussed in Section 8.3.2. If we substitute the approximate values of the hyperbolic functions into Eqs. (9-14), we obtain

$$\text{(9-19a)} \qquad \alpha = \frac{I_C}{I_E} = \frac{\text{csch } W_b/L_p}{\text{ctnh } W_b/L_p} = \text{sech } \frac{W_b}{L_p} \simeq 1 - \frac{1}{2}\left(\frac{W_b}{L_p}\right)^2$$

$$\text{(9-19b)} \qquad \beta = \frac{I_C}{I_B} = \frac{\text{csch } W_b/L_p}{\text{tanh } W_b/2L_p} \simeq \frac{2L_p{}^2}{W_b{}^2}$$

In approximating β we have used only the first term in the expansion of csch (W_b/L_p). This is a fairly good approximation when $W_b \ll L_p$. The value of α calculated in Eq. (9-19a) is actually the base transport factor B, since $\alpha = B\gamma$, and we have assumed the emitter injection efficiency $\gamma = 1$ in this section. This approximation of perfect hole injection is violated in some transistors.

The straight-line approximation of the excess hole density (Fig. 9-3b) is fairly accurate in calculating the base current. On the other hand, it does not give a valid picture of I_E and I_C. If the distribution were perfectly straight, the slope would be the same at each end of the base region. This would imply zero base current, which is not the case. There must be some "droop" to the distribution, as in the more accurate curve of Fig. 9-2. This slight deviation from linearity gives a steeper slope at $x_n = 0$ than at $x_n = W_b$, and the value of I_E is larger than I_C by the amount I_B. The reason we can use the straight-line approximation in the charge control calculation of base current is that the area under the hole distribution curve is essentially the same in the two cases.

9.2 Generalized Biasing; Switching

The expressions derived in Section 9.1 describe the terminal currents of the transistor, if the device geometry and other factors are consistent with the assumptions. Real transistors may deviate from these approximations, as we shall see in Section 9.3. The collector and emitter junctions may differ in area, saturation current, and other parameters, so that the proper description of the terminal currents may be more complicated than Eqs. (9-11) through (9-13) suggest. For example, if the roles of emitter and collector are reversed, these equations predict that the behavior of the transistor is symmetrical. Real transistors, on the other hand, are generally not symmetrical between emitter and collector. This is a particularly important consideration when the transistor is not biased in the usual way. We have discussed normal biasing (sometimes called the *normal active* mode), in which the emitter junction is forward biased and the collector is reverse biased. In some applications, particularly in switching, this normal biasing rule is violated. In these cases it is important to account for the differences in injection and collection properties of the two junctions. In this section we shall develop a generalized approach which accounts for transistor operation in terms of a coupled-diode model, valid for all combinations of emitter and collector bias. This model involves four measurable parameters which can be related to the geometry and material properties of the device. Using this model in conjunction with the charge control approach, we can describe the physical operation of a transistor in switching circuits and in other applications.

9.2.1 The Coupled-Diode Model. If the collector junction of a transistor is forward biased, we cannot neglect Δp_C; instead, we must use a more general hole distribution in the base region. Figure 9-4a illustrates a situation in which the emitter and collector junctions are each forward biased, so that Δp_E and Δp_C are positive numbers. We can handle this situation with Eqs. (9-11) through (9-13) for the symmetrical transistor. It is interesting to note that these equations can be considered as linear superpositions of the effects of injection by each junction. For example, the straight-line hole distribution of Fig. 9-4a can be broken into the two components of Figs. 9-4b and 9-4c. One component (Fig. 9-4b) accounts for the holes injected

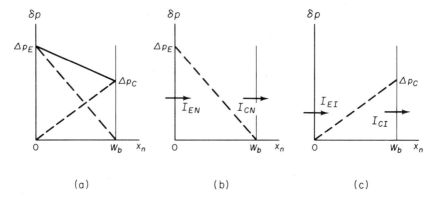

FIGURE 9-4. Evaluation of a hole distribution in terms of components due to normal and inverted modes: (a) approximate hole distribution in the base with emitter and collector junctions forward biased; (b) component due to injection and collection in the normal mode; (c) component due to the inverted mode.

by the emitter and collected by the collector. We can call the resulting currents (I_{EN} and I_{CN}) the *normal mode* components, since they are due to injection from emitter to collector. The component of the hole distribution illustrated by Fig. 9-4c results in currents I_{EI} and I_{CI}, which describe injection in the *inverted mode* of injection from collector to emitter.[†] Of course, these inverted components will be negative, since they account for hole flow opposite to our original definitions of I_E and I_C. *For the symmetrical transistor,* these various components are described by Eqs. (9-11) and (9-12). Defining $a \equiv (qAD_p/L_p) \operatorname{ctnh}(W_b/L_p)$ and $b \equiv (qAD_p/L_p) \operatorname{csch}(W_p/L_p)$ we have

(9-20a) $I_{EN} = a\,\Delta p_E$ and $I_{CN} = b\,\Delta p_E$, with $\Delta p_C = 0$

(9-20b) $I_{EI} = -b\,\Delta p_C$ and $I_{CI} = -a\,\Delta p_C$, with $\Delta p_E = 0$

The four components are combined by linear superposition in Eqs. (9-11) and (9-12)

(9-21a) $$I_E = I_{EN} + I_{EI} = a\,\Delta p_E - b\,\Delta p_C$$
$$= A(e^{qV_{EB}/kT} - 1) - B(e^{qV_{CB}/kT} - 1)$$

(9-21b) $$I_C = I_{CN} + I_{CI} = b\,\Delta p_E - a\,\Delta p_C$$
$$= B(e^{qV_{EB}/kT} - 1) - A(e^{qV_{CB}/kT} - 1)$$

where $A \equiv ap_n$ and $B \equiv bp_n$.

[†]Here the words *emitter* and *collector* refer to physical regions of the device rather than to the functions of injection and collection of holes.

We can see from these equations that a linear superposition of the normal and inverted components does give the result we derived previously for the symmetrical transistor. To be more general, however, we must relate the four components of current by factors which allow for asymmetry in the two junctions. For example, the emitter current in the normal mode can be written

(9-22) $\qquad\qquad I_{EN} = I_{ES}(e^{qV_{EB}/kT} - 1), \qquad \Delta p_C = 0$

where I_{ES} is the magnitude of the emitter saturation current in the normal mode. Since we specify $\Delta p_C = 0$ in this mode, we imply $V_{CB} = 0$ in Eq. (9-1b). Thus we shall consider I_{ES} to be the magnitude of the emitter saturation current with the collector junction short circuited. Similarly, the collector current in the inverted mode is

(9-23) $\qquad\qquad I_{CI} = -I_{CS}(e^{qV_{CB}/kT} - 1), \qquad \Delta p_E = 0$

where I_{CS} is the magnitude of the collector saturation current with $V_{EB} = 0$. As before, the minus sign associated with I_{CI} simply means that in the inverted mode holes are injected opposite to the definition of I_C.

The corresponding collected currents for each mode of operation can be written by defining a new α for each case

(9-24a) $\qquad\qquad I_{CN} = \alpha_N I_{EN} = \alpha_N I_{ES}(e^{qV_{EB}/kT} - 1)$

(9-24b) $\qquad\qquad I_{EI} = \alpha_I I_{CI} = -\alpha_I I_{CS}(e^{qV_{CB}/kT} - 1)$

where α_N and α_I are the ratios of collected current to injected current in each mode. We notice that in the inverted mode the injected current is I_{CI} and the collected current is I_{EI}.

The total currents can again be obtained by superposition of the components

(9-25a) $\quad I_E = I_{EN} + I_{EI} = I_{ES}(e^{qV_{EB}/kT} - 1) - \alpha_I I_{CS}(e^{qV_{CB}/kT} - 1)$

(9-25b) $\quad I_C = I_{CN} + I_{CI} = \alpha_N I_{ES}(e^{qV_{EB}/kT} - 1) - I_{CS}(e^{qV_{CB}/kT} - 1)$

These relations were derived by J. J. Ebers and J. L. Moll and are referred to as the *Ebers–Moll equations*.[†] While the general form is the same as Eqs. (9-21) for the symmetrical transistor, these equations allow for variations in I_{ES}, I_{CS}, α_I, and α_N due to asymmetry between the junctions. Although

---

[†] J. J. Ebers and J. L. Moll, "Large-Signal Behavior of Junction Transistors," *Proc. IRE*, vol. 42, pp. 1761–1772, December 1954. In the original paper and in many texts, the terminal currents are all defined as flowing *into* the transistor. This introduces minus signs into the expressions for I_C and I_B as we have developed them here.

we shall not prove it here, it is possible to show by reciprocity arguments that

(9-26) $$\alpha_N I_{ES} = \alpha_I I_{CS}$$

even for nonsymmetrical transistors.

An interesting feature of the Ebers–Moll equations is that I_E and I_C are described by terms resembling diode relations (I_{EN} and I_{CI}), plus terms which provide coupling between the properties of the emitter and collector (I_{EI} and I_{CN}). This *coupled-diode* property is illustrated by the equivalent circuit of Fig. 9-5. In this figure we take advantage of Eq. (9-1) to write the Ebers–Moll equations in the form

(9-27a) $$I_E = I_{ES}\frac{\Delta p_E}{p_n} - \alpha_I I_{CS}\frac{\Delta p_C}{p_n} = \frac{I_{ES}}{p_n}(\Delta p_E - \alpha_N \Delta p_C)$$

(9-27b) $$I_C = \alpha_N I_{ES}\frac{\Delta p_E}{p_n} - I_{CS}\frac{\Delta p_C}{p_n} = \frac{I_{CS}}{p_n}(\alpha_I \Delta p_E - \Delta p_C)$$

It is often useful to relate the terminal currents to each other as well as to the saturation currents. We can eliminate the saturation current from the coupling term in each part of Eq. (9-25). For example, multiplying Eq. (9-25a) by α_N and subtracting the resulting expression from Eq. (9-25b), we have

(9-28) $$I_C = \alpha_N I_E - (1 - \alpha_N \alpha_I)I_{CS}(e^{qV_{CB}/kT} - 1)$$

Similarly, the emitter current can be written in terms of the collector current

(9-29) $$I_E = \alpha_I I_C + (1 - \alpha_N \alpha_I)I_{ES}(e^{qV_{EB}/kT} - 1)$$

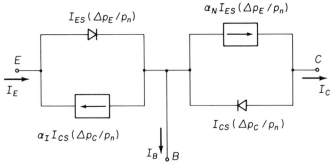

$$I_B = (1 - \alpha_N)I_{ES}\frac{\Delta p_E}{p_n} + (1 - \alpha_I)I_{CS}\frac{\Delta p_C}{p_n}$$

FIGURE 9-5. An equivalent circuit synthesizing the Ebers-Moll equations.

The terms $(1 - \alpha_N\alpha_I)I_{CS}$ and $(1 - \alpha_N\alpha_I)I_{ES}$ can be abbreviated as I_{CO} and I_{EO}, respectively, where I_{CO} is the magnitude of the collector saturation current with the emitter junction *open* ($I_E = 0$), and I_{EO} is the magnitude of the emitter saturation current with the collector open (Prob. 9.11). The Ebers–Moll equations then become

(9-30a) $$I_E = \alpha_I I_C + I_{EO}(e^{qV_{EB}/kT} - 1)$$

(9-30b) $$I_C = \alpha_N I_E - I_{CO}(e^{qV_{CB}/kT} - 1)$$

and the equivalent circuit is shown in Fig. 9-6a. In this form the equations describe both the emitter and collector currents in terms of a simple diode characteristic plus a current generator proportional to the other current. For example, under normal biasing the equivalent circuit reduces to the form shown in Fig. 9-6b. The collector current is α_N times the emitter current plus the collector saturation current, as expected. The resulting collector character-

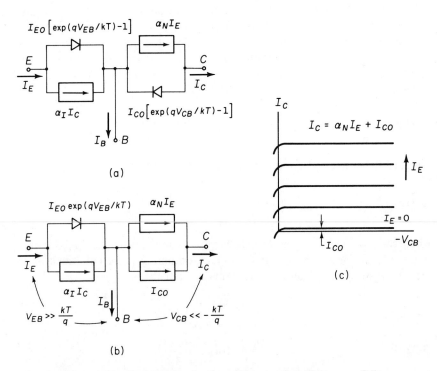

FIGURE 9-6. Equivalent circuits of the transistor in terms of the terminal currents and the open-circuit saturation currents: (a) synthesis of Eqs. (9-30); (b) equivalent circuit with normal biasing; (c) collector characteristics with normal biasing.

istics of the transistor appear as a series of reverse diode curves, displaced by increments proportional to the emitter current (Fig. 9-6c).

9.2.2 Charge Control Analysis.

The charge control approach is useful in analyzing the transistor terminal currents, particularly in a-c applications. Considerations of transit time effects and charge storage are revealed easily by this method. Following the techniques of the previous section, we can separate an arbitrary excess hole distribution in the base into the normal and inverted distributions of Fig. 9-4. The charge stored in the normal distribution will be called Q_N and the charge under the inverted distribution will be called Q_I. Then we can evaluate the currents for the normal and inverted modes in terms of these stored charges. For example, the collected current in the normal mode I_{CN} is simply the charge Q_N divided by the mean time required for this charge to be collected. This time is the transit time for the normal mode τ_{tN}. On the other hand, the emitter current must support not only the rate of charge collection by the collector but also the recombination rate in the base Q_N/τ_{pN}. Here we use a subscript N with the transit time and recombination lifetime in the normal mode in contrast to the inverted mode, to allow for possible asymmetries due to imbalance in the transistor structure. With these definitions, the normal components of current become

(9-31a)
$$I_{CN} = \frac{Q_N}{\tau_{tN}}, \qquad I_{EN} = \frac{Q_N}{\tau_{tN}} + \frac{Q_N}{\tau_{pN}}$$

Similarly, the inverted components are

(9-31b)
$$I_{EI} = -\frac{Q_I}{\tau_{tI}}, \qquad I_{CI} = -\frac{Q_I}{\tau_{tI}} - \frac{Q_I}{\tau_{pI}}$$

where the I subscripts on the stored charge and on the transit and recombination times designate the inverted mode. Combining these equations as in Eq. (9-25) we have the terminal currents for general biasing

(9-32a)
$$I_E = Q_N\left(\frac{1}{\tau_{tN}} + \frac{1}{\tau_{pN}}\right) - \frac{Q_I}{\tau_{tI}}$$

(9-32b)
$$I_C = \frac{Q_N}{\tau_{tN}} - Q_I\left(\frac{1}{\tau_{tI}} + \frac{1}{\tau_{pI}}\right)$$

It is not difficult to show that these equations correspond to the Ebers–Moll relations [Eq. (9-27) and Prob. 9.12], where

$$(9\text{-}33) \qquad \alpha_N = \frac{\tau_{pN}}{\tau_{tN} + \tau_{pN}}, \qquad \alpha_I = \frac{\tau_{pI}}{\tau_{tI} + \tau_{pI}}$$

$$I_{ES} = q_N\left(\frac{1}{\tau_{tN}} + \frac{1}{\tau_{pN}}\right), \qquad I_{CS} = q_I\left(\frac{1}{\tau_{tI}} + \frac{1}{\tau_{pI}}\right)$$

$$Q_N = q_N \frac{\Delta p_E}{p_n}, \qquad Q_I = q_I \frac{\Delta p_C}{p_n}$$

The base current in the normal mode supports recombination, and the base-to-collector current amplification factor β_N takes the form predicted by Eq. (8-32)

$$(9\text{-}34) \qquad I_{BN} = \frac{Q_N}{\tau_{pN}}, \qquad \beta_N = \frac{I_{CN}}{I_{BN}} = \frac{\tau_{pN}}{\tau_{tN}}$$

This expression for β_N is also obtained from $\alpha_N/(1 - \alpha_N)$. Similarly, I_{BI} is Q_I/τ_{pI}, and the total base current is

$$(9\text{-}35) \qquad I_B = I_{BN} + I_{BI} = \frac{Q_N}{\tau_{pN}} + \frac{Q_I}{\tau_{pI}}$$

This expression for the base current is substantiated by $I_E - I_C$ from Eq. (9-32).

The effects of time dependence of stored charge can be included in these equations by the methods introduced in Section 5.5.1. We can include the proper dependencies by adding a rate of change of stored charge to each of the injection currents I_{EN} and I_{CI}

$$(9\text{-}36a) \qquad i_E = Q_N\left(\frac{1}{\tau_{tN}} + \frac{1}{\tau_{pN}}\right) - \frac{Q_I}{\tau_{tI}} + \frac{dQ_N}{dt}$$

$$(9\text{-}36b) \qquad i_C = \frac{Q_N}{\tau_{tN}} - Q_I\left(\frac{1}{\tau_{tI}} + \frac{1}{\tau_{pI}}\right) - \frac{dQ_I}{dt}$$

$$(9\text{-}36c) \qquad i_B = \frac{Q_N}{\tau_{pN}} + \frac{Q_I}{\tau_{pI}} + \frac{dQ_N}{dt} + \frac{dQ_I}{dt}$$

We shall return to these equations in Section 9.4, when we discuss the use of transistors at high frequencies.

9.2.3 Switching.

In a switching operation a transistor is usually controlled in two conduction states which can be referred to loosely as the "on" state and the "off" state. Ideally, a switch should appear as a short circuit when turned on and an open circuit when turned off. Furthermore, it is desirable to switch the device from one state to the other with no lost time in between. Transistors do not fit this ideal description of a switch, but they can serve as a useful approximation in practical electronic circuits. The two states of a transistor in switching can be seen in the simple common-emitter example of Fig. 9-7. In this figure the collector current i_C is controlled by the base current i_B over most of the family of characteristic curves. The load line specifies the locus of allowable $(i_C, -v_{CE})$ points for the circuit, in analogy with Fig. 8-2c for triodes. If i_B is such that the operating point (Q-point) lies somewhere between the two end points of the load line (Fig. 9-7b), the transistor operates in the normal active mode. That is, the emitter junction is forward biased and the collector is reverse biased, with a reasonable value of i_B flowing out of the base. On the other hand, if the base current is zero or negative, the point C is reached at the bottom end of the load line, and the collector current is negligible. This is the "off" state of the transistor, and the device is said to be operating in the *cutoff* regime. If the base current is positive and sufficiently large, the device is driven to the *saturation* regime, marked S. This is the "on" state of the transistor, in which a large value of

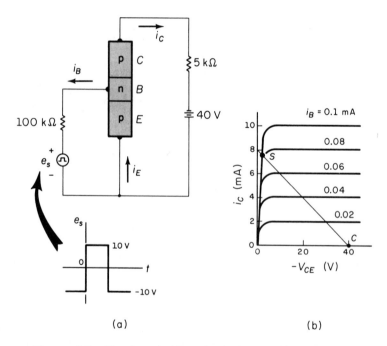

(a)   (b)

FIGURE 9-7. Simple switching circuit for a transistor in the common-emitter configuration: (a) biasing circuit; (b) collector characteristics and load line for the circuit, with cutoff and saturation indicated.

i_C flows with only a very small voltage drop v_{CE}. As we shall see below, the beginning of the saturation regime corresponds to the loss of reverse bias across the collector junction. In a typical switching operation the base current swings from positive to negative, thereby driving the device from saturation to cutoff, and vice versa. In this section we shall explore the nature of conduction in the saturation and cutoff regimes; also we shall investigate the factors affecting the speed with which the transistor can be switched between the two states.

Cutoff:

If the emitter junction is reverse biased in the cutoff regime, (negative i_B), we can approximate the excess hole densities at the edges of the reverse-biased emitter and collector junctions as

$$(9\text{-}37) \qquad \frac{\Delta p_E}{p_n} \simeq \frac{\Delta p_C}{p_n} \simeq -1$$

With a straight-line approximation, the excess hole density distribution in the base appears constant at $-p_n$, as shown in Fig. 9-8a. Actually, there will be some slope to the distribution at each edge to account for the reverse saturation current in the junctions, but Fig. 9-8a is approximately correct.

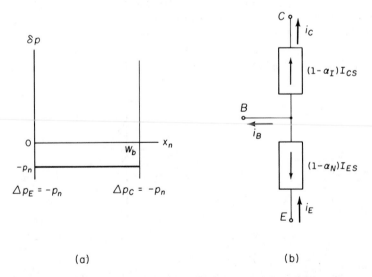

(a)   (b)

FIGURE 9-8. The cutoff regime of a p-n-p transistor: (a) excess hole distribution in the base region with emitter and collector junctions reverse biased; (b) equivalent circuit corresponding to Eqs. (9-38).

The base current i_B can be approximated for a symmetrical transistor on a charge storage basis as $-qAp_nW_b/\tau_p$. In this calculation a negative excess hole density corresponds to *generation* in the same way that a positive distribution indicates recombination. This expression is also obtained by applying Eq. (9-37) to Eq. (9-13) with an approximation from Table 9-1. Physically, a small saturation current flows from n to p in each reverse-biased junction, and this current is supplied by the base current i_B (which is negative when flowing into the base of a p-n-p device according to our definitions). A more general evaluation of the currents can be obtained from the Ebers–Moll equations by applying Eq. (9-37) to Eq. (9-27)

(9-38a) $i_E = -I_{ES} + \alpha_I I_{CS} = -(1 - \alpha_N)I_{ES}$

(9-38b) $i_C = -\alpha_N I_{ES} + I_{CS} = (1 - \alpha_I)I_{CS}$

(9-38c) $i_B = i_E - i_C = -(1 - \alpha_N)I_{ES} - (1 - \alpha_I)I_{CS}$

If the short-circuit saturation currents I_{ES} and I_{CS} are small and α_N and α_I are both near unity, these currents will be negligible and the cutoff regime will closely approximate the "off" condition of an ideal switch. The equivalent circuit corresponding to Eq. (9-38) is illustrated in Fig. 9-8b.

Saturation:

The saturation regime begins when the reverse bias across the collector junction is reduced to zero, and it continues as the collector becomes forward biased. The excess hole distribution in this case is illustrated in Fig. 9-9. The device is saturated when $\Delta p_C = 0$, and forward bias of the collector junction (Fig. 9-9b) leads to a positive Δp_C, driving the device further into

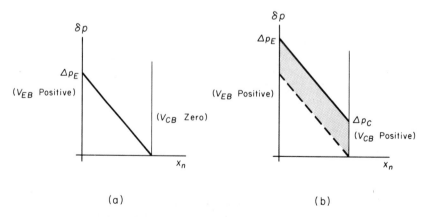

(a)   (b)

FIGURE 9-9.   Excess hole distributions in the base of a saturated transistor: (a) the beginning of saturation; (b) oversaturation.

saturation. With the load line fixed by the battery and the 5-kΩ resistor in
Fig. 9-7, saturation is reached by increasing the base current i_B. We can see
how a large value of i_B leads to saturation by applying the reasoning of charge
control to Fig. 9-9. Since a certain amount of stored charge is required to
accommodate a given i_B (and vice versa), an increase in i_B calls for an increase
in the area under the $\delta p(x_n)$ distribution.

In Fig. 9-9a the device has just reached saturation, and the collector
junction is no longer reverse biased. The implication of this condition for
the circuit of Fig. 9-7 is easy to state. Since the emitter junction is forward
biased and the collector junction has zero bias, very little voltage drop appears
across the device from collector to emitter. The magnitude of $-v_{CE}$ is only
a fraction of a volt, due to the forward-biased emitter voltage v_{EB}. Therefore,
almost all of the battery voltage appears across the resistor, and the collector
current is approximately 40 V/5 kΩ = 8 mA. As the device is driven deeper
into saturation (Fig. 9-9b), the collector current stays essentially constant
while the base current increases. In this saturation condition the transistor
approximates the "on" state of an ideal switch.

Whereas the degree of "oversaturation" (indicated by the shaded area
in Fig. 9-9b) does not affect the value of i_C significantly, it is important in
determining the time required to switch the device from one state to the other.
For example, from previous experience we expect the turn-off time (from
saturation to cutoff) to be longer for larger values of stored charge in the
base. We can calculate the various charging and delay times from Eq. (9-36).
Detailed calculations are somewhat involved, but we can simplify the problem
greatly with approximations of the type used in Chapter 5 for transient
effects in p-n junctions.

The Switching Cycle:

The various mechanisms of a switching cycle are illustrated in Fig.
9-10. If the device is originally in the cutoff condition, a step increase of base
current to I_B causes the hole distribution to increase approximately as
illustrated in Fig. 9-10b. As in the transient analysis of Chapter 5, we assume
for simplicity of calculation that the distribution maintains a simple form in
each time interval of the transient. At time t_s the device enters saturation,
and the hole distribution reaches its final state at t_2. As the stored charge in
the base Q_b increases, there is an increase in the collector current i_C. The
collector current does not increase beyond its value at the beginning of
saturation t_s, however. We can call this saturated collector current $I_C \simeq
E_{CC}/R_L$, where E_{CC} is the value of the collector circuit battery and R_L is the
load resistor ($I_C \simeq$ 8 mA for the example of Fig. 9-7). There is an essentially
exponential increase in the collector current while Q_b rises to its value Q_s at
t_s; this rise time serves as one of the limitations of the transistor in a switching application. Similarly, when the base current is switched negative (e.g.,

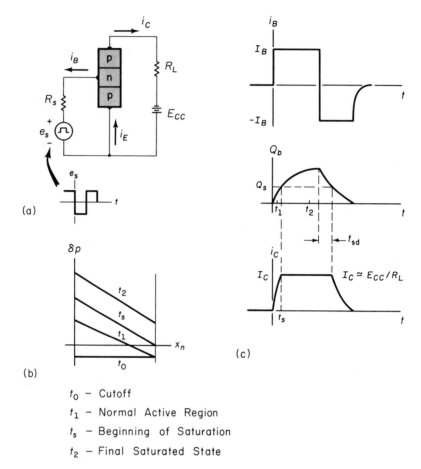

t_0 – Cutoff
t_1 – Normal Active Region
t_s – Beginning of Saturation
t_2 – Final Saturated State

FIGURE 9-10. Switching effects in a common-emitter transistor circuit: (a) circuit diagram; (b) approximate hole distributions in the base during switching from cutoff to saturation; (c) base current, stored charge, and collector current during a turn-on and a turn-off transient.

to the value $-I_B$), the stored charge must be withdrawn from the base before cutoff is reached. While Q_b is larger than Q_s, the collector current remains at the value I_C, fixed by the battery and resistor. Thus there is a storage delay time t_{sd} after the base current is switched and before i_C begins to fall toward zero. After the stored charge is reduced below Q_s, i_C drops exponentially with a characteristic fall time. Once the stored charge is withdrawn, the base current cannot be maintained any longer at its large negative value and must decay to the small cutoff value described by Eq. (9-38c).

Turn-On Transient:

We can calculate the various times involved in the switching transient of a symmetrical device by relating the base current to the stored charge and the rate of change of stored charge

$$(9\text{-}39) \qquad i_B(t) = \frac{Q_b(t)}{\tau_p} + \frac{dQ_b(t)}{dt}$$

This equation is analogous to Eq. (5-52) in the diode transient analysis. In calculating the turn-on time, we can simplify the problem if we neglect the small negative excess hole distribution in cutoff and assume the stored charge Q_b increases from zero to its final value $I_B\tau_p$. Thus, when the base current switches from essentially zero to I_B, the Laplace transform of the stored charge relation [Eq. (9-39)] is

$$(9\text{-}40) \qquad \frac{I_B}{s} = \frac{Q_b(s)}{\tau_p} + s\,Q_b(s)$$

$$Q_b(s) = \frac{I_B}{s(s + 1/\tau_p)}$$

with the solution

$$(9\text{-}41) \qquad Q_b(t) = I_B\tau_p(1 - e^{-t/\tau_p})$$

While the stored charge increases with time, the collector current follows according to $i_C = Q_b(t)/\tau_t$, where τ_t is the transit time. However, this increase in i_C continues only until saturation is reached at t_s, when i_C reaches its maximum value I_C. Therefore, the turn-on time t_s can be calculated by solving for the time at which $Q_b(t)/\tau_t$ reaches I_C

$$(9\text{-}42) \qquad \frac{I_B\tau_p}{\tau_t}(1 - e^{-t_s/\tau_p}) = I_C, \qquad \text{when } t_s = \tau_p \ln \frac{1}{1 - I_C/\beta I_B}$$

We have neglected an important delay time resulting from the necessity of charging the emitter junction capacitance, which we shall discuss below. It is clear from Eq. (9-42), however, that the time t_s will be short if (1) the lifetime τ_p is short and (2) the limiting value of collector current I_C is small compared with the product of β and the base current drive I_B. The latter condition is met by driving the device into oversaturation.

Turn-Off Transient:

If the turn-off transient involves simply the reduction of the base current from its "on" value I_B to zero, the decays of the stored charge and collector

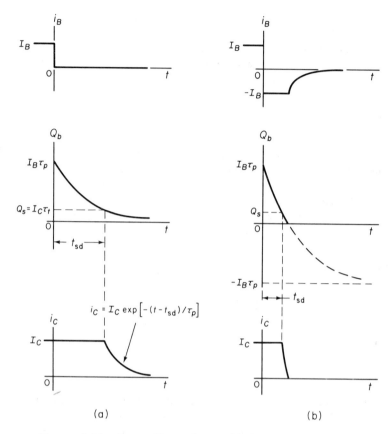

FIGURE 9-11. Turn-off transients: (a) base current switched from a large positive value to zero; (b) base current switched from positive to negative.

current are particularly easy to calculate (Fig. 9-11a). The stored charge at the beginning of the turn-off transient is simply $Q_b = I_B \tau_p$. The stored charge falls exponentially from this value to zero with the time constant τ_p. The collector current remains at its saturated value I_C until Q_b falls to the value for minimum saturation $Q_s = I_C \tau_t$. Therefore, the storage delay time t_{sd} over which the collector current remains constant after the base current has been switched is given by

$$(9\text{-}43) \qquad\qquad t_{sd} = \tau_p \ln \frac{I_B \tau_p}{I_C \tau_t} = \tau_p \ln \frac{\beta I_B}{I_C}$$

After the storage delay time, the collector current falls to zero with the time constant τ_p. We notice that in this case the delay time is increased by the condition of oversaturation ($\beta I_B > I_C$).

The decay time can be shortened by driving the device to cutoff, that is, by switching the base current to a negative value instead of simply reducing it to zero. For example, Fig. 9-11b illustrates a case in which the base current is switched from I_B to an equal negative value $-I_B$. In this case the stored charge decays according to (Prob. 9.13)

$$(9\text{-}44) \qquad\qquad Q_b(t) = I_B \tau_p (2e^{-t/\tau_p} - 1)$$

Thus Q_b falls from its oversaturation value of $I_B \tau_p$ toward a negative value of equal magnitude. Of course, the stored charge does not reach the final negative value but stops at the negligibly small negative stored charge corresponding to cutoff. The effect, however, is to drive the stored charge (and therefore the collector current) to the cutoff value in a shorter time.

Specifications for Switching Transistors:

As mentioned above, we have simplified this discussion greatly so the basic principles of switching times could be understood apart from secondary effects. Thus we have neglected the effects of asymmetry in the transistor, typified by differing values of τ_{pN}, τ_{pI}, and the other parameters defined in the Ebers–Moll model. The main discrepancy of this analysis with the switching of real devices is the neglect of the charging time of the emitter junction capacitance in going from cutoff to saturation. Since the emitter junction is reverse biased in cutoff, it is necessary for the emitter space charge layer to be charged to the forward bias condition before collector current can flow. Therefore, we should include a *delay time* t_d as in Fig. 9-12 to account for

t_d – Delay Time While Junction
        Capacitance is Charging

t_r – Rise Time from 0.1-$0.9 I_c$

t_f – Fall Time from 0.9-$0.1 I_c$

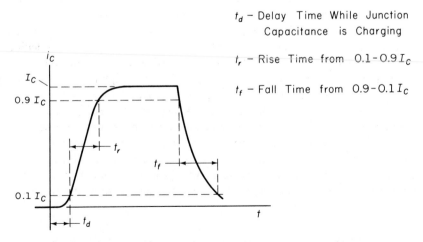

FIGURE 9-12. Collector current during switching transients, including the delay time required for charging the junction capacitance; definitions of the rise time and fall time.

this effect. Typical values of t_d are given in the specification information of most switching transistors, along with a *rise time* t_r defined as the time required for the collector current to rise from 10 to 90 per cent of its final value. A third specification is the *fall time* t_f required for i_C to fall through a similar fraction of its turn-off excursion.

In fabricating transistors for switching applications it is desirable to make the lifetime τ_p in the base region as small as possible, as suggested by Eqs. (9-42), (9-43), and (9-44). One common technique is to increase the density of recombination centers in the base—for example, by doping a Si device with Au. This reduction of τ_p can be tolerated while maintaining the β of the device at an acceptable value by making the base region very narrow.

9.3 Secondary Effects

The approach we have taken in analyzing the properties of transistors has involved a number of simplifying assumptions. Some of the assumptions must be modified in dealing with practical devices. In this section we investigate some common deviations from the basic theory and indicate situations in which each effect is important. Since the various effects discussed here involve modifications of the more straightforward theory, they are labeled "secondary effects." This does not imply that they are unimportant; in fact, the effects described in this section can dominate the conduction in transistors under certain conditions of device geometry and circuit application.

In this section we shall consider the effects of nonuniform doping in the base region of the transistor. In particular, we shall find that graded doping can lead to a drift component of charge transport across the base, adding to the diffusion of carriers from emitter to collector. We shall discuss the effects of large reverse bias on the collector junction, in terms of widening the space charge region about the junction and avalanche multiplication. We shall see that transistor parameters are affected at high current levels by the degree of injection and by heating effects. We shall consider several structural effects which are important in practical devices, such as asymmetry in the areas of the emitter and collector junctions, series resistance between the base contact and the active part of the base region, and nonuniformity of injection at the emitter junction. All these effects are important in understanding the operation of transistors, and proper consideration of their interactions can contribute greatly to the usefulness of practical transistor circuits.

9.3.1 Drift in the Base Region. The assumption of uniform doping in the base is generally valid for an alloyed transistor, in which the base region is made up of a thin layer of the starting material sandwiched between two abrupt alloyed junctions. On the other hand, diffused junction transistors usually involve an appreciable amount of impurity grading; for example, the double-diffused transistor of Fig. 8-26 has a doping profile similar to that sketched in Fig. 9-13. In this example there is a fairly sharp discontinuity in the doping profile, when the donor density in the base region becomes smaller than the constant p-type background doping in the collector. Similarly, the emitter is assumed to be a heavily doped (p^+) shallow region, providing a second rather sharp boundary for the base. Within the base region itself, however, the net doping density ($N_d - N_a \equiv N$) varies along a profile which decreases from the emitter edge to the collector edge. The most likely profile for this grading of the doping density is essentially a portion of a gaussian distribution (see Section 5.1.3); however, it is often a good approximation to assume that $N(x_n)$ varies exponentially within the base region (Fig. 9-13b).

One important result of a graded base region is that a built-in electric field exists from emitter to collector (for a p-n-p), thereby adding a drift component to the transport of holes across the base. We can demonstrate this effect very simply by considering the required balance of drift and diffusion in the base at equilibrium. If the net donor doping of the base is large enough to allow the usual approximation $n(x_n) \simeq N(x_n)$, the balance of electron drift and diffusion currents at equilibrium requires

$$(9\text{-}45) \qquad I_n(x_n) = qA\mu_n N(x_n)\,\mathscr{E}(x_n) + qAD_n\frac{dN(x_n)}{dx_n} = 0$$

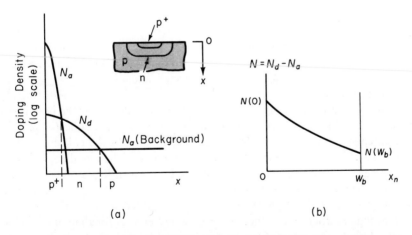

FIGURE 9-13. Graded doping in the base region of a p-n-p transistor: (a) typical doping profile; (b) approximate exponential distribution of the net donor density in the base region.

Therefore, the built-in electric field is

$$(9\text{-}46) \qquad \mathscr{E}(x_n) = -\frac{D_n}{\mu_n}\frac{1}{N(x_n)}\frac{dN(x_n)}{dx_n} = -\frac{kT}{q}\frac{1}{N(x_n)}\frac{dN(x_n)}{dx_n}$$

For a doping profile $N(x_n)$ which decreases in the positive x_n direction, this field is positive, directed from emitter to collector.

For the example of an exponential doping profile, the electric field $\mathscr{E}(x_n)$ turns out to be constant with position in the base. We can represent an exponential distribution as

$$(9\text{-}47) \qquad N(x_n) = N(0)e^{-ax_n/W_b}, \qquad \text{where } a \equiv \ln\frac{N(0)}{N(W_b)}$$

Taking the derivative of this distribution and substituting in Eq. (9-46), we obtain the constant field

$$(9\text{-}48) \qquad\qquad\qquad \mathscr{E}(x_n) = \frac{kT}{q}\frac{a}{W_b}$$

Since this field aids the transport of holes across the base region from emitter to collector, the transit time τ_t is reduced below that of a comparable uniform base transistor. This shortening of the transit time can be very important in high-frequency devices (Section 9.4.2). Since the base transport factor B for a graded junction is even closer to unity than we indicated for the uniform device, the current transfer ratio α is often determined almost entirely by the emitter injection efficiency.

9.3.2 Base Narrowing.
In the discussion of transistors thus far, we have assumed that the effective base width W_b is essentially independent of the bias voltages applied to the collector and emitter junctions. This assumption is not always valid; for example, the p^+-n-p^+ transistor of Fig. 9-14 is affected by the reverse bias applied to the collector. If the base region is lightly doped, the depletion region at the reverse-biased collector junction can extend significantly into the n-type base region. As the collector voltage is increased, the space charge layer takes up more of the metallurgical width of the base L_b, and as a result, the effective base width W_b is decreased. This effect is variously called *base narrowing*, *base–width modulation*, and the *Early effect* after J. M. Early, who first interpreted it. The effects of base narrowing are apparent in the collector characteristics for the common-base and common-emitter configurations (Fig. 9-14b and c). In each case the current transfer ratio increases with increased reverse bias on the collector junction. As a result, I_C is not independent of $-V_{CB}$ as predicted earlier, but instead increases as α and β increase.

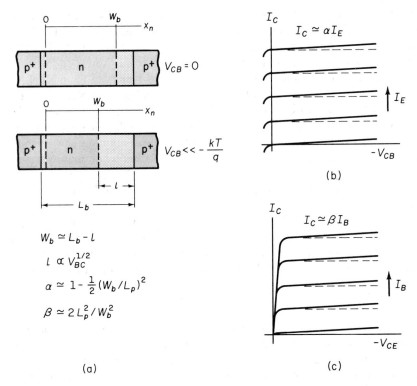

FIGURE 9-14. The effects of base narrowing on the characteristics of a p^+-n-p^+ transistor: (a) decrease in the effective base width as the reverse bias on the collector junction is increased; (b) common-base and (c) common-emitter characteristics showing the increase in I_C with increased collector voltage. The dashed lines in (b) and (c) indicate the constant I_C curves which would result if base narrowing were not present.

For the p^+-n-p^+ device of Fig. 9-14 we can approximate the length l of the collector junction depletion region in the n material from Eq. (5-23b) with V_0 replaced by $V_0 - V_{CB}$ and V_{CB} taken to be large and negative

$$(9\text{-}49) \qquad\qquad l = \left(\frac{2\epsilon V_{BC}}{qN_d}\right)^{1/2}$$

Therefore, the effective base width $W_b = L_b - l$ decreases approximately with the square root of the reverse bias on the collector junction. The approximations of Eq. (9-19) indicate the effect this variation in W_b will have on α and β. These approximate relations for the current transfer ratios are repeated in Fig. 9-14 for convenience. It is clear that a decrease in W_b drives α closer to unity, and that β increases as W_b becomes smaller. The effects of base narrowing are particularly noticeable in the common-emitter configuration, since β can continue to increase as W_b decreases. Since the voltage drop across the forward-biased emitter junction is small, most of $-V_{CE}$ appears across the reverse-biased collector junction.

If the reverse bias on the collector junction is increased far enough, it is possible to decrease W_b to the extent that the collector depletion region essentially fills the entire base. This effect is called *punch-through* as in the narrow base diode of Section 6.1.1; in this condition holes are swept directly from the emitter region to the collector, and transistor action is lost. Punch-through is a breakdown effect which is generally avoided in circuit design. In most cases, however, avalanche breakdown of the collector junction occurs before punch-through is reached. We shall discuss the effects of avalanche multiplication in the following section.

In devices with graded base doping, base narrowing is of less importance. For example, if the donor density in the base region of a p-n-p increases with position from the collector to the emitter, the intrusion of the collector space charge region into the base becomes less important with increased bias as more donors are available to accommodate the space charge.

9.3.3 Avalanche Breakdown.

9.3.3 Avalanche Breakdown. Before punch-through occurs in most transistors, avalanche multiplication at the collector junction becomes important (see Section 5.4.2). As Fig. 9-15 indicates, the collector current increases sharply at a well-defined breakdown voltage BV_{CBO} for the common-base configuration. For the common-emitter case, however, there is a strong influence of carrier multiplication over a fairly broad range of collector voltage. Furthermore, the breakdown voltage in the common-emitter case BV_{CEO} is significantly smaller than BV_{CBO}. We can understand these effects by considering breakdown for the condition $I_E = 0$ in the common-base case

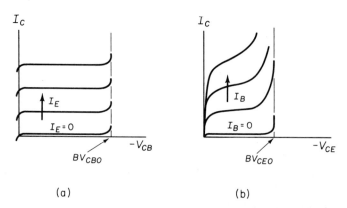

(a)   (b)

FIGURE 9-15. Avalanche breakdown in a transistor: (a) common-base configuration; (b) common-emitter configuration.

and for $I_B = 0$ in the common-emitter case. These conditions are implied by the O in the subscripts of BV_{CEO} and BV_{CBO}. In each case the terminal current I_C is the current entering the collector depletion region multiplied by the factor M. Including multiplication due to impact ionization, Eq. (9-30b) becomes

$$(9\text{-}50) \qquad I_C = (\alpha_N I_E + I_{CO})M = (\alpha_N I_E + I_{CO})\frac{1}{1 - (V_{BC}/BV_{CBO})^n}$$

where for M we have used the empirical expression given in Eq. (5-49).

For the limiting common-base case of $I_E = 0$ (the lowest curve in Fig. 9-15a), I_C is simply MI_{CO}, and the breakdown voltage is well defined, as in an isolated junction. The term BV_{CBO} signifies the collector junction breakdown voltage in common-base with the emitter open. In the common-emitter case the situation is somewhat more complicated. Setting $I_B = 0$, and therefore $I_C = I_E$ in Eq. (9-50), we have

$$(9\text{-}51) \qquad\qquad\qquad I_C = \frac{MI_{CO}}{1 - M\alpha_N}$$

We notice that in this case the collector current increases indefinitely when $M\alpha_N$ approaches unity. By contrast, M must approach infinity in the common-base case before BV_{CBO} is reached. Since α_N is close to unity in most transistors, M need be only slightly larger than unity for Eq. (9-51) to approach breakdown. Avalanche multiplication thus dominates the current in a common-emitter transistor well below the breakdown voltage of the isolated collector junction. The sustaining voltage for avalanching in the common-emitter case BV_{CEO} is therefore smaller than BV_{CBO}.

We can understand physically why multiplication is so important in the common-emitter case by considering the effect of M on the base current. When an ionizing collision occurs in the collector junction depletion region, a secondary hole and electron are created. The primary and secondary holes are swept into the collector, but the electron is swept into the base by the junction field. Therefore, the supply of electrons to the base is increased, and from our charge control analysis we conclude that hole injection at the emitter must increase to maintain space charge neutrality. This is a regenerative process, in which an increased injection of holes from the emitter causes an increased multiplication current at the collector junction; this in turn increases the rate at which secondary electrons are swept into the base, calling for more hole injection. Because of this regenerative effect, it is easy to understand why the multiplication factor M need be only slightly greater than unity to start the avalanching process.

9.3.4 Injection Level; Thermal Effects.

In discussions of transistor characteristics we have assumed that α and β are independent of carrier injection level. Actually, the parameters of a practical transistor may vary considerably with injection level, which is determined by the magnitude of I_E or I_C. For very low injection, the assumption of negligible recombination in the junction depletion regions is invalid (see Section 5.6.2). This is particularly important in the case of recombination in the emitter junction, where any recombination tends to degrade the emitter injection efficiency γ. Thus we might expect that α and β can decrease for low values of I_C, causing the curves of the collector characteristics to be spaced more closely for low currents than for higher currents.

As I_C is increased beyond the low injection level range, α and β increase but fall off again at very high injection. The primary cause of this fall-off is the increase of majority carriers at high injection levels (see Section 5.6.1). For example, as the density of excess holes injected into the base becomes large, the matching excess electron density can become greater than the background density n_n. This conductivity modulation effect results in a decrease in γ as more electrons are injected across the emitter junction into the emitter region.

Large values of I_C may be accompanied by significant power dissipation in the transistor and therefore heating of the device. In particular, the product of I_C and the collector voltage V_{BC} is a measure of the power dissipated at the collector junction. This dissipation is due to the fact that carriers swept through the collector junction depletion region are given increased kinetic energy, which in turn is given up to the lattice in scattering collisions. It is very important that the transistor be operated in a range such that $I_C V_{BC}$ does not exceed the maximum power rating of the device. In devices designed for high power capability, the transistor is mounted on a copper stud or other efficient heat sink, so that thermal energy can be transferred away from the junction.

If the temperature of the device is allowed to increase due to power dissipation or thermal environment, the transistor parameters change. The most important parameters dependent on temperature are the carrier lifetimes and diffusion constants. In Si or Ge devices the lifetime τ_p increases with temperature for most cases (see Fig. 4-12), due to thermal reexcitation from recombination centers. This increase in τ_p tends to increase β for the transistor. On the other hand, the mobility decreases with increasing temperature in the lattice-scattering range, varying approximately as $T^{-3/2}$ (see Fig. 3-23). Thus from the Einstein relation, we expect D_p to decrease as the temperature increases, thereby causing a drop in β due to an increasing transit time τ_t. Of these competing processes, the effect of increasing lifetime with temperature usually dominates, and β becomes larger as the device is heated. It is clear from this effect that *thermal runaway* can occur if the circuit is not designed to prevent it. For example, a large power dissipation in the device can cause an increase in T; this results in a larger β and therefore a larger

I_C for a given base current; the larger I_C causes more collector dissipation and the cycle continues. This type of runaway of the collector current can result in overheating and destruction of the device.

9.3.5 Base Resistance and Emitter Crowding. A number of structural effects are important in determining the operation of a transistor. For example, the emitter and collector areas are considerably different in the diffused transistor of Fig. 9-16a. This and most other structural effects can be accounted for by differences in α_N, α_I, and the other parameters in the Ebers–Moll model. Several effects caused by the structural arrangement of

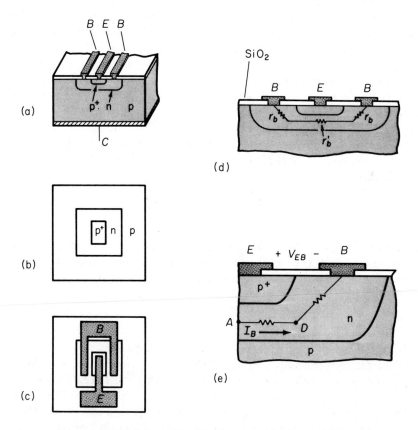

FIGURE 9-16. Effects of base resistance: (a) cross section of a diffused transistor; (b) and (c) top view, showing emitter and base areas and metallized contacts; (d) illustration of base resistance; (e) expanded view of distributed resistance in the active part of the base region.

real transistors deserve special attention, however. One of the most important of these effects is the fact that base current must pass from the active part of the base region to the base contacts B. Thus, to be accurate, we should include a resistance r_b in equivalent models for the transistor to account for voltage drops which may occur between B and the active part of the base. Because of r_b, it is common to contact the base with the metallization pattern on both sides of the emitter, as in Fig. 9-16c.

If the transistor is designed so that the n-type regions leading from the base to the contacts are large in cross-sectional area, the base resistance r_b may be negligible. On the other hand, the distributed resistance r_b' along the thin base region is almost always important.† Since the width of the base between emitter and collector is very narrow, this distributed resistance is usually quite high. Therefore, as base current flows from points within the base region toward each end, a voltage drop occurs along r_b'. This effect is similar to that discussed in Section 8.2.1 regarding the current in the channel of an FET. In this case the forward bias across the emitter–base junction is not uniform but varies with position according to the voltage drop in the distributed base resistance. In particular, the forward bias on the emitter junction is largest at the corner of the emitter region near the base contact. We can see that this is the case by considering the simplified example of Fig. 9-16e. Neglecting variations in the base current along the path from point A to the contact B, the forward bias on the emitter above point A is approximately

$$(9\text{-}52) \qquad V_{EA} = V_{EB} - I_B(R_{AD} + R_{DB})$$

Actually, the base current is not uniform along the active part of the base region, and the distributed resistance of the base is more complicated than we have indicated (Prob. 9.15). But this example does illustrate the point of nonuniform injection. Whereas the forward bias at A is approximately described by Eq. (9-52), the emitter bias voltage at point D is

$$(9\text{-}53) \qquad V_{ED} = V_{EB} - I_B R_{DB}$$

which can be significantly closer to the applied voltage V_{EB}.

Since the forward bias is largest at the edge of the emitter, it follows that the injection of holes is also greatest there. This effect is called *emitter crowding*, and it can strongly affect the behavior of the device. The most important result of emitter crowding is that high-injection effects described in the previous section can become dominant locally at the corners of the emitter before the overall emitter current is very large. In transistors designed to handle appreciable current, this is a problem which must be dealt with by

---

†The distributed resistance r_b' is often called the *base spreading resistance*.

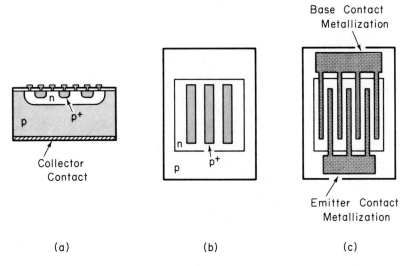

FIGURE 9-17. An interdigitated geometry to compensate for the effects of emitter crowding in a power transistor: (a) cross section; (b) top view of diffused regions; (c) top view with metallized contacts. The metal interconnections are isolated from the device by an oxide layer except where they contact the appropriate base and emitter regions at "windows" in the oxide.

proper structural design. The most effective approach to the problem of emitter crowding is to distribute the emitter current along a relatively large emitter edge, thereby reducing the current density at any one point. Clearly, what is needed is an emitter region with a large perimeter compared with its area. A likely geometry to accomplish this is a long thin stripe for the emitter, with base contacts on each side (Fig. 9-16b and c). With this geometry the total emitter current I_E is spread out along a rather long edge on each side of the stripe. An even better geometry is several emitter stripes, connected electrically by the metallization and separated by interspersing base contacts (Figs. 9-17 and 9-18). Many such thin emitter and base contact "fingers" can be interlaced to provide for handling large current in a power transistor. This is often called very descriptively an *interdigitated* geometry.

9.4 Frequency Limitations of Transistors

In this section we discuss the properties of transistors under high-frequency operation. Some of the frequency limitations are junction capacitance, charging times required when excess carrier distributions are altered, and

(a)

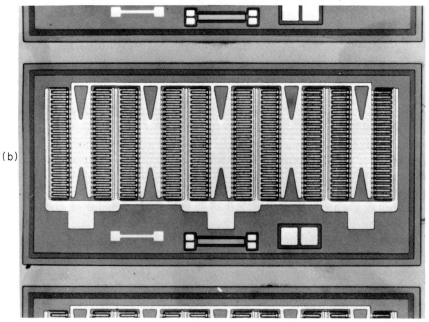

(b)

FIGURE 9-18. Interdigitated power transistors: (a) a Si power transistor mounted on a header designed for efficient heat dissipation. The collector is alloyed to a metallized region on a ceramic substrate, and the base and emitter contact "fingers" are connected to two other metallized regions by Al wires. Photograph courtesy of Texas Instruments, Inc. (b) A Si transistor with many interlaced base and emitter regions. This device is rated at 25 W and 175 MHz. Photograph courtesy of Motorola Semiconductor Products, Inc.

transit time of carriers across the base region. Our aim here is not to attempt
a complete analysis of high-frequency operation, but rather to consider the
physical basis of the most important effects. Therefore, we shall include
the dominant capacitances and charging times and discuss the effects of the
transit time on high-frequency devices. In conclusion, we shall consider
appropriate design features of a microwave transistor.

9.4.1 Capacitance and Charging Times.

The most obvious fre-
quency limitation of transistors is the presence of junction capacitance at
the emitter and collector junctions. We have considered this type of capacitance in Chapter 5, and we can include junction capacitors C_{je} and C_{jc} in
circuit models for the transistor (Fig. 9-19a). If there is some equivalent
resistance r_b between the base contact and the active part of the base region,
we can also include it in the model, along with r_c to account for a series
collector resistance. Clearly, the combinations of r_b with C_{je} and r_c with C_{jc}
can introduce important time constants into a-c circuit applications of the
device.

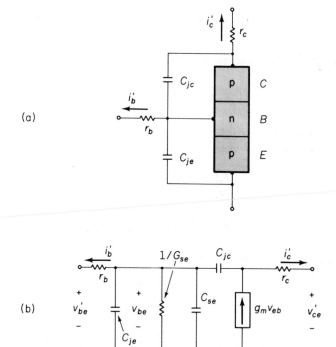

(a)

(b)

FIGURE 9-19. Models for a-c operation: (a) inclusion of base
and collector resistances and junction capacitances; (b) hybrid-pi
model synthesizing Eqs. (9-58) and (9-59).

From Section 5.5.3 we recall that capacitive effects can arise from the requirements of altering the carrier distributions during time-varying injection. In a-c circuits the transistor is usually biased to a certain steady state operating point (Q-point) characterized by the d-c quantities V_{BE}, V_{CE}, I_C, I_B, and I_E; then a-c signals are superimposed upon these steady state values. We shall call the a-c terms v_{be}, v_{ce}, i_c, i_b, and i_e. Total (a-c + d-c) quantities will be lower case with capitalized subscripts.

If a small a-c signal is applied to the emitter p-n junction along with a d-c level, the analysis leading to Eq. (5-75) is valid

$$(9\text{-}54) \qquad \Delta p_E(t) \simeq \Delta p_E(\text{d-c})\left(1 + \frac{qv_{eb}}{kT}\right)$$

We can relate this time-varying excess hole density to the stored charge in the base region, and then use Eq. (9-36) to determine the resulting currents. For simplicity we shall assume the device is biased in the normal active mode and use only $Q_N(t)$. Assuming an essentially triangular excess hole distribution in the base, Eq. (9-17) gives

$$(9\text{-}55) \qquad Q_N(t) = \frac{1}{2}qAW_b\,\Delta p_E(t) = \frac{1}{2}qAW_b\,\Delta p_E(\text{d-c})\left[1 + \frac{qv_{eb}}{kT}\right]$$

The terms outside the brackets constitute the d-c stored charge $I_B\tau_p$

$$(9\text{-}56) \qquad Q_N(t) = I_B\tau_p\left(1 + \frac{qv_{eb}}{kT}\right)$$

Now that we have a simple relation for the time-dependent stored charge, we can use Eq. (9-36c) to write the total base current as

$$(9\text{-}57) \qquad i_B(t) = \frac{Q_N(t)}{\tau_p} + \frac{dQ_N(t)}{dt}$$

$$= I_B + \frac{q}{kT}I_Bv_{eb} + \frac{q}{kT}I_B\tau_p\frac{dv_{eb}}{dt}$$

The a-c component of the base current is

$$(9\text{-}58) \qquad i_b = G_{se}v_{eb} + C_{se}\frac{dv_{eb}(t)}{dt}$$

where

$$G_{se} \equiv \frac{q}{kT}I_B \quad \text{and} \quad C_{se} \equiv \frac{q}{kT}I_B\tau_p = G_{se}\tau_p$$

Thus, as in the case of the simple diode, an a-c conductance and capacitance are associated with the emitter–base junction due to charge storage effects. From Eq. (9-36b) we have

$$i_C(t) = \frac{Q_N(t)}{\tau_t} = \beta I_B + \frac{q}{kT}\beta I_B v_{eb}$$

(9-59) $$i_c = g_m v_{eb}, \qquad \text{where } g_m \equiv \frac{q}{kT}\beta I_B = \frac{C_{se}}{\tau_t}$$

The quantity g_m is an a-c *transconductance*, which is evaluated at the quiescent value of collector current $I_C = \beta I_B$. We can synthesize Eqs. (9-58) and (9-59) in an equivalent a-c circuit as in Fig. 9-19b. In this equivalent circuit the voltage v_{be} used in the calculations appears "inside" the device, so that a new applied voltage v'_{be} must be used external to r_b to refer to the voltage applied between the contacts, and similarly for v'_{ce}. This equivalent model is discussed in detail in most electronic circuits texts; it is often called a *hybrid-pi* model.

From Fig. 9-19b it is clear that several charging times are important in the a-c operation of a transistor; the most important are the time required to charge the emitter and collector depletion regions and the delay time in altering the charge distribution in the base region. Other delay times included in a complete analysis of high-frequency transistors are the transit time through the collector depletion region and the charge storage time in the collector region. If all of these are included in a single delay time τ_d, we can estimate the upper frequency limit of the device. This is usually defined as the *cutoff frequency* for the transistor $f_T \equiv (2\pi\tau_d)^{-1}$. It is possible to show that f_T represents the frequency at which the a-c amplification for the device $[\beta(\text{a-c}) \equiv h_{fe} = \partial i'_c / \partial i'_B]$ drops to unity.

9.4.2 Transit Time Effects.

In high-frequency transistors the ultimate limitation is often the transit time across the base. For example, in a p-n-p device the time τ_t required for holes to diffuse from emitter to collector can determine the maximum frequency of operation for the device. We can calculate τ_t for a transistor with normal biasing from Eq. (9-19b) and the relation $\beta = \tau_p / \tau_t$

$$\beta = \frac{2L_p{}^2}{W_b{}^2} = \frac{2D_p \tau_p}{W_b{}^2} = \frac{\tau_p}{\tau_t}$$

(9-60) $$\tau_t = \frac{W_b{}^2}{2D_p}$$

Another instructive way of calculating τ_t is to consider that the diffusing holes *seem* to have an average velocity $\langle v(x_n) \rangle$ (actually the individual hole motion is completely random, as discussed in Section 4.4.1). The hole current $i_p(x_n)$ is then given by

(9-61) $$i_p(x_n) = qAp(x_n)\langle v(x_n)\rangle$$

The transit time is

(9-62) $$\tau_t = \int_0^{W_b} \frac{dx_n}{\langle v(x_n)\rangle} = \int_0^{W_b} \frac{qAp(x_n)}{i_p(x_n)}\, dx_n$$

For a triangular distribution as in Fig. 9-3b, the diffusion current is almost constant at $i_p = qAD_p\,\Delta p_E/W_b$, and τ_t becomes

(9-63) $$\tau_t = \frac{qA\,\Delta p_E W_b/2}{qAD_p\,\Delta p_E/W_b} = \frac{W_b^2}{2D_p}$$

as before. The average velocity concept should not be pushed too far in the case of diffusion, but it does serve to illustrate the point that a delay time exists between the injection and collection of holes.

We can estimate the transit time for a typical device by choosing a value of W_b, say one micron (10^{-4} cm). For Si, a typical number for D_p is about 10 cm^2/sec; then for this transistor $\tau_t = 0.5 \times 10^{-9}$ sec. Approximating the upper frequency limit as $1/\tau_t$, we can use the transistor to about 2 GHz. Actually, this estimate is too optimistic because of other delay times. A factor of four improvement can be obtained in τ_t by reducing W_b to one-half of a micron. By careful diffusion or elaborate fabrication processing such as ion implantation, it is possible to reduce W_b to about one-tenth of a micron. The transit time can also be reduced by making use of field-driven currents in the base. For the diffused transistor of Fig. 9-13, the holes drift in the built-in field from emitter to collector over most of the base region. By increasing the doping gradient in the base, we can reduce the transit time and thereby increase the maximum frequency of the transistor.

9.4.3 High-Frequency Transistors.
The most obvious generality we can make about the fabrication of high-frequency transistors is that the physical size of the device must be kept small. The base width must be narrow to reduce the transit time, and the emitter and collector areas must be small to reduce junction capacitance. Unfortunately, the requirement of small size generally works against the requirements of power rating for the device. Since we usually require a trade-off between frequency and power, the dimensions and other design features of the transistor must be tailored to the specific circuit requirements. On the other hand, many of the fabrication techniques useful for power devices can be adapted to increase the frequency

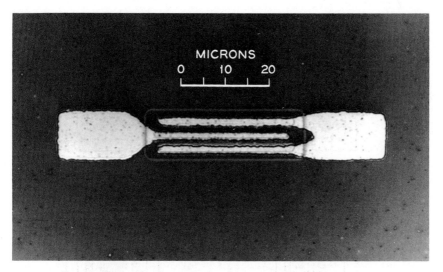

FIGURE 9-20. A Si transistor with a cutoff frequency of more than 7 GHz, illustrating the interdigital arrangement of the two base contacts flanking the center emitter contact. The light rectangular outline around the Al contact fingers is the boundary of the diffused base region. Photograph courtesy of Bell Telephone Laboratories, Incorporated.

range. For example, the method of interdigitation (Fig. 9-17) provides a means of increasing the useful emitter edge length while keeping the overall emitter area to a minimum. Therefore, some form of interdigitation is often used in transistors designed for high frequency and reasonable power requirements (Fig. 9-20).

A second fabrication technique which is very useful in increasing the emitter perimeter-to-area ratio is the *overlay* method (Fig. 9-21). This method employs the insulation provided by an oxide layer between the semiconductor and the emitter contact metallization. Many small emitter regions are diffused into the base region (Fig. 9-21c); with an oxide layer protecting the base, the emitter metallization contacts all of the small emitter regions and connects them in parallel. We notice the emitter contact metallization overlays part of the base, separated from it by the oxide. The base and emitter regions are contacted in the same metallization step by opening appropriate windows before the metal is deposited.

Another set of parameters which must be considered in the design of a high-frequency device is the effective resistance associated with each region of the transistor. Since the emitter, base, and collector resistances affect the various *RC* charging times, it is important to keep them to a minimum. Therefore, the metallization patterns contacting the emitter and base regions

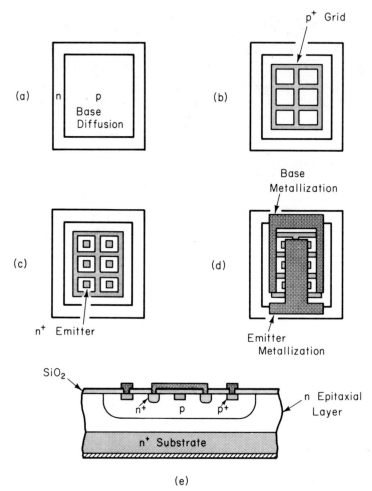

FIGURE 9-21. An overlay transistor: (a) p-type base diffusion into an n-type Si epitaxial layer on an n+ substrate; (b) p+ diffusion in a grid pattern to interlace the base with low-resistance paths to the base contact; (c) n+ emitter regions diffused into the p material; (d) base and emitter metallization; (e) cross section of the device.

must not present significant series resistance. Furthermore, the semiconductor regions themselves must be designed to reduce resistance. For example, the series base resistance r_b of an n-p-n device can be reduced greatly by performing a p+ diffusion between the contact area on the surface and the active part of the base region. The base region surrounding the small emitter areas of the overlay device can be interlaced with a grid of p$^+$ material to provide low-resistance paths to the base contact metallization (Fig. 9-21b).

In Si, n-p-n transistors are usually preferred, since the electron mobility and diffusion constant are higher than for holes. It is common to fabricate n-p-n transistors on n-type epitaxial material grown on an n$^+$ substrate. The heavily doped substrate provides a low-resistance contact to the collector region, while maintaining low doping in the collector to ensure a high breakdown voltage of the collector junction. It is important, however, to keep the collector depletion region as small as possible to reduce the transit time of carriers drifting through the collector junction. This can be accomplished by making the lightly doped collector region narrow so that the depletion region under bias extends to the n$^+$ substrate.

In addition to the various parameters of the device itself, the transistor must be packaged properly to avoid parasitic resistance, inductance, or capacitance at high frequencies. We shall not attempt to describe the many techniques for mounting and packaging transistors here, since methods vary greatly among manufacturers. We should mention, however, that the sealed junction, beam–lead method to be described in Chapter 10 offers many advantages not only for IC's but also for high-frequency discrete transistors.

READING LIST

J. L. MOLL, "Junction Transistor Electronics," *Proceedings of the IRE*, vol. 43, pp. 1807–1819, December 1955.

J. F. GIBBONS, *Semiconductor Electronics*. New York: McGraw-Hill, Inc., 1966, pp. 315–371.

P. E. GRAY and C. L. SEARLE, *Electronic Principles; Physics, Models, and Circuits*. New York: John Wiley & Sons, Inc., 1969, pp. 245–311.

S. M. SZE, *Physics of Semiconductor Devices*. New York: John Wiley & Sons, Inc., 1969, pp. 261–318.

J. R. HAUSER, "Bipolar Transistors," Sec. I in *Fundamentals of Silicon Integrated Device Technology, Vol. II: Bipolar and Unipolar Transistors*, eds. R. M. BURGER and R. P. DONOVAN. Englewood Cliffs, N.J.: Prentice-Hall, Inc., 1968, pp. 3–262.

A. S. GROVE, *Physics and Technology of Semiconductor Devices*. New York: John Wiley & Sons, Inc., 1967, pp. 208–241.

J. L. MOLL, *Physics of Semiconductors*. New York: McGraw-Hill, Inc., 1964, pp. 141–166.

A. VAN DER ZIEL, *Solid State Physical Electronics* (2nd ed.). Englewood Cliffs, N.J.: Prentice-Hall, Inc., 1968, pp. 337–400.

PROBLEMS

9.1 Calculate and plot the excess hole distribution $\delta p(x_n)$ in the base of a p-n-p transistor from Eq. (9-7), assuming $W_b/L_p = 0.5$. The calculations are simplified if the vertical scale is measured in units of $\delta p/\Delta p_E$ and the horizontal scale in units of x_n/L_p. In good transistors, W_b/L_p is much smaller than 0.5; however, $\delta p(x_n)$ is quite linear even for this rather large base width.

9.2 Derive Eq. (9-13) from the charge control approach by integrating Eq. (9-4) across the base region and applying Eq. (9-6).

9.3 Extend Eq. (9-14a) to include the effects of nonunity emitter injection efficiency ($\gamma < 1$). Find an expression for γ. Assume the emitter region is long compared with an electron diffusion length. See Prob. 5.10.

9.4 Apply the charge control approach to the linear distribution in Fig. 9-3b to find α and β; use Eq. (9-63) for the transit time. Is the result the same as Eqs. (9-19)?

9.5 The symmetrical p⁺-n-p⁺ transistor of Fig. P9-5 is connected as a diode in the four configurations shown. Assume $V \gg kT/q$. Sketch $\delta p(x_n)$ in the base region for each case. Which connection seems most appropriate for use as a diode? Why?

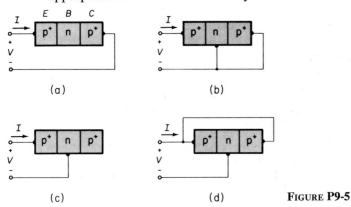

FIGURE P9-5

9.6 For the transistor connection in Fig. P9-5a, (a) show that $V_{EB} = (kT/q) \ln 2$; (b) find the expression for I when $V \gg kT/q$ and sketch I vs. V.

9.7 (a) Find the expression for the current I for the transistor connection of Fig. P9-5b; compare the result with the narrow base diode problem (Prob. 6.5).

(b) How does the current I divide between the base lead and the collector lead?

9.8 Suppose V is negative in Fig. P9-5c.

(a) Find I from the Ebers–Moll equations.

(b) Find the expression for V_{CB}.

(c) Sketch $\delta p(x_n)$ in the base.

9.9 For the transistor connection of Fig. P9-5d, (a) find the expression for $\delta p(x_n)$ in the base region; (b) find the current I.

9.10 Assume a symmetrical p^+-n-p^+ Ge alloyed transistor has $W_b = 5 \ \mu m$, $A = 10^{-3} \ cm^2$, base doping $N_d = 5 \times 10^{15} \ cm^{-3}$, and $\tau_p = 10 \ \mu sec$.

(a) Calculate the saturation current $I_{ES} = I_{CS}$.

(b) Assume $V_{EB} = 0.26 \ V$ and $V_{CB} = -50 \ V$; calculate I_B from Eq. (9-13) and from Eq. (9-18).

(c) Calculate α and β.

9.11 It is obvious from Eqs. (9-28) and (9-29) that I_{EO} and I_{CO} are the saturation currents of the emitter and collector junctions, respectively, with the opposite junction open circuited.

(a) Show that this is true from Eq. (9-25).

(b) Find expressions for the following excess densities: Δp_C with the emitter junction forward biased and the collector open; Δp_E with the collector junction forward biased and the emitter open.

(c) Sketch $\delta p(x_n)$ in the base for the two cases of part (b).

9.12 (a) Show that the definitions of Eq. (9-33) are correct; what does q_N represent?

(b) Show that Eqs. (9-32) correspond to Eqs. (9-27), using the definitions of Eq. (9-33).

9.13 Show that the stored charge $Q_b(t)$ decays according to Eq. (9-44) when the base current is driven from I_B to $-I_B$ at $t = 0$.

9.14 Assume the base region of a p-n-p transistor is doped exponentially as in Fig. 9-13b. For low injection levels we can assume the electric field remains constant at the value given by Eq. (9-48).

(a) Neglecting recombination, show that the hole distribution in the base with a hole current I_p flowing is

$$p(x_n) = \frac{I_p W_b}{q A D_p a}[1 - e^{a(x_n/W_b - 1)}]$$

In the derivation, assume the collector junction is reverse biased, such that $p(W_b) = 0$. (Continued)

(Continued 9.14)

    (b) Sketch the hole distribution for $a = 5$ and discuss the relative influences of hole drift and diffusion across the base.

9.15 To calculate the base spreading resistance r_b', we must find an average value $\langle V_B \rangle$ for the lateral voltage drop in the base region. For simplicity, assume the stripe geometry of Fig. P9-15, in which the base region is contacted on one side and all base current must flow along the base region to this contact. Assume low injection, so that emitter crowding is negligible (I_E is uniform across the emitter).

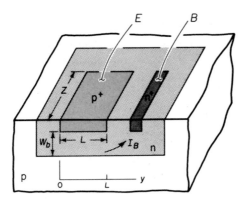

FIGURE P9-15

    (a) Show that the base current at point y is $I_B(y) = I_B y/L$. *Hint:* Write an expression for the differential base current $dI_B(y)$ supplying an element dy of the base, and integrate current and distance from the point $y = 0$ to the point y.

    (b) Find the differential voltage drop $dV_B(y)$ in the element dy and then show that the voltage drop from $y = 0$ to y is $V_B(y) = \rho_b y^2 I_B / 2ZW_b L$, where ρ_b is the average resistivity of the base region.

    (c) Find the average voltage drop $\langle V_B \rangle = L^{-1} \int_0^L V_B(y)\, dy$ and the average base resistance $r_b' = \langle V_B \rangle / I_B$.

INTEGRATED CIRCUITS 10

Just as the transistor revolutionized electronics by offering more flexibility, convenience, and better reliability than the vacuum tube, the integrated circuit enables new applications for electronics which were not possible with discrete devices. Integration allows complex circuits consisting of many transistors, diodes, resistors, and capacitors to be included in a chip of semiconductor or on a small insulating substrate. This means that sophisticated circuitry can be miniaturized for use in space vehicles, in large-scale computers, and in other applications where a large collection of discrete components would be impractical. In addition to offering the advantages of miniaturization, however, the simultaneous fabrication of hundreds of IC's on a single Si wafer greatly reduces the cost and increases the reliability of each of the finished circuits. Certainly discrete components are still used in great numbers in many electronic circuits; however, more and more circuits are being fabricated on the Si chip rather than with a collection of individual components. Because of this trend, the traditional distinctions between the roles of circuit and system designers do not apply to IC fabrication.

In this chapter we shall discuss various types of IC's and the fabrication steps used in their production. We shall investigate techniques for building large numbers of transistors, diodes, and resistors on a single chip of Si, as well as the interconnection, contacting, and packaging of these circuits in usable form. All the processing techniques discussed here are very basic and general. There would be no purpose in attempting a comprehensive review of all the subtleties of device fabrication in a book of this type. In fact, the only way to keep up with such an expanding field is to study the current literature. Many good reviews are suggested in the reading list at the end of this chapter; more important, current issues of those periodicals cited can be consulted for up-to-date information regarding IC technology. Having the background of this chapter, one should be able to read the current literature and thereby keep abreast of the present trends in this very important field of electronics.

10.1 Background

In this section we provide an overview of the nature of integrated circuits and the motivation for using them. It is important to realize the reasons, both technical and economic, for the dramatic rise of IC's to their present role in electronics. We shall discuss several main types of IC's and point out some of the applications of each. More specific fabrication techniques will be presented in later sections.

10.1.1 Advantages of Integration. It might appear that building complicated circuits, involving many interconnected components on a single tiny substrate, would be risky both technically and economically. In fact, however, modern techniques allow this to be done reliably and relatively inexpensively; in most cases an entire circuit on a Si chip can be produced more inexpensively and with greater reliability than a similar circuit built up from individual components. The basic reason is that hundreds of identical circuits can be built simultaneously on a single Si wafer (Fig. 10-1); this process is called *batch fabrication*. Although the processing steps for the wafer may be involved and expensive, the large number of resulting circuits makes the ultimate cost of each fairly low. In fact, the main cost of most IC's can be traced to the bonding and packaging of the individual chips once they are separated from the wafer. This means that the number of components in each circuit is relatively unimportant in terms of the ultimate cost of the circuit. The implications of this principle are tremendous for circuit designers; it greatly increases the flexibility of design criteria. Unlike circuits with individual transistors and other components wired together or placed on a circuit board, IC's allow many "extra" components to be included economically. Thus, redundancy and "back-up" circuitry can be included without greatly raising the cost of the final product. Reliability is also improved since all devices and interconnections are made on a single rigid substrate, greatly minimizing failures due to the soldered interconnections of discrete component circuits.

The advantages of IC's in terms of miniaturization are obvious. Since many circuits can be packed into a small space, complex electronic equipment can be employed in many applications where weight and space are critical, such as in aircraft or space vehicles. In large-scale computers it is now possible not only to reduce the size of the overall unit but also to facilitate maintenance by allowing for the replacement of entire circuits quickly and easily. As IC techniques develop, more and more applications are found in such consumer products as automobiles, telephones, television, and appli-

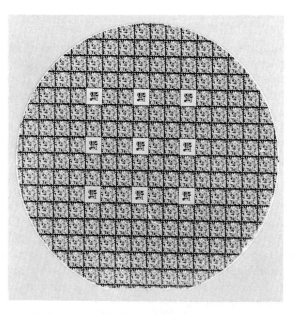

Figure 10-1. A wafer of integrated circuits: (a) overall view of the Si wafer, containing about 180 complete circuits; (b) an expanded view of one of the circuits. Each circuit, about 75 mils square, contains 65 elements (transistors, diodes, and resistors). The nine noncircuit squares are used to check diffusions and other processing parameters during fabrication. Photographs courtesy of Delco Electronics Division, General Motors Corporation.

ances. Miniaturization and the cost reduction provided by IC's mean that we all have much more sophisticated electronics at our disposal than would ever be possible through discrete components.

Some of the most important advantages of miniaturization pertain to response time and the speed of signal transfer between circuits. For example, in high-frequency circuits it is necessary to keep the separation of various components small to reduce time delay of signals. Similarly, in very high-speed computers it is important that the various logic and information storage circuits be placed close together. Since electrical signals are ultimately limited by the speed of light (about one foot per nanosecond), physical separation of the circuits can be an important limitation. As we shall see in Section 10.4, *large-scale integration* of many circuits on a Si wafer can lead to major reductions in computer size, thereby tremendously increasing the potential for speed and function density. In addition to decreasing the signal transfer time, integration can reduce parasitic capacitance and inductance between circuits. Reduction of these parasitics can provide significant improvement in the operating speed of the system.

We have discussed several advantages of reducing the size of each unit in the batch fabrication process, such as miniaturization, high-frequency and switching speed improvements, and cost reduction due to the large number of circuits fabricated on a single wafer. Another important advantage has to do with the percentage of usable devices (often called the *yield*) which results from batch fabrication. Faulty devices usually occur because of some defect in the Si wafer or in the fabrication steps. Defects in the Si can occur because of lattice imperfections and strains introduced in the crystal growth, cutting, and handling of the wafers. Usually such defects are extremely small, but their presence can ruin devices built on or around them. Reducing the size of each device greatly increases the chance for a given device to be free of such defects. The same is true for fabrication defects, such as the presence of a dust particle on a photographic mask. For example, a lattice defect or dust particle one-half micron in diameter can easily ruin a circuit which includes the damaged area. If a fairly large circuit is built around the defect it will be faulty; however, if the device size is reduced so that four circuits occupy the same area, chances are good that only the one containing the defect will be faulty and the other three will be good. Therefore, the percentage yield of usable circuits increases over a certain range of decreasing chip area. There is an optimum area for each circuit, above which defects are needlessly included and below which the elements are spaced too closely for reliable fabrication.

10.1.2 Types of Integrated Circuits. There are several ways of categorizing IC's as to their use and method of fabrication. The most common categories are *linear* or *digital* according to application, and *monolithic* or *hybrid* according to fabrication.

A linear IC is one which performs amplification or other essentially linear operations on signals. Examples of linear circuits are simple amplifiers, operational amplifiers, and other signal-processing circuits. Digital circuits involve logic and switching, primarily for computer applications. By far the greatest volume of IC's has been in the digital field, since large numbers of such circuits are required in computers. Since digital circuits generally require only "on–off" operation of transistors, the design requirements for integrated digital circuits are often less stringent than for linear circuits. Although transistors can be fabricated as easily in integrated form as in discrete form, passive elements (resistors and capacitors) are usually more difficult to produce to close tolerances in IC's.

Integrated circuits which are included entirely on a single chip of semiconductor (usually Si) are called *monolithic* circuits (Fig. 10-1). The word monolithic literally means "one stone" and implies that the entire circuit is contained in a single piece of semiconductor. Any additions to the semiconductor sample, such as insulating layers and metallization patterns, are intimately bonded to the surface of the chip. A *hybrid* circuit may contain one or more monolithic circuits or individual transistors bonded to an insulating substrate with resistors, capacitors, or other circuit elements, with appropriate interconnections (Fig. 10-2). Monolithic circuits have the advantage that all components are contained in a single rigid structure which can be batch fabricated; that is, hundreds of identical circuits can be built simultaneously on a Si wafer. On the other hand, hybrid circuits offer excellent isolation between components and allow the use of more precise resistors and capacitors. Furthermore, hybrid circuits are often less expensive to build in small numbers.

10.1.3 Passive Components for Hybrid Circuits. When resistors and capacitors are made external to the monolithic Si chip, basically two types of technology are used; the passive elements are fabricated and interconnected by *thick-film* or *thin-film* processes. Although the dividing line between thin and thick films is not precise, they are fairly well separated in application to IC's: "thin" films are typically 0.1 to 0.5 micron, and "thick" films are about 25 microns thick.

The processing steps for the two techniques are quite different. In thick-film circuits the resistors and interconnection patterns are "printed" on a ceramic substrate by silk-screen or similar processes. Conductive and resistive pastes consisting of metal powders in organic binders are printed on the substrate and cured in an oven. One advantage of this process is that resistors can be made below the rated values and then trimmed by abrasion, or by

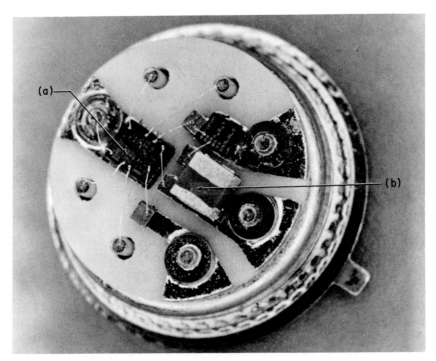

FIGURE 10-2. A hybrid integrated circuit mounted on a ceramic substrate. This device includes (a) a monolithic circuit, and (b) a ceramic chip capacitor. The various elements are interconnected by small Al wires. Photograph courtesy of Motorola Semiconductor Products, Inc.

selective evaporation using a pulsed laser. These corrections can be made quickly with automated procedures while the resistance values are under test. Small ceramic chip capacitors can be bonded into place in the interconnection pattern, along with monolithic circuits or individual transistors.

Thin-film technology allows for greater miniaturization and is generally preferred when space is an important limitation. Thin-film interconnection patterns and resistors can be vacuum deposited on a glass or glazed ceramic substrate. The resistive films are usually made of tantalum or other resistive metal, and the conductors are often aluminum or gold. In general, the resistive materials must be deposited by sputtering. Pattern definition for the resistors and conductor paths can be achieved by depositing the films through metal shields which contain appropriate apertures. Better definition is obtained by metallizing the entire substrate, or large parts of it, and using photolithographic methods to remove the metal except in the desired pattern. This technique is similar to that used in forming metal contact patterns on monolithic circuits. Capacitors can be fabricated by thin-film techniques by depositing an insulating layer between two metal films or by oxidizing the surface of one film and then depositing a second film on top. Whether thin-film or thick-film techniques are used, the object is to fabricate passive components of greater precision than could be obtained on the Si substrate.

An important passive component which is missing from the discussion thus far is the inductor. Inductors are difficult to integrate into small substrates and are usually avoided in the circuit design or placed external to the IC. This limitation is serious but seldom disastrous, since the function of an inductor can often be obtained by proper design of an active circuit. Since active devices can be made easily in IC's, it is usually possible to avoid the use of inductors altogether.

10.2 Fabrication of Monolithic Circuits

The most important element of IC technology is the monolithic chip, which may contain several or dozens of individual transistors, diodes, resistors, and capacitors, all properly interconnected. Such monolithic circuits are batch fabricated on a Si wafer one or two inches in diameter (Fig. 10-1); each wafer may contain hundreds of the monolithic circuits. These circuits are then separated by selective etching or by scribing the wafer with a diamond stylus and breaking it apart into small squares or rectangles containing individual circuits. Each circuit is then mounted on an appropriate substrate, contacted, and packaged.

The fabrication process for making the wafer of monolithic circuits is the subject of this section. We shall review the processes of masking and selective diffusion and shall point out some of the steps in obtaining the various monolithic components. Many of the techniques discussed here are similar to those described earlier for discrete components, but there are also many other methods which are peculiar to monolithic circuits. We shall discuss the problem of isolation between components, special techniques for fabricating monolithic transistors, and the various ways of obtaining diodes, resistors, and capacitors on a monolithic chip. We shall also discuss techniques for protecting the device surface and indicate some of the current approaches to problems of reliability. Design techniques for the various steps in the fabrication process will be discussed qualitatively and use of computer-aided design will be described.

10.2.1 Masking and Selective Diffusion. Although the process of selective diffusion has been discussed in previous chapters, it is worth-

while to review it here and investigate the extensions to monolithic IC's. As with discrete components, the object of the process is to selectively dope certain regions of the semiconductor and to properly interconnect the resulting components with a metallization pattern. Including oxidation steps, the number of operations involved in fabricating a monolithic circuit can be quite large. As an example, let us review the simple diffused transistor of Fig. 8-26. The basic steps in the process are as follows:

1. Grow the first oxide layer.
2. Open a window in the SiO_2 for the base diffusion.
3. Perform a boron diffusion.
4. Grow a second oxide layer.
5. Open a window for the emitter diffusion.
6. Perform a phosphorus diffusion.
7. Grow a third oxide layer.
8. Open windows for the base and emitter contacts.
9. Evaporate Al on the surface.
10. Remove the Al except in the desired metallization pattern.

In this simple example there are three oxidation steps, two diffusions, and one metallization. The number of contact masks required is four: two for diffusions, one for the contacting windows, and one for the metallization definition. More steps and therefore more masks are required for transistors suitable for monolithic circuits. The important point, however, is that many identical circuits are made simultaneously on the wafer, thus making the process economically feasible. Therefore, it is important to reduce the size of each circuit to multiply the number of usable devices resulting from the batch fabrication. This means that the various masks must be extremely accurate and well aligned during each photolithographic step.

Usually, the original artwork for each mask is made on large plastic (Mylar) sheets. The sheet is a laminate of a transparent layer and a red layer of Mylar. The desired pattern (including reference marks for alignment purposes) is cut into the red layer which is selectively stripped off, leaving transparent windows (Fig. 10-3). This original sheet usually contains the mask pattern for only one circuit; duplication into the multiple-pattern contact mask is done at a later stage. Since small errors in master artwork can ruin every device on the wafer, the patterns must be extremely precise. The patterns are cut in the Mylar sheet on a precision drafting table called a *coordinatograph*, which provides for calibrated positioning of the knife point in the plane of the table. In another method, mask patterns are generated on a photographic film by a precision light pen which can be controlled from a computer output.

After the original artwork is completed, it is photographed and reduced in size. A *step-and-repeat* camera is used to photograph the reduced pattern,

FIGURE **10-3.** Fabrication of master artwork for a monolithic contact mask. This photograph illustrates the precision cutting of transparent windows in a Mylar sheet, using a coordinatograph. Photograph courtesy of Motorola Semiconductor Products, Inc.

perform the final reduction, and repeat the process for every rectangle in the final array. The overall reduction in the pattern size is a factor of several hundred, and the final array may contain hundreds of identical patterns. The final pattern array is printed on a glass contact mask to be placed over the Si wafer in the photolithographic process.

Each contact mask is positioned over the wafer on a precision alignment apparatus, and the reference marks are viewed through a microscope to align the mask with previous patterns on the wafer. Ultraviolet light is then passed through the contact mask to selectively expose the photoresist emulsion (Fig. 5-6). Definition in the pattern lines on the wafer can be one micron or less.† To achieve this kind of accuracy, the original artwork and every photographic step in the reduction and step-and-repeat processes must be controlled very carefully. To reduce the total number of masks required, as many functions as possible are performed by each mask. For example, patterns for diffused resistors may be included in the same mask used to open windows for the base diffusion. We shall discuss the fabrication of resistors and other components in Section 10.2.3.

10.2.2 Isolation. An important step in the fabrication of monolithic circuits is the provision of electrical isolation between components. For

---

†For smaller dimensions than can be obtained by contact masks, the pattern is projected onto the wafer through a precision microscope lens. Another alternative is to use a computer-controlled electron beam to selectively expose the photoresist.

example, if several transistors were made on a monolithic chip by the technique of Fig. 8-26, all collector regions would be in common. While this is a useful connection in some logic circuits, it is generally necessary to isolate most components and then interconnect them with the metallization pattern. Several techniques are available for isolation; we shall discuss two of the most common methods here and reserve discussion of a third method until after beam–lead technology has been presented in Section 10.3.2.

For typical n-p-n monolithic devices, the most convenient method of isolation involves diffusing a pattern of p-type "moats" into an n-type epitaxial layer on a p-type substrate (Fig. 10-4a). The p-type substrate provides mechanical support for the structure and, in conjunction with the diffused p-type pattern, defines isolated regions of n-type material. The epitaxial layer is only about ten microns thick in most cases, which is thick

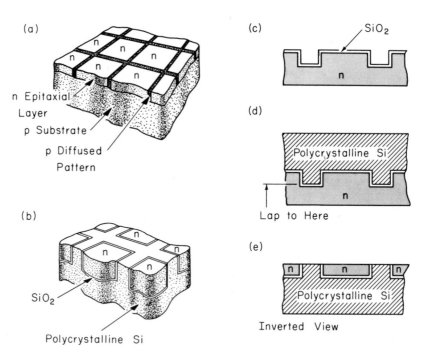

FIGURE 10-4. Isolation techniques for monolithic circuits: (a) junction isolation achieved by diffusing a p-type grid through the n-type epitaxial layer; (b) SiO_2 isolation. Steps in the SiO_2 isolation process are as follows: (c) etch grooves in the n-type wafer and grow an oxide layer on the surface; (d) deposit polycrystalline Si; (e) remove the n-type material until the isolated islands are exposed. The wafer in (e) is inverted to indicate that the polycrystalline layer becomes the substrate.

enough to accommodate subsequent diffusion of n-p-n transistors and other elements. Since each component can be contained in an island of n-type material, good isolation results if the substrate p material is held at the most negative potential in the circuit. In effect, each component is surrounded by a reverse-biased p-n junction. The room temperature saturation current for Si junctions is typically in the nanoampere range. Thus the isolation between circuit elements is good, except for capacitance effects at high frequencies.

Among the more elaborate schemes for providing isolation is the use of SiO$_2$ layers surrounding each n-type island. One method for accomplishing this is illustrated in Fig. 10-4b. An n-type wafer is masked and etched to remove Si in a grid pattern, leaving grooves in the surface of the wafer. The entire surface, including the grooves, is coated with about one micron of SiO$_2$ (Fig. 10-4c). Then a thick layer of Si is deposited on the oxide by a process similar to epitaxial growth. However, since growth occurs on the amorphous oxide, the deposited Si grows in polycrystalline form rather than as a single-crystal epitaxial layer (Fig. 10-4d). Now the original n-type Si material is lapped away until the oxide-isolated regions are exposed (Fig. 10-4e). This step requires great precision to obtain uniformity across the surface. In the final wafer to be used for the monolithic diffusion processes, the polycrystalline layer serves as the substrate, and the n-type islands are insulated by the surrounding SiO$_2$ layers. This method is obviously much more difficult than the p-n junction isolation described above. Such complex methods are used only in special circumstances, such as in devices designed to function in nuclear radiation environments (radiation-hardened devices).

In subsequent discussions of monolithic circuit fabrication, we shall assume that the diffused epitaxial method of p-n junction isolation is used unless otherwise stated. Another important method of isolation will be discussed in conjunction with the beam–lead technology of Section 10.3.2.

10.2.3 Monolithic Device Elements.

10.2.3 Monolithic Device Elements. Now we shall consider the various steps in fabricating device elements for monolithic circuits. The techniques for making monolithic transistors, diodes, resistors, capacitors, and interconnections are discussed here for fairly standard cases, without the complications which may arise in special circumstances. We shall consider double-diffused epitaxial (DDE) transistors with p-n junction isolation.

Transistors:

One important difference between monolithic circuit transistors and the discrete transistors we have discussed earlier is that in the monolithic case all three terminals must be available on the top surface of the chip. For example, interconnection of several isolated transistors requires that collector contacts be made at the surface. As a result, collector current must pass along a high-resistance path in the lightly doped n-epitaxial material while flowing from the active part of the collector to the contact. The resulting series collector resistance can be detrimental to the transistor properties and should be avoided if possible. One common method of decreasing the collector resistance is to include a heavily doped n layer just below the collector. Usually, this n^+ *buried layer* is included by diffusion of n^+ regions into the p-type substrate before the epitaxial layer is grown (Fig. 10-5a). After the n-epitaxial layer is grown and the p-type isolation diffusion is made, the wafer has a pattern typified by Fig. 10-5c. During the isolation diffusion and in subsequent diffusions, there is some movement of the n^+ buried layer into the epitaxial region. To minimize this spreading of the buried layer, it is usually doped with Sb, which diffuses much more slowly in Si than does phosphorus.

After the isolation pattern has been diffused through the epitaxial layer, the wafer is masked for the base diffusion (boron) and then again for the emitter diffusion (phosphorus). During the shallow n^+ emitter diffusion, it is common to diffuse n^+ stripes above the collector areas to improve the collector contact. The resulting transistor structure is shown in Fig. 10-5d (the oxide and metallization layers are omitted for clarity). If very low collector resistance is required, deep n^+ diffusions can be made from the surface to the buried layer (Fig. 10-5e). This step is done before the base diffusion. After all diffusions have been completed, the surface is passivated and the interconnection pattern is included, as discussed below.

We have presented the diffusion processes for fabrication of monolithic n-p-n transistors, since this is the most common type in IC's. If p-n-p transistors are to be included as well, special techniques must be used. One way to make a p-n-p device in the n-type island of Fig. 10-5c would be to employ a triple diffusion: a p-layer, then an n-layer, and finally another p-layer. This method is undesirable because of the extra diffusion step required and because the ultimate emitter region has been compensated three times. When p-n-p transistors are required in an otherwise n-p-n circuit, they are usually made by the method shown in Fig. 10-5f. This is called a *lateral transistor*, since the current flows laterally from the emitter to the base to the collector. The base width and other parameters are more difficult to control in the lateral transistor than in *planar* transistors, in which the emitter and collector junctions are parallel. In fact, the β of lateral transistors is often near unity; therefore, the lateral p-n-p transistor is used most often in conjunction with a planar n-p-n such that the n-p-n boosts the gain of the overall unit (Fig. 10-5g). This is called a *composite transistor;* the overall gain β is essentially the product of the gains of the individual transistors, but the terminals of the composite transistor have the properties of a p-n-p device (Prob. 10.5).

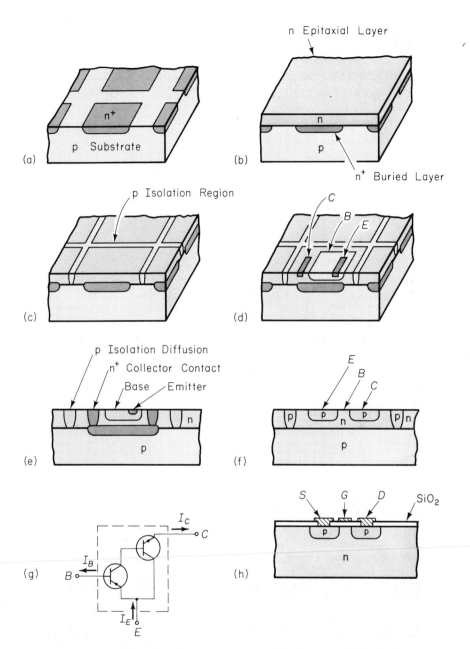

FIGURE 10-5. Transistors for integrated circuits: (a) buried layer diffusion; (b) growth of epitaxial layer; (c) isolation diffusion; (d) double-diffused epitaxial transistor; (e) improved collector contact by deep n^+ diffusion; (f) lateral p-n-p transistor; (g) interconnection of a p-n-p with an n-p-n in a composite transistor; (h) induced-channel FET. The oxide layers in (c)—(f) and metallizations in (d)—(f) are omitted for clarity.

Field-effect transistors can be used successfully in IC's if care is taken to control surface effects at the oxide–semiconductor interface. Induced-channel FET's (Fig. 10-5h) are convenient for integrated fabrication; they require fewer processing steps than BJT's and can be made smaller, with high packing densities on the surface of the chip. In some cases the gate electrode is made of deposited polycrystalline Si instead of Al (*silicon gate technology*). This lowers the threshold voltage and can decrease the overall transistor size.

Diodes:

It is simple to build p-n junction diodes in a monolithic circuit. It is also common practice to use transistors to perform diode functions. Since many transistors are included in a monolithic circuit, no special diffusion step is required to fabricate the diode elements. There are a number of ways in which a transistor can be connected as a diode (Prob. 9.5). Perhaps the most common method is to use the emitter junction as the diode, with the collector and base shorted. This configuration is essentially the narrow base diode structure, which has high switching speed with little charge storage. Since all the transistors can be made simultaneously, the proper connections can be included in the metallization pattern to convert some of the transistors into diodes.

Resistors:

Diffused resistors can be obtained in monolithic circuits by using the shallow diffusions for the transistor base and emitter regions (Fig. 10-6a). For example, during the base diffusion, a resistor can be diffused which is made up of a thin p-type layer within one of the n-type islands. Alternatively,

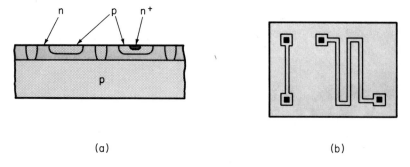

(a)   (b)

Figure 10-6. Monolithic resistors: (a) cross section showing use of base and emitter diffusions for resistors; (b) top view of two resistor patterns.

a p region can be made during the base diffusion, and an n-type resistor channel can be included within the resulting p region during the emitter diffusion step. In either case, the resistance channel can be isolated from the rest of the circuit by proper biasing of the surrounding material. For example, if the resistor is a p-type channel obtained during the base diffusion, the surrounding n material will be connected to the most positive potential in the circuit to provide reverse-bias isolation. The resistance of the channel depends on its length, width, depth of the diffusion, and resistivity of the diffused material. Since the depth and resistivity are determined by the requirements of the base or emitter diffusion, the variable parameters are the length and width. Two typical resistor geometries are shown in Fig. 10-6b. In each case the resistor is long compared with its width, and a provision is made on each end for making contact to the metallization pattern.

Design of diffused resistors begins with a quantity called the *sheet resistance* of the diffused layer. If the average resistivity of a diffused region is ρ, the resistance of a given length L is $R = \rho L/w\mathrm{t}$, where w is the width and t is the thickness of the layer. Now if we consider one square of the material, such that $L = w$, we have the sheet resistance $R_s \equiv \rho/\mathrm{t}$ in units of ohms per square. We notice that R_s measured for a given layer is numerically the same for any size square. This quantity is simple to measure for a thin diffused layer by a four-point probe technique.[†] Therefore, for a given diffusion, the sheet resistance is generally known with good accuracy. The resistance then can be calculated from the known value of R_s and the ratio L/w (the *aspect ratio*) for the resistor. We can make the width w as small as possible within the requirements of heat dissipation and other limitations and then calculate the required length from w and R_s. These are a few of the design criteria for diffused resistors; other criteria include geometrical factors, such as the presence of high current density at the inside corner of a sharp turn. In some cases it is necessary to round corners slightly in a folded or zig-zag resistor (Fig. 10-6b) to reduce this problem.

If precision resistors are required, it is often necessary to use thin-film techniques. We have discussed the formation of precision resistors on an insulating substrate for hybrid circuits. Much the same technology can be used on the chip; tantalum or other thin resistive films can be deposited on the oxide layer and shaped by photolithography to give resistors of reasonable precision. Although this process involves placing the resistor above the Si chip, it is considered a monolithic process because the film is solidly bound to the oxide layer.

[†]This is a very useful method, in which current is introduced into a wafer at one probe, collected at another probe, and the voltage is measured by two probes in between. Special formulas are required to calculate resistivity or sheet resistance from these measurements. Good reviews of this method are given in the texts by Burger and Donovan and by Ghandhi in the reading list for this chapter.

Capacitors:

When capacitors are included on the monolithic chip, either junction capacitance or an MOS structure is used. As in the cases of diodes and resistors, it is desirable to form junction capacitors during one of the transistor diffusion steps. The base–collector junction can be used without an emitter diffusion, the emitter–base junction can be used, or an n^+ region can be diffused into one of the p-isolation regions during an emitter diffusion. All of these junction capacitors have the disadvantage that the value of capacitance varies with voltage. A good approximation to the usual parallel plate capacitor can be made by MOS techniques without increasing the fabrication steps. A relatively large n^+ region is diffused into the n-type island during the emitter diffusion step. After an oxide is grown on the surface, an Al layer is deposited on top during the metallization step (Fig. 10-7a). Although the resulting MOS capacitor takes up more room than the junction type, it has several advantages, including that of constant capacitance with applied voltage.

Interconnections:

During the metallization step, the various regions of each circuit element are contacted, MOS devices are completed, and proper interconnection of the circuit elements is made. Aluminum is commonly used for the metallization, since it adheres well to Si and to SiO_2 if the temperature is raised

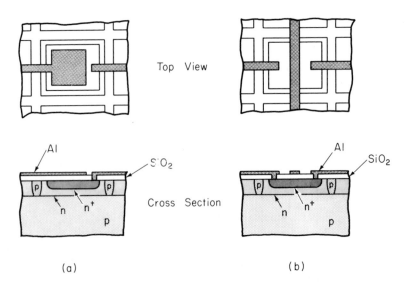

FIGURE 10-7. Elements formed during metallization: (a) MOS capacitor; (b) conductor crossover.

briefly to about 550°C after deposition. Gold is used on some monolithic devices, but the adherence properties of Au to Si and SiO_2 are poor. Therefore, intermediate metallizations must be made to bond the Au to the surface, as we shall discuss with regard to Fig. 10-14.

The steps involved in metallization have been mentioned previously. The wafer is oxidized and windows are opened over the regions to be contacted. These windows are made smaller than the region to be contacted to ensure making only the desired contact. After the Al is deposited on the surface, the wafer is coated again with photoresist and exposed to define the appropriate metallization pattern. The Al is etched away everywhere except under the polymerized emulsion, and the wafer is then heated to bond the Al to the Si and the SiO_2. After heat treatment, the Al–Si bond forms a shallow p^+ region, which serves as a good ohmic contact to the p regions of the device. Contact with the n^+ emitter-diffused regions is essentially ohmic because of properties of the heavily doped junction. In the case of the n-type collector contact, it is important that an n^+ shallow diffusion be made during the emitter diffusion to provide a heavily doped region for contacting with the Al.

In designing the layout of elements for a monolithic circuit, topological problems must be solved to provide efficient interconnection without *crossovers*—points at which one conductor crosses another conductor. If crossovers must be made, they can be accomplished easily at a resistor. Since the diffused resistor is a channel in the Si covered by SiO_2, a conductor can be deposited crossing the insulated channel. In cases requiring crossovers where no resistor is available, a low-value diffused resistor can be inserted in one of the conductor paths. For example, a short n^+ region can be diffused during the emitter step and contacted at each end by one of the conductors. The other conductor can then cross over the oxide layer above the n^+ region (Fig. 10-7b). Usually, this can be accomplished without appreciable increase in resistance, since the n^+ region is heavily doped and its length can be made small.

During the metallization step, appropriate points in the circuit are connected to relatively large *pads* to provide for external contacts. These metal pads are visible in photographs of monolithic circuits (Fig. 10-1b) as rectangular or circular areas spaced around the periphery of the device. In the mounting and packaging process, these pads are contacted by small Au or Al wires or by special techniques such as those discussed in Section 10.3.

10.2.4 Passivation.

One of the most important fabrication steps in terms of long-term stability and reliability of the monolithic circuit is *passivation* of the surface. By passivation we mean the protection of the surface, particularly the intersections of the various junctions with the surface, from outside contaminants. For example, if water vapor or small amounts of metal contamination are allowed to migrate to the junction area of a component, the characteristics of the device can be degraded seriously.

The SiO_2 layer used in the diffusion-masking process is a good protection for the surface against most contaminants. Since the final step before opening the contact windows is an oxidation of the surface, all junctions will be protected by at least one layer of oxide. In diffusion through an oxide window, the junction meets the surface under the oxide instead of at the edge of the oxide, since the diffusion takes place laterally as well as normal to the surface (Fig. 10-8a). This means that the junction–surface boundary occurs automatically under an oxide layer. Further SiO_2 protection is provided by one or more additional oxide layers grown in the subsequent fabrication steps; when the final window is opened for the contact metallization, it is made smaller than the original diffusion mask window.

Whereas SiO_2 provides good protection from water vapor and other contaminants, an important problem exists due to the possibility of migration of certain metal ions, particularly Na^+ ions, through the SiO_2 layer. This problem can be reduced greatly by depositing a thin layer of phosphorus glass $(P_2O_5 + SiO_2)$ on top of the SiO_2. The Na^+ migration occurs via oxygen vacancies in the SiO_2, and the presence of excess oxygen in the phosphorus–glass layer reduces the vacancies available for this migration.

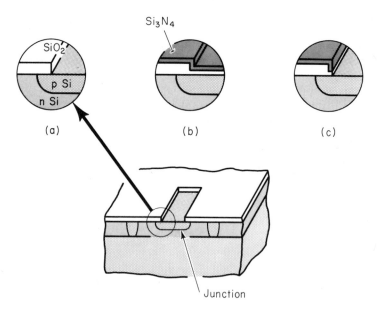

FIGURE 10-8. Silicon nitride passivation of a Si surface: (a) oxidation and diffusion through the window; (b) thin oxidation and nitride deposition; (c) window opened in Si_3N_4 and SiO_2 layers for contact metallization.

A particularly effective passivation technique is the use of a thin silicon nitride (Si_3N_4) layer on top of the SiO_2. This type of protective layer eliminates the migration of all important impurities to the surface of the Si; it is often called a *sealed junction* method, indicating that each junction on the surface is effectively sealed against contamination. The use of Si_3N_4 passivation requires several extra processing steps, but the resulting structure is sufficiently protected to permit greatly simplified packaging techniques. After the final diffusion step, an additional thin oxide layer is grown to include the diffusion windows. After this oxide growth, the silicon nitride layer is deposited (Fig. 10-8b). Special etching techniques must be used to open windows in the silicon nitride; therefore, another layer of SiO_2 is deposited over the Si_3N_4. After opening windows in this top oxide layer with an HF etch, the underlying nitride layer can be etched in the windows by a boiling phosphoric acid etch. A final rinse in HF removes the thin oxide layer in the contact windows and the top oxide layer over the silicon nitride. Therefore, the final passivation medium is a sandwich structure of SiO_2 and Si_3N_4 (Fig. 10-8c). Now metallization can be performed over the nitride layer and in the windows.

10.2.5 Use of Computers in Fabrication and Testing.

Much of the time-consuming design and testing work of monolithic circuit fabrication can be reduced greatly with the aid of computers. In fact, for very complex circuits involving large numbers of components and interconnections, the use of a computer is mandatory if the design and testing is to be economically feasible.

In circuit layout, various design criteria including allowable dimensions, tolerances, and other data are stored in the computer memory and are recalled during the design of a particular circuit. The computer-aided design attempts to find the best layout for maximum utilization of area with a minimum of crossovers in the interconnection pattern. In some cases the computer can be programmed to test a trial circuit layout so that the expense and time of actually building the circuit can be postponed until a proper design is established. Once a circuit layout is determined, the output of the computer can be printed in coordinate form for each mask, or the output information can be fed directly to an automatic mask generator. One of the most intricate designs is the generation of a metallization mask for large-scale integrated circuits, to be discussed in Section 10.4. In this case the computer must make very complex decisions regarding the optimum interconnection pattern.

Another important use of the computer is in testing and evaluation. After the wafer has been processed and the final metallization pattern defined, the wafer of monolithic circuits is placed in a holder under a microscope and is aligned for testing by a multiple-point probe (Fig. 10-9). The probe contacts

FIGURE 10-9. A wafer of monolithic circuits under test by a multiple-point probe. Photograph courtesy of Texas Instruments, Incorporated.

the various pads on an individual circuit, and a series of tests are made of the electrical properties of the device. The various tests are programmed to be made automatically in a very short time (usually milliseconds). The information from these tests is fed into a computer, which compares the results with information stored in its memory, and a decision is made regarding the acceptability of the circuit. If there is some defect so that the circuit falls below specifications, the computer instructs the test probe to mark the circuit with a dot of ink. The probe automatically steps the prescribed distance to the next circuit on the wafer and repeats the process. The time between measurements on adjacent circuits can be 1 sec or less, and the entire wafer can be tested in a relatively short time. After all of the circuits have been tested and the substandard ones marked, the wafer is removed from the testing machine, scribed between the circuits, and broken apart. Then the individual chips are mounted on their substrates for packaging, after the inked circuits have been discarded. In the testing process, information from tests on each circuit can be printed out to facilitate analysis of the rejected circuits or to evaluate the fabrication process for possible changes.

10.3 Bonding and Packaging

After the preceding discussions of rather dramatic fabrication steps in monolithic circuit technology, the processes of attaching leads and packaging the devices could seem rather mundane. Such an impression would be far from accurate, however, since the techniques discussed in this section are crucial to the overall fabrication process. In fact, the handling and packaging of individual circuits can be the most critical steps of all from the viewpoints of cost and reliability. The individual IC chip must be connected properly to outside leads and packaged in a way that is convenient for use in a larger circuit or system. Since the devices are handled individually once they are separated from the wafer, bonding and packaging are expensive processes. Considerable work has been done to reduce the steps required in bonding. We shall discuss the most straightforward technique first, which involves bonding individual leads from the contact pads on the circuit to terminals in the package. Then we shall consider two important methods for making all bonds simultaneously. Finally, we shall discuss a few typical packaging methods for IC's.

10.3.1 Wire Bonding.

The earliest method used for making contacts from the monolithic chip to the package was the bonding of fine Au wires. Later techniques expanded wire bonding to include Al wires and several types of bonding processes. Here we shall outline only a few of the most important aspects of wire bonding.

If the chip is to be wire bonded, it is first mounted solidly on a header or on a metallized region ("land") of an insulating substrate. In this process a thin layer of Au (perhaps combined with Ge or other elements to improve the metallurgy of the bond) is placed between the bottom of the chip and the substrate; heat and a slight scrubbing motion are applied, forming an alloyed bond which holds the chip firmly to the substrate. This process is called *die bonding* ("die," the singular form of "dice," is used interchangeably with "chip" to refer to the individual monolithic device). Once the chip is mounted, the interconnecting wires are attached from the various contact pads to posts on the header (or metallized regions on the substrate).

In Au wire bonding, a spool of fine Au wire (about 0.0007 to 0.002 in. diameter) is mounted in a *lead bonder* apparatus, and the wire is fed through a glass or tungsten carbide *capillary* (Fig. 10-10a). A hydrogen gas flame jet is swept past the wire to form a ball on the end. In *thermocompression bonding* the chip (or in some cases the capillary) is heated to about 360°C, and the capillary is brought down over the contact pad. When pressure is exerted by the capillary on the ball, a bond is formed between the Au ball and the Al pad (Fig. 10-10b). Then the capillary is raised and moved to a header post (or a metal land). The capillary is brought down again, and the combination of force and temperature bonds the wire to the post. After raising the capillary again, the hydrogen flame is swept past, forming a new ball (Fig. 10-10c); then the process is repeated for the other pads on the chip.

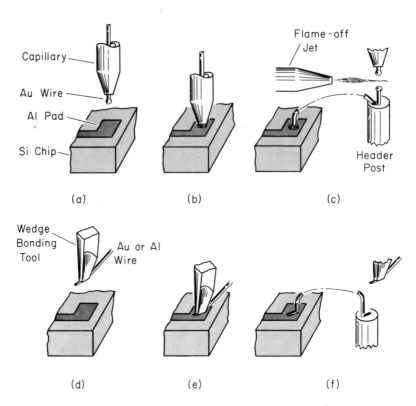

FIGURE 10-10. Wire bonding techniques: (a) capillary positioned over one of the contact pads for a ball (nail-head) bond; (b) pressure exerted to bond the wire to the pad; (c) post bond and flame-off; (d) wedge bonding tool; (e) pressure and ultrasonic energy applied; (f) post bond completed and wire broken or cut for next bond.

There are many variations in this basic method. For example, the substrate heating can be eliminated by *ultrasonic bonding*. In this method a tungsten carbide capillary is held by a tool connected to an ultrasonic transducer. When it is in contact with a pad or a post, the wire is vibrated under pressure to form a bond. Other variations include techniques for automatically removing the "tail" which is left on the post in Fig. 10-10c. When the bond to the chip is made by exerting pressure on a ball at the end of the Au wire, it is called a *ball bond* or a *nail-head bond*, because of the shape of the deformed ball after the bond is made.

Aluminum wire can be used in ultrasonic bonding; it has several advantages over Au, including the absence of possible metallurgical problems in bonds between Au and Al pads. When Al wire is used, the flame-off step is replaced by cutting or breaking the wire at appropriate points in the process. In forming a bond, the wire is bent under the edge of a wedge-shaped bonding tool (Fig. 10-10d). The tool then applies pressure and ultrasonic vibration, forming the bond (Fig. 10-10e and f). The resulting flat bond, formed by the bent wire wedged between the tool and the bonding surface, is called a *wedge bond*.

Whether thermocompression or ultrasonic techniques are used, the fine wires can be bonded fairly rapidly under a microscope by an experienced operator of the lead bonder. However, this method suffers from the fact that each connection is made separately, and labor costs escalate quickly in circuits with many terminals. A further disadvantage of wire bonding is that the chip must be firmly mounted on a header or substrate, and the contacted circuit must be packaged carefully to protect the fragile wires (Fig. 10-11). Finally, the heating and/or pressure involved in wire bonding can be undesirable for some IC chips. All these disadvantages suggest the use of other bonding techniques whenever possible. Two methods in common use are discussed below.

10.3.2 Flip–Chip and Beam–Lead Techniques.

The disadvantages of bonding wires individually to each pad on the chip can be overcome by several methods of simultaneous bonding. The *flip–chip* and the *beam–lead* approaches are typical of these methods. In each case, relatively thick metal is deposited on the contact pads before the devices are separated from the wafer. After separation, the deposited metal is used to contact a matching metallized pattern on the substrate.

In the flip–chip method, "bumps" of solder or special metal alloys are deposited on each contact pad. These metal bumps rise about two mils above the surface of the monolithic chip (Fig. 10-12). After separation from the wafer, each chip is turned upside down, and the bumps are properly aligned with the metallization pattern on the substrate. At this point, ultrasonic bonding or solder alloying attaches each bump to its corresponding connector on the substrate. An obvious advantage of this method is that all connections are made simultaneously. Disadvantages include the fact that the bonds are made under the chip and therefore cannot be inspected visually. Furthermore, it is necessary to heat and/or exert pressure on the chip.

In beam–lead technology, bonds to the substrate pattern are made external to the Si chip. The process is more complicated than flip–chip methods, but there are important compensations. Basically, the beam–lead technique calls for thick (about ten microns) metal tabs on the wafer, leading away from the circuit at each contact pad. These relatively large ("beam") leads are commonly made by electroplating Au onto the wafer in regions defined by a photoresist pattern. The beam leads extend from the metallized

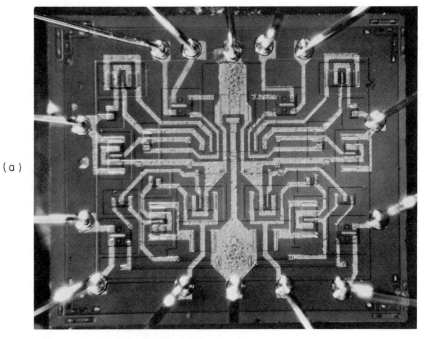

(a)

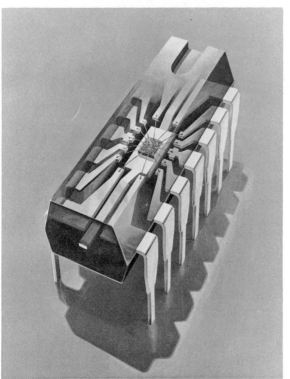

FIGURE 10-11. Illustration of Au wire ball bonding: (a) Au wires bonded to Al pads on a monolithic circuit; (b) drawing of the circuit after encapsulation. Illustrations courtesy of Texas Instruments, Incorporated.

(b)

FIGURE 10-12. Portion of a wafer of flip–chip circuits. After separation, each circuit is approximately sixty mils square. The metal bumps on this wafer are plated Ag, about two mils high. Photograph courtesy of Delco Electronics Division, General Motors Corporation.

interconnection pattern to outside the active area of each circuit (Fig. 10-13a). Then the wafer is mounted face down in wax on a flat disk and is lapped from the back side to a total wafer thickness of about fifty microns. A photoresist pattern is used on the back surface to delineate the circuits, and the Si is then etched away from between the individual chips.† After the Si has been removed in a rectangular pattern, the beam leads are left protruding from each circuit. Then the circuit can be mounted face down on the substrate, and the protruding beam leads can be bonded simultaneously to the substrate metallization pattern (Fig. 10-13b). Alternatively, the chip can be placed face up into a cavity in the substrate, with the beam leads over the

---

†To align a contact mask on the back side of a Si wafer with circuits on the front side, infrared light ($hv < E_g$) is shone through the wafer from the front and is viewed with a microscope fitted with an infrared-to-visible converter tube.

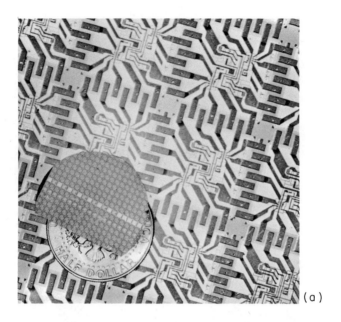

(a)

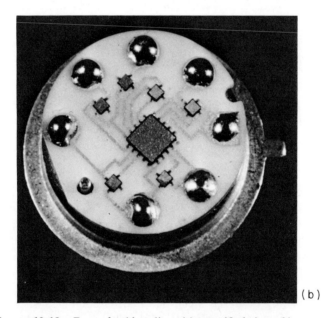

(b)

FIGURE 10-13. Beam–lead bonding: (a) magnified view of beam–lead circuits compared with a wafer of the same circuits and a coin; (b) beam–lead circuits mounted on a ceramic substrate with a metallized interconnection pattern. Photograph (a) courtesy of Bell Telephone Laboratories, Incorporated; (b) courtesy of Texas Instruments, Incorporated.

cavity edge. In either case, all bonds to the connector pattern can be made simultaneously without heating or exerting pressure on the chip itself.

In electroplating Au on the surface of the wafer it is necessary to use interspersing layers of other metals to ensure a good bond. In one common method titanium and platinum are used between the Si (and the oxide or nitride layers) and the Au. In the example of Fig. 10-14, contact to the Si in the window is made in several steps: (1) A thin layer of platinum is sputtered onto the surface, and the wafer is heated to form a platinum silicide layer, which is a good ohmic contact to the Si; (2) successive layers of titanium and platinum are deposited; (3) the interconnection pattern is defined in the platinum, and a thin layer of Au is electroplated on the pattern by masking the rest of the wafer with photoresist; (4) using a new photoresist pattern, the thick beam leads are electroplated on.

Although the beam–lead process requires many extra fabrication steps, the accessibility of the leads and other advantages make it worthwhile in many applications. In addition to the advantages of one-step bonding, beam–lead technology provides excellent isolation between devices. By

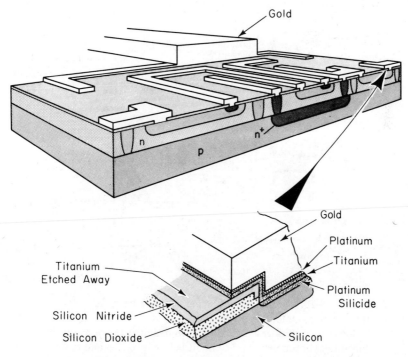

FIGURE 10-14. Gold metallization for beam–lead circuits. From S. S. Hause and R. A. Whitner, "Manufacturing Beam–Lead Sealed-Junction Monolithic Integrated Circuits," *The Western Electric Engineer*, vol. 11, no. 4, pp. 2–15, December 1967.

(a)

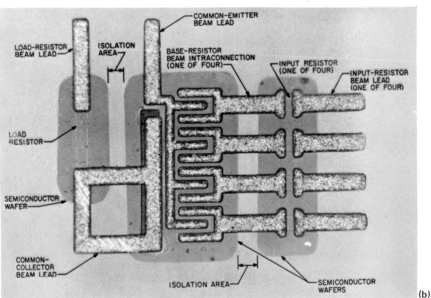

(b)

Figure 10-15. Circuit element separation in a beam–lead device: (a) circuit mounted on a ceramic substrate after separation by etching; the beam leads are alloyed to metallized regions on the substrate; (b) top view of the circuit before inverting and mounting; this view shows the Au beam leads and illustrates the separation of the transistor section from the resistor sections. Photographs courtesy of Bell Telephone Laboratories, Incorporated.

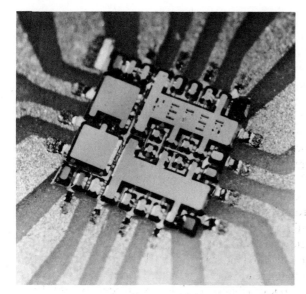

FIGURE 10-16. Example of air isolation in beam–lead circuits. This view of the inverted and mounted structure illustrates the precision of preferential etching in separating circuit elements. This beam–lead logic circuit contains 6 transistors, 48 diodes, and 18 resistors in an area 38 mils square. After etching, the circuit consists of 24 individual chips; the distance between adjacent chips is 0.2 mil. Photograph courtesy of Bell Telephone Laboratories, Incorporated.

using beam leads within individual circuits, transistors and other circuit elements can be separated by etching from the back side (Fig. 10-15). Utilizing the property of preferential etching along certain crystallographic directions, air gaps as small as 0.2 mil can be etched between circuit elements (Fig. 10-16). This procedure results in isolation between elements typical of discrete components, while maintaining batch fabrication and other advantages of IC's.

10.3.3 Packaging. The final step in IC fabrication is packaging the device in a suitable medium which can protect it from the environment of its intended application. In most cases this means the surface of the device must be isolated from moisture and contaminants and the bonds and other elements must be protected from corrosion and mechanical shock. The problems of surface protection are greatly minimized by modern passivation techniques, but it is still necessary to provide some protection in the package. In every case, the choice of package type must be made within the requirements of the application and cost considerations. There are many techniques for encapsulating devices, and the various methods are constantly refined and changed. Here we shall consider just a few general methods for the purpose of illustration.

In the early days of IC technology, all devices were packaged in metal headers. In this method the device is alloyed to the surface of the header, wire bonds are made to the header posts, and a metal lid is welded over the device and wiring. In the case of a hybrid circuit, a ceramic disk with proper metallization is mounted on the header and is contacted to the posts and to the circuit elements (Fig. 10-2). Although this method has several drawbacks, it does provide complete sealing of the unit from the outside environment. This is often called a *hermetically sealed* device. After the chip is mounted on the header and bonds are made to the posts, the header cap can be welded shut in a controlled environment (e.g., an inert gas), which maintains the device in a prescribed atmosphere.

The excessive processing steps and the bulky package resulting from header mounting can be avoided by several techniques, one of which is *plastic encapsulation*. A passivated device can be mounted on an appropriate substrate with projecting leads, and an epoxy, silicone, or other plastic material can be molded around the entire unit (Fig. 10-11b). This method is well suited for assembly-line processing and is less expensive than header mounting. Since the plastic encapsulation provides mechanical support, flip–chip or beam–lead circuits need not be mounted on a substrate but can be placed directly on a metal lead pattern instead. Such a metal pattern can be stamped out of a continuous ribbon, leaving a metal edge to hold the pattern together; then this edge can be trimmed off after plastic encapsulation, to leave the leads protruding from the package.

A combination of sealed junction passivation and beam–lead technology (Fig. 10-14) results in a device which may require little further encapsulation. Such a device can be mounted on a ceramic substrate as one "functional block" of a larger system, much of which may be attached to the same substrate. This is an excellent illustration of the assertion made earlier that IC design overlaps the design of the overall electronic system. A further illustration of this view is provided in the following section.

10.4 Large-Scale Integration

As the use of IC's grew and the technology became more sophisticated, it was inevitable that integration should expand to include larger and more complicated circuits. For example, the success of building several computer logic gates on a monolithic chip leads naturally to the following questions: "Why separate the chips for interconnection on a substrate? Why not leave the circuits on the wafer and interconnect them by appropriate metallization?" These questions are now being answered by the technology of *large-scale integration* (*LSI*).

There is no exact definition of LSI; it can include separated chips containing many circuits, or it can be a complete Si wafer of interconnected circuits. As a general rule for computer circuits, LSI refers to a chip or wafer

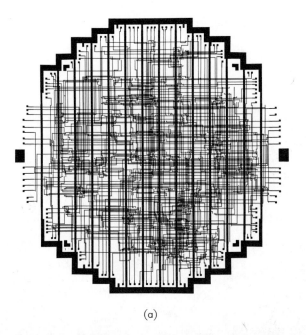

(a)

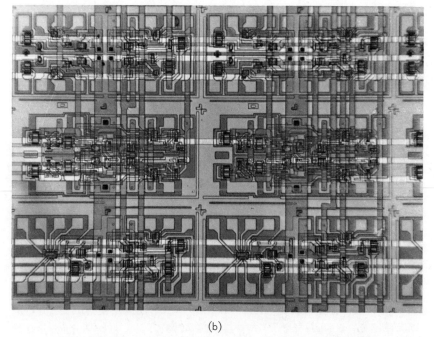

(b)

FIGURE 10-17. Example of an LSI computer memory module:
(a) computer-generated discretionary wiring pattern; (b) closeup
view of individual integrated circuits on the wafer, interconnected by three layers of metallization;   (Continued)

Figure 10-17 (Continued). (c) LSI wafer mounted in a ceramic
package with protruding leads; (Continued)

FIGURE **10-17** (Continued). (d) computer memory bank consisting of 34 LSI wafer modules compared with a similar memory
bank utilizing conventional integrated circuits. Photographs
courtesy of Texas Instruments, Incorporated.

with more than 100 properly interconnected logic gates. The element density in these circuits can be more than 50,000 transistors, diodes, and resistors per square inch. As an example of LSI technology, we shall discuss the basic methods for incorporating the circuits on a Si wafer into a functional system.

An obvious problem in LSI fabrication is the complex pattern of interconnections required. For example, if several hundred logic circuits are to be interconnected on a wafer, many crossings will be required in the metallization pattern. Even if a computer is used to determine the optimum interconnection, crossings will be inevitable. This problem can be solved by using *multilayer metallization.* This involves laying down the interconnection pattern in several steps, with insulating layers in between. An important difficulty arises, however, from the presence of faulty circuits on the wafer. In the usual IC, this difficulty is overcome simply by testing individual circuits and rejecting the faulty ones. If a predetermined interconnection pattern were used on an entire wafer, however, faulty circuits would automatically be "wired" into the system. Therefore, some process of *discretionary wiring* must be used, in which only the good circuits are interconnected. It is clear that this task cannot be approached without the aid of a computer. Basically, discretionary wiring requires building more circuits on the wafer than are needed in the completed system, testing each circuit with a multiple-point probe, and designing a special interconnection pattern for each wafer of circuits.

In one method, a discretionary wiring pattern is generated by computer and is displayed on a special precision CRT screen. From this pattern a photographic mask is prepared for the metallization (Fig. 10-17a). After several layers of metallization, the acceptable circuits on the wafer are interconnected to perform the desired operation (Fig. 10-17b). The resulting LSI wafer is then mounted in a package (Fig. 10-17c) and encapsulated to serve as a functional module in a larger system (Fig. 10-17d). This technology is particularly useful in computer systems, in which a collection of LSI modules can serve as the memory.†

READING LIST

W. C. HITTINGER and M. SPARKS, "Microelectronics," *Scientific American,* vol. 213, no. 5, pp. 56–70, November 1965.

---

†Bipolar and FET integrated circuit modules compete favorably with more traditional computer elements, such as magnetic cores. Recently, these relatively new IC techniques have been challenged by even newer developments, such as charge-coupled devices (p. 300) and *magnetic bubbles.* For an introduction to magnetic bubbles, see A. H. Bobeck and H. E. D. Scovil, *Scientific American,* Vol. 224, No. 6, pp. 78-90, June 1971.

F. G. HEATH, "Large-Scale Integration in Electronics," *Scientific American*, vol. 222, no. 2, pp. 22–31, February 1970.

Special issue on integrated circuits, *The Western Electric Engineer*, vol. 11, no. 4, December 1967. This issue contains excellent articles on the fabrication and assembly of beam–lead sealed junction devices, mask manufacture, thin-film technology, and the economics of IC fabrication.

Special feature, "Dielectric Isolation," *Electronics*, vol. 40, no. 6, pp. 91–108, 20 March 1967.

H. JOHNSON, "The Anatomy of Integrated–Circuit Technology," *IEEE Spectrum*, vol. 7, no. 2, pp. 56–66, February 1970.

J. J. SURAN, "A Perspective on Integrated Electronics," *IEEE Spectrum*, vol. 7, no. 1, pp. 67–79, January 1970.

H. H. HACKMAN, "The Art of Building LSI's," *IEEE Spectrum*, vol. 6, no. 9, pp. 29–36, September 1969.

R. M. WARNER, JR., "Comparing MOS and Bipolar Integrated Circuits," *IEEE Spectrum*, vol. 4, no. 6, pp. 50–58, June 1967.

Special issues of the *IEEE Journal of Solid State Circuits:* "Large Scale Integration," vol. SC-2, no. 4, December 1967; "Microwave Integrated Circuits," vol. SC-3, no. 2, June 1968; "Linear Integrated Circuits," vol. SC-3, no. 4, December 1968; "Technology for Integrated-Circuit Design," vol. SC-5, no. 1, February 1970.

Special issues of the *Proceedings of the IEEE:* "Integrated Electronics," vol. 52, no. 12, December 1964; "Materials and Processes in Integrated Electronics," vol. 57, no. 9, September 1969.

Special issue on device photolithography, *Bell System Technical Journal*, vol. 49, no. 9, pp. 1955–2220, November 1970.

R. M. BURGER and R. P. DONOVAN, *Fundamentals of Silicon Integrated Device Technology*, vols. 1 and 2. Englewood Cliffs, N.J.: Prentice-Hall, Inc., 1967.

S. K. GHANDHI, *The Theory and Practice of Microelectronics*. New York: John Wiley & Sons, Inc., 1968.

PROBLEMS

10.1 (a) Assume boron is diffused into a uniform n-type Si sample, resulting in a net doping profile $N_a(x) - N_d$. Set up an expression relating the sheet resistance of the diffused layer to the acceptor profile $N_a(x)$ and the junction depth x_j. Assume $N_a(x)$ is much greater than the background doping N_d over most of the diffused layer and that the hole mobility is constant. (Continued)

(Continued 10.1)

    (b) A typical sheet resistance of a base diffusion layer is 100 Ω/square. What should be the aspect ratio of a 1 kΩ resistor, using this diffusion?

10.2 A 10-μm n-type epitaxial layer ($N_d = 10^{16}$ cm^{-3}) is grown on a p-type Si substrate. Areas of the n layer are to be junction isolated by a boron diffusion at 1200°C ($D = 2.5 \times 10^{-12}$ cm^2/sec). The surface boron concentration is held constant at 5×10^{20} cm^{-3} (see Prob. 5.22).

    (a) What time is required for this isolation diffusion?

    (b) How far does an Sb-doped buried layer ($D = 2 \times 10^{-13}$ cm^2/sec) diffuse into the epitaxial layer during this time, assuming the concentration at the substrate–epitaxial boundary is constant at at 10^{20} cm^{-3}?

10.3 In Section 10.2.1 ten steps are listed for the fabrication of a double-diffused transistor. Make a list of this type for the double-diffused epitaxial transistor illustrated in Fig. 10-5d, including metallization.

10.4 Repeat Prob. 10.3 for the beam–lead transistor of Fig. 10–14.

10.5 We wish to find the overall values of α and β for the composite p-n-p transistor of Fig. 10-5g. Assume the terminal currents of the lateral p-n-p device are related by α_1 and β_1 (neglect saturation currents); similarly, let α_2 and β_2 describe the n-p-n.

    (a) Find α and β for the composite p-n-p.

    (b) Show that when $\alpha_2 = 1$, the overall $\alpha = 1$ and $\beta = \beta_1 \beta_2$.

10.6 Figure P10-6 shows the metallization pattern and a cross section of a monolithic circuit consisting of a transistor and a resistor. Each element is junction isolated.

    (a) Identify the various regions in the two drawings.

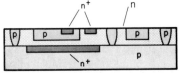

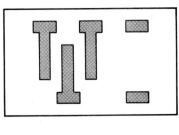

Figure P10-6

(Continued 10.6)

    (b) Draw and identify a set of six contact masks required in the fabrication of this device. You may draw the masks for a single circuit instead of for an array.

10.7 The circuit shown in Fig. P10-7 consists of a p-n junction diode, an n-p-n double-diffused epitaxial transistor (with buried layer), and three diffused resistors. Each element is junction isolated.

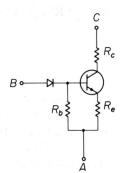

FIGURE **P10-7**

    (a) Sketch the processing steps in fabricating this circuit, using Fig. 10-5 (a–d) as an example; you may omit oxide layers and metallization in these illustrations. In forming a layout of the circuit, show a cross section through the transistor flanked by the diode on one side and the collector resistor R_c on the other side; the other two resistors can be placed conveniently.

    (b) Show a final metallization pattern, including contact pads for points A, B, and C.

SWITCHING DEVICES **11**

One of the most common applications of electronic devices is in switching, which requires the device to change from an "off" or *blocking* state to an "on" or *conducting* state. We have discussed the use of transistors in this application, in which base current drives the device from cutoff to saturation. Similarly, diodes and other devices can be used to serve as certain types of switches. There are a number of important switching applications which require that a device remain in the blocking state under forward bias until switched to the conducting state by an external signal. Several devices which fulfill this requirement have been developed, and we shall discuss two of the most important ones in this chapter: the *unijunction transistor* (*UJT*) and the *semiconductor controlled rectifier* (*SCR*). These devices are typified by a high impedance ("off" condition) under forward bias until a switching signal is applied; after switching they exhibit low impedance ("on" condition). In each case the signal required for switching can be varied externally. Therefore, each of these devices can be used to block or pass currents at predetermined levels. In this chapter we shall discuss the physical operation of these two devices, as well as certain other switching devices related to the SCR.

11.1 Unijunction Transistor

In this section we consider a switching device which operates by *conductivity modulation*. The unijunction transistor switches from a blocking state to a conducting state when the applied voltage reaches a critical value; furthermore, this critical switching voltage can be altered by a second bias voltage. This device is useful in various switching applications, including timing circuits, and in triggering other devices.

The unijunction transistor is also known as a *filamentary transistor* and a *double-base diode.* First we shall discuss the operation of this device in its simplest geometry and then include several variations on the basic structure.

11.1.1 Basis of Operation.

In its simplest form the UJT is a bar of high-resistivity semiconductor with ohmic contacts on each end and a junction placed at an intermediate position along its length (Fig. 11-1a). For

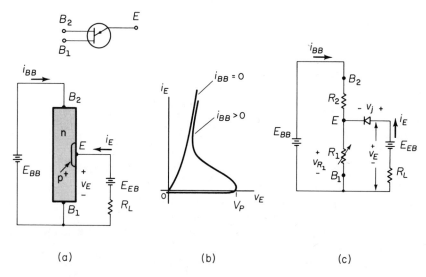

(a)   (b)   (c)

FIGURE 11-1. Schematic representation of a UJT: (a) UJT biasing circuit; (b) *I-V* characteristics with B_2 open and also with i_{BB} present; (c) equivalent circuit.

illustration we shall discuss an n-type bar with a p^+ contact. The p^+ region is called the emitter (*E*), and the two ends of the n region (B_1 and B_2) are base contacts. In effect, the device is a p^+-n junction in which there are two n-type base regions (thus the alternate name "double-base diode"). If the p^+ emitter region is forward biased with respect to the n material, hole injection occurs. For example, if B_2 is left open and a forward bias is applied between *E* and B_1, the *I-V* characteristic of a normal diode results (Fig. 11-1b, $i_{BB} = 0$). There is some slope to the forward characteristic because of series resistance in the n region between *E* and B_1, but in general, the curve is similar to those we have considered for diodes.

The situation is changed considerably if a current i_{BB} is passed through the high-resistivity n material. For small values of v_E, the diode cannot turn on because of the voltage $v_{R_1} = i_{BB}R_1$ across the lower half of the base region (Fig. 11-1c); the voltage v_E must exceed v_{R_1} before the junction can inject minority carriers into the bar. For simplicity of illustration we have divided the n region into two sections at the emitter contact in the equivalent circuit of Fig. 11-1c; we approximate the equivalent resistance of each section by R_1 and R_2. The voltage drop in the n material between E and B_1 is described by the voltage divider formula

$$(11\text{-}1) \qquad\qquad v_{R_1} = E_{BB}\frac{R_1}{R_1 + R_2}, \qquad \text{for } i_E = 0$$

The voltage v_{R_1} tends to reverse bias the junction; as long as v_{R_1} is greater than v_E, the junction voltage v_j is negative. Therefore, the current I_E is restricted to the small reverse saturation current of the junction. This condition persists until v_E has been increased to equal v_{R_1}. At this point v_j and i_E are zero. As v_E becomes slightly larger, the junction is forward biased and hole injection begins from p^+ into n. As the holes are injected into the base, they are swept by the electric field toward B_1. The presence of the excess holes reduces R_1 slightly which in turn reduces v_{R_1}. This is an example of conductivity modulation, since the conductivity of the material between E and B_1 increases as carriers are injected into it. This is obviously a regenerative process, since a smaller v_{R_1} results in a stronger forward bias v_j at the junction; more holes are injected, and v_{R_1} is reduced further. As a result, the junction current i_E increases, while the total voltage $v_E = v_j + v_{R_1}$ decreases. This produces a negative resistance region in the I–V characteristic as the device switches from the "off" state to the "on" state. Ultimately, i_E follows a curve near that of the $i_{BB} = 0$ case. There is a slight offset in voltage for large i_E, since a small component of v_{R_1} remains due to the product of i_{BB} and R_1' (the reduced resistance from E to B_1). After the junction is forward biased, the value of v_E is approximately

$$(11\text{-}2) \qquad\qquad v_E = E_{BB}\frac{R_1'}{R_2 + R_1'} + i_E R_1' + v_j$$

where v_j is the small voltage drop across the forward-biased junction and R_1' is small compared with R_2.

The voltage v_{R_1} before switching is controlled by the battery E_{BB}, according to Eq. (11-1). In turn, the critical value of v_E required to initiate the switching action (V_P) is determined by the magnitude of v_{R_1}. Therefore, the UJT is a switching device in which the required turn-on voltage V_P can be varied by changing an external voltage source E_{BB}. This property of the UJT switch is very useful in pulse circuits and in other applications. Another advantage of the UJT is that its switching properties are relatively independent of temperature. This results from the fact that the resistances of the upper and lower parts of the bar change together with temperature.

11.1.2 *UJT Geometries.* There are several ways of fabricating UJT devices, utilizing alloying or impurity diffusion. In one early technique, the two ohmic base contacts are formed by alloying a small bar of n-type Si to Au metallized regions on a ceramic substrate. The p^+ emitter contact is then formed by alloying an Al wire to the top surface (Fig. 11-2a). Alternatively, a Si chip can be Au-alloyed to a header to form the B_1 contact; then the B_2 and E contacts can be alloyed to the top surface of the chip with Au and Al wires, respectively. In these alloying processes, it is common to use Au containing a small percentage of Sb to form ohmic base contacts to the n-type Si.

In diffused UJT structures all three contacts usually appear at the top surface. This type of planar geometry offers the advantages of passivation, better design capability, and possible integration with other devices. In the

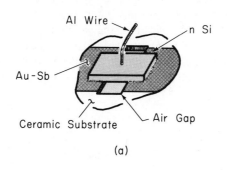

(a)

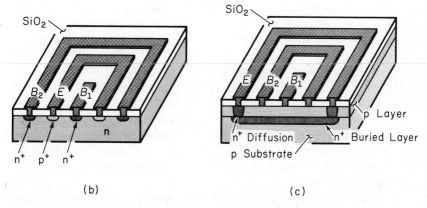

(b)   (c)

FIGURE 11-2. UJT geometries: (a) alloyed bar; (b) planar diffused device; (c) planar diffused-epitaxial device with junction isolation.

diffused planar structure of Fig. 11-2b the three contact regions appear on the surface of an n-type Si substrate. The current i_{BB} flows laterally from the B_2 contact to B_1, with E in between. This geometry does not provide for isolation; therefore, if several devices are to be integrated into a chip, a diffused epitaxial approach must be used. An example of a passivated, junction-isolated device is shown in Fig. 11-2c. In this geometry an n^+ buried layer in a p substrate serves as the emitter region. A p-type epitaxial layer is grown, and n diffusions through this epitaxial layer provide surface contact to the emitter and define the isolated device. The base contacts do not require diffusions; Al metallization defines B_1 and B_2 and provides ohmic contact to the p-layer.

The most critical parameters in UJT design are the *interbase resistance* $R_{BB} = R_1 + R_2$, the *intrinsic stand-off ratio* $\eta = R_1/R_{BB}$ [Eq. (11-1)], and the transit time and charge storage properties of the region between E and B_1. It is desirable that R_{BB} be fairly large so that i_{BB} can be small. The ratio η should be large to provide a turn-on voltage V_P which is a significant fraction of E_{BB} (Prob. 11.1). This implies that the junction E should be closer to B_2 than to B_1. On the other hand, switching time requirements may limit the distance from E to B_1. The injected hole transit time depends on this distance, and the storage delay time depends on the volume of the bar between E and B_1.

11.2 Four-Layer (p-n-p-n) Devices

One of the most important semiconductor devices in the field of power switching is the *semiconductor controlled rectifier (SCR)*.† This device is a four-layer (p-n-p-n) structure which effectively blocks current through two terminals until it is turned on by a small signal at a third terminal. There are many varieties of the basic p-n-p-n structure, and we shall not attempt to cover all of them; however, we can discuss the basic operation and physical mechanisms involved in these devices. We shall begin by investigating the current flow in a two-terminal p-n-p-n device and then extend the discussion to include triggering by a third terminal. We shall see that the p-n-p-n structure can be considered for many purposes as a combination of p-n-p and n-p-n transistors, and the analysis in Chapter 9 can be used as an aid in understanding its behavior.

The SCR is useful in many applications, such as in power switching and in various control circuits. This device can handle currents from a few

---

†Since Si is the material commonly used for this device, it is often called a *silicon controlled rectifier*.

milliamperes to hundreds of amperes. Since it can be turned on externally, the SCR can be used to regulate the amount of power delivered to a load simply by passing current only during selected portions of the line cycle. A common example of this application is the light-dimmer switch used in many homes. At a given setting of this switch, an SCR is turned on and off repetitively, such that all or only part of each power cycle is delivered to the lights. As a result, the light intensity can be varied continuously from full intensity to dark. The same control principle can be applied to motors, heaters, and many other systems. We shall discuss this type of application in this section, after first establishing the fundamentals of device operation.

11.2.1 The p-n-p-n Diode. First we consider a four-layer diode structure with an *anode* terminal A at the outside p region and a *cathode* terminal K at the outside n region (Fig. 11-3a). We shall refer to the junction nearest the anode as j_1, the center junction as j_2, and the junction nearest

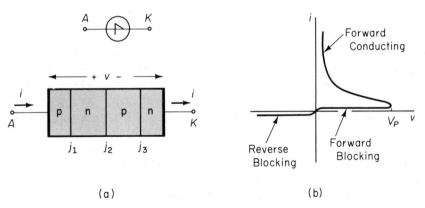

FIGURE 11-3. A two-terminal p-n-p-n device: (a) basic structure and common circuit symbol; (b) *I-V* characteristic.

the cathode as j_3. When the anode is biased positively with respect to the cathode (v positive), the device is forward biased. However, as the *I-V* characteristic of Fig. 11-3b indicates, the forward-biased condition of this diode can be considered in two separate states, the high-impedance or *forward-blocking* state and the low-impedance or *forward-conducting* state. In the device illustrated here the forward *I-V* characteristic switches from the blocking to the conducting states at a critical peak forward voltage V_P.

We can anticipate the discussion of conduction mechanisms to follow by noting that an initial positive voltage v places j_1 and j_3 under forward bias and the center junction j_2 under reverse bias. As v is increased, most of the forward voltage in the blocking state must appear across the reverse-biased junction j_2. After switching to the conducting state, the voltage from A to K is very small (less than 1 V), and we conclude that in this condition all three junctions must be forward biased. The mechanism by which j_2 switches from reverse bias to forward bias is the subject of much of the discussion to follow.

In the *reverse-blocking state* (v negative), j_1 and j_3 are reverse biased and j_2 is forward biased. Since the supply of electrons and holes to j_2 is restricted by the reverse-biased junctions on either side, the device current is limited to a small saturation current arising from thermal generation of EHP near j_1 and j_3. The current remains small in the reverse-blocking condition until avalanche breakdown occurs at a large reverse bias. In a properly designed device, with guards against surface breakdown (see Fig. 6-3), the reverse breakdown voltage can be several thousand volts.

We shall now consider the mechanism by which this device, often called a *Shockley diode*, switches from the forward-blocking state to the forward-conducting state.

The Two-Transistor Analogy:

The four-layer configuration of Fig. 11-3a suggests that the p-n-p-n diode can be considered as two coupled transistors: j_1 and j_2 form the emitter and collector junctions, respectively, of a p-n-p transistor; similarly, j_2 and j_3 form the collector and emitter junctions of an n-p-n (note the emitter of the n-p-n is on the right, which is the reverse of what we usually draw). In this analogy, the collector region of the n-p-n is in common with the base of the p-n-p, and the base of the n-p-n serves as the collector region of the p-n-p. The center junction j_2 serves as the collector junction for both transistors.

This two-transistor analogy is illustrated by Fig. 11-4. The collector current i_{C1} of the p-n-p transistor drives the base of the n-p-n, and the base current i_{B1} of the p-n-p is dictated by the collector current i_{C2} of the n-p-n. If we associate an emitter-to-collector current transfer ratio α with each transistor, we can use the analysis in Chapter 9 to solve for the current i. Using Eq. (9-30b) with $\alpha_1 \equiv \alpha_N$ for the p-n-p, $\alpha_2 \equiv \alpha_N$ for the n-p-n, and with I_{CO1} and I_{CO2} for the respective collector saturation currents, we have

(11-3a) $$i_{C1} = \alpha_1 i + I_{CO1} = i_{B2}$$
(11-3b) $$i_{C2} = \alpha_2 i + I_{CO2} = i_{B1}$$

But i_{B1} can also be written as $i - i_{C1}$ by a current summation in the p-n-p device

(11-4) $$i_{B1} = i - i_{C1} = i(1 - \alpha_1) - I_{CO1}$$

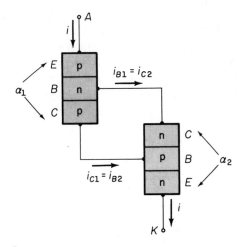

FIGURE 11-4. Two-transistor analogy of the p-n-p-n diode.

Equating (11-4) and (11-3b) we have

$$i(1 - \alpha_1) - I_{co1} = \alpha_2 i + I_{co2}$$

(11-5)
$$i = \frac{I_{co1} + I_{co2}}{1 - (\alpha_1 + \alpha_2)}$$

As Eq. (11-5) indicates, the current i through the device is small (approximately the combined collector saturation currents of the two equivalent transistors) as long as the sum $\alpha_1 + \alpha_2$ is small compared with unity. As the sum of the alphas approaches unity, the current i increases rapidly. The current does not increase without limit as Eq. (11-5) implies, however, because the derivation is not valid as $\alpha_1 + \alpha_2$ approaches unity. Since j_2 becomes forward biased in the forward-conducting region, both transistors become saturated after switching. The two transistors remain in saturation while the device is in the forward-conducting state, being held in saturation by the device current.

Variation of α with Injection:

Since the two-transistor analogy implies that switching involves an increase in the alphas to the point that $\alpha_1 + \alpha_2$ approaches unity, it may be helpful to review how alpha varies with injection for a transistor. The emitter-to-collector current transfer ratio α is given in Section 8.3.2 as the product of the emitter injection efficiency γ and the base transport factor B. An increase in α with injection can be caused by increases in either of these factors, or both. At very low currents (such as in the forward-blocking state of the p-n-p-n diode), γ is usually dominated by recombination in the transition region of the emitter junction (Section 9.3.4). As the current is increased, injection across the junction begins to dominate over recombination within the transition region (Section 5.6.2) and γ increases. There are several mechanisms by which the base transport factor B increases with injection, including the saturation of recombination centers as the excess carrier density becomes large. Whichever mechanism dominates, the increase in $\alpha_1 + \alpha_2$ required for switching of the p-n-p-n diode is automatically accomplished. In general, no special design is required to maintain $\alpha_1 + \alpha_2$ smaller than unity during the forward-blocking state; this requirement is usually met at low currents by the dominance of recombination within the transition regions of j_1 and j_3.

Forward-Blocking State:

When the device is biased in the forward-blocking state (Fig. 11-5a), the applied voltage v appears primarily across the reverse-biased junction j_2. Although j_1 and j_3 are forward biased, the current is small. The reason for this becomes clear if we consider the supply of electrons available to j_1 and holes to j_3. Focusing attention first upon j_1, let us assume a hole is injected from p_1 into n_1. If the hole recombines with an electron in n_1 (or in the j_1 transition region), that electron must be resupplied to the n_1 region to maintain space charge neutrality. The supply of electrons in this case is severely

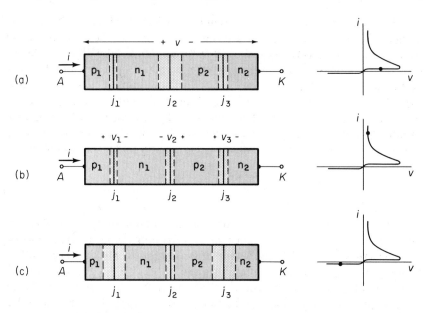

FIGURE 11-5. Three bias states of the p-n-p-n diode: (a) the forward-blocking state; (b) the forward-conducting state; (c) the reverse-blocking state.

restricted, however, by the fact that n_1 is terminated in j_2, a reverse-biased junction. In a normal p-n diode the n region is terminated in an ohmic contact, so that the supply of electrons required to match recombination (and injection into p) is unlimited. In this case, however, the electron supply is restricted essentially to those electrons generated thermally within a diffusion length of j_2. As a result, the current passing through the j_1 junction is approximately the same as the reverse saturation current of j_2. A similar argument holds for the current through j_3; holes required for injection into n_2 and to feed recombination in p_2 must originate in the saturation current of the center junction j_2. The applied voltage v divides appropriately among the three junctions to accommodate this small current throughout the device.

In this discussion we have tacitly assumed that the current crossing j_2 is strictly the thermally generated saturation current. This implies that electrons injected by the forward-biased junction j_3 do not diffuse across p_2 in any substantial numbers, to be swept across the reverse-biased junction into n_1 by transistor action. This is another way of saying that α_2 (for the "n-p-n transistor") is small. Similarly, the supply of holes to p_2 is primarily thermally generated, since few holes injected at j_1 reach j_2 without recombination (i.e., α_1 is small for the "p-n-p"). Now we can see physically why Eq. (11-5) implies a small current while $\alpha_1 + \alpha_2$ is small: Without the transport of charge provided by transistor action, the thermal generation of carriers is the only significant source of electrons to n_1 and holes to p_2.

Conducting State:

The charge transport mechanism changes dramatically when transistor action begins. As $\alpha_1 + \alpha_2$ approaches unity by one of the mechanisms described above, many holes injected at j_1 survive to be swept across j_2 into p_2. This helps to feed the recombination in p_2 and to support the injection of holes into n_2. Similarly, the transistor action of electrons injected at j_3 and collected at j_2 supplies electrons for n_1. Obviously, the current through the device can be much larger once this mechanism begins. The transfer of injected carriers across j_2 is regenerative, in that a greater supply of electrons to n_1 allows greater injection of holes at j_1 while maintaining space charge neutrality; this greater injection of holes further feeds p_2 by transistor action, and the process continues to repeat itself.

If $\alpha_1 + \alpha_2$ is large enough, so that many electrons are collected in n_1 and many holes are collected in p_2, the depletion region at j_2 begins to decrease. Finally the reverse bias disappears across j_2, in analogy with a transistor biased deep in saturation. When this occurs, the three small forward-bias voltages appear as shown in Fig. 11-5b. Two of these voltages essentially cancel in the overall v, so that the forward voltage drop of the device from anode to cathode in the conducting state is not much greater than that of a single p-n junction. For Si this forward drop is less than 1 V, until ohmic losses become important at high current levels.

We have discussed the current transport mechanisms in the forward-blocking and forward-conducting states, but we have not indicated how switching is initiated from one state to the other. Basically, the requirement is that the carrier injection at j_1 and j_3 must somehow be increased so that significant transport of injected carriers across j_2 occurs. Once this transport begins, the regenerative nature of the process takes over and switching is completed.

Triggering Mechanisms:

There are several methods by which a p-n-p-n diode can be switched (or *triggered*) from the forward-blocking state to the forward-conducting state. For example, an increase in the device temperature can cause triggering, by sufficiently increasing the carrier generation rate and the carrier lifetimes (see Fig. 4-12). These effects cause a corresponding increase in device current and in the alphas discussed above. Similarly, optical excitation can be used to trigger a device by increasing the current through EHP generation.† The most common method of triggering a two-terminal p-n-p-n, however, is simply to raise the bias voltage to the peak value V_P. This type of *voltage triggering* results in a breakdown (or significant leakage) of the reverse-biased junction j_2; the accompanying increase in current provides the injection at j_1 and j_3 and transport required for switching to the conducting state. The breakdown mechanism commonly occurs by a combination of *base-width narrowing* and *avalanche multiplication*.

When carrier multiplication occurs in j_2, many electrons are swept into n_1 and holes into p_2. This process provides the majority carriers to these regions needed for increased injection by the emitter junctions. Because of transistor action, the full breakdown voltage of j_2 need not be reached. As we showed in Eq. (9-51), breakdown occurs in the collector junction of a transistor with $i_B = 0$ when $M\alpha = 1$. In the coupled transistor case of the p-n-p-n diode, breakdown occurs at j_2 when

$$(11\text{-}6) \qquad M_p\alpha_1 + M_n\alpha_2 = 1$$

where M_p is the hole multiplication factor and M_n is the multiplication factor for electrons.

As the bias v increases in the forward-blocking state, the depletion region about j_2 spreads to accommodate the increased reverse bias on the center junction. This spreading means that the neutral base regions on either side

---

†Four-layer devices which can be triggered by a pulse of light are useful in many optoelectronic systems. This type of device is often called a *light-activated SCR*, or *LASCR*.

(n_1 and p_2) become thinner. Since α_1 and α_2 increase as these base regions decrease, triggering can occur by the effect of base-width narrowing. A true punch-through of the base regions is seldom required, since moderate narrowing of these regions can increase the alphas enough to cause switching. Furthermore, breakdown may be the result of a combination of avalanche and base-width narrowing, along with possible leakage current through j_2 at high voltage. From Eq. (11-6) it is clear that with avalanche multiplication present, the sum $\alpha_1 + \alpha_2$ need not approach unity to initiate breakdown of j_2. Once breakdown begins, the increase of carriers in n_1 and p_2 drives the device to the forward-conducting state by the regenerative process of coupled transistor action. As switching proceeds, the reverse bias is lost across j_2 and the junction breakdown mechanisms are no longer active. Therefore, base narrowing and avalanche multiplication serve only to start the switching process.

If a forward-bias voltage is applied rapidly to the device, switching can occur by a mechanism commonly called *dv/dt triggering*. Basically, this type of triggering occurs as the depletion region of j_2 adjusts to accommodate the increasing voltage. As the depletion width of j_2 increases, electrons are removed from the n_1 side and holes are removed from the p_2 side of the junction. For a slow increase in voltage, the resulting flow of electrons toward j_1 and holes toward j_3 does not constitute a significant current. If dv/dt is large, however, the rate of charge removal from each side of j_2 can cause the current to increase significantly. In terms of the junction capacitance (C_{j_2}) of the reverse-biased junction, the transient current is given by

$$(11\text{-}7) \qquad i(t) = \frac{dC_{j_2}v_{j_2}}{dt} = C_{j_2}\frac{dv_{j_2}}{dt} + v_{j_2}\frac{dC_{j_2}}{dt}$$

where v_{j_2} is the instantaneous voltage across j_2. This type of current flow is often called *displacement current*. The rate of change of C_{j_2} must be included in calculating the current, since the capacitance varies with time as the depletion width changes.

The increase in current due to a rapid rise in voltage can cause switching well below the steady state triggering voltage V_P. Therefore, a *dv/dt* rating is usually specified along with V_P for p-n-p-n diodes. Obviously, *dv/dt* triggering can be a disadvantage in circuits subjected to unpredictable voltage transients.

The various triggering mechanisms discussed in this section apply to the two-terminal p-n-p-n diode. As we shall see in the following section, the semiconductor controlled rectifier is triggered by an external signal applied to a third terminal.

11.2.2 The Semiconductor Controlled Rectifier. The most important four-layer device in power circuit applications is the three-terminal

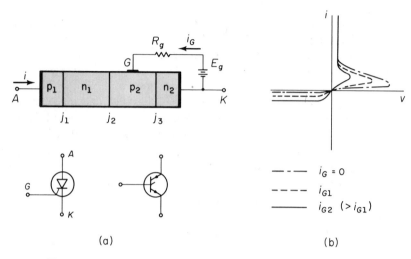

FIGURE 11-6. A semiconductor controlled rectifier: (a) four-layer geometry and common circuit symbols; (b) *I-V* characteristics.

SCR† (Fig. 11-6). This device is similar to the p-n-p-n diode, except that a third lead (*gate*) is attached to one of the base regions. When the SCR is biased in the forward-blocking state, a small current supplied to the gate can initiate switching to the conducting state. As a result, the anode switching voltage V_P decreases as the current i_G applied to the gate is increased (Fig. 11-6b). This type of turn-on control makes the SCR a useful and versatile device in switching and control circuits.

To visualize the *gate triggering* mechanism, let us assume the device is in the forward-blocking state, with a small saturation current flowing from anode to cathode. A positive gate current causes holes to flow from the gate into p_2, the base of the n-p-n transistor. This added supply of holes and the accompanying injection of electrons from n_2 into p_2 initiates transistor action in the n-p-n. After a transit time τ_{t2}, the electrons injected by j_3 arrive at the center junction and are swept into n_1, the base of the p-n-p. This causes an increase of hole injection by j_1, and these holes diffuse across the base n_1 in a transit time τ_{t1}. Thus, after a delay time of approximately $\tau_{t1} + \tau_{t2}$, transistor action is established across the entire p-n-p-n and the device is driven into the forward-conducting state. In most SCR's the delay time is less than a few microseconds, and the required gate current for turn-on is only a few

---

†This device is often called a *thyristor* to indicate its function as a solid state analogue of the *gas thyratron*. The thyratron is a gas-filled tube which passes current when an arc discharge occurs at a critical firing voltage. Analogous to the gate current control of the SCR, this firing voltage can be varied by a voltage applied to a third electrode.

milliamperes. Therefore, the SCR can be turned on by a very small amount of power in the gate circuit. On the other hand, the device current i can be many amperes, and the power controlled by the device may be very large.

It is not necessary to maintain the gate voltage once the SCR switches to the conducting state; in fact, the gate essentially loses control of the device after regenerative transistor action is initiated. For most devices a gate current pulse lasting a few microseconds is sufficient to ensure switching. Ratings of minimum gate pulse height and duration are generally provided for particular SCR devices.

Turn-off of the SCR from the conducting state to the blocking state can be accomplished by reducing the load current i below a critical value (called the *holding current*) required to maintain the $\alpha_1 + \alpha_2 = 1$ condition. In some SCR devices, gate turn-off can be used to reduce the alpha sum below unity. For example, if the gate voltage is reversed in Fig. 11-6, holes are extracted from the p_2 base region. If the rate of hole extraction by the gate is sufficient to remove the n-p-n transistor from saturation, the device turns off. However, there are often problems involving the lateral flow of current in p_2 to the gate; nonuniform biasing of j_3 can result from the fact that the bias on this emitter junction varies with position when a lateral current flows. Therefore, SCR devices must be specifically designed for turn-off control; at best, this turn-off capability can be utilized only over a limited range for a given device.

Some four-layer devices have two gate leads, one attached to n_1 and the other to p_2 (Fig. 11-7). This type of device is often called a *semiconductor controlled switch* (SCS). The availability of the second gate electrode provides additional flexibility in circuit design. The SCS biased in the forward-blocking state can be switched to the conducting state by a positive current pulse applied to the *cathode gate* (at p_2) or by a negative pulse at the *anode gate*

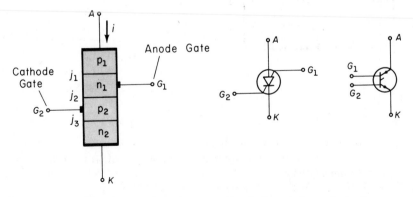

FIGURE 11-7. Semiconductor controlled switch: schematic configuration and common circuit symbols.

(n_1). If the device is designed for turn-off capability, separate circuits can be employed for turn-on at one gate and turn-off at the other. Other advantages of the SCS configuration include the possibility of minimizing unwanted dv/dt switching; for example, the gate not used for triggering can be capacitively coupled to the nearest current terminal (G_1 to A or G_2 to K) to allow a charging path for j_2 during a voltage transient without increasing i to the switching condition.†

11.2.3 Bilateral Devices. In many applications it is useful to employ devices which switch symmetrically with forward and reverse bias. This type of *bilateral* device is particularly useful in a-c circuits in which sinusoidal signals are switched on and off during positive and negative portions of the cycle. A typical bilateral p-n-p-n diode configuration is shown in Fig. 11-8a. This device differs from the p-n-p-n diode of Fig. 11-3 in that the n_2 region extends over only half the width of the cathode, and a new region n_3 is diffused into half of the anode region. In effect, this *bilateral diode switch* consists of two separate p-n-p-n diodes: the p_1-n_1-p_2-n_2 section and the p_2-n_1-p_1-n_3 section. We notice that the device shown in Fig. 11-8a is symmetrical. With the anode A biased positively with respect to the cathode K, junction j_1 is forward biased, while j_2 is reverse biased. Junction j_3 is shorted at

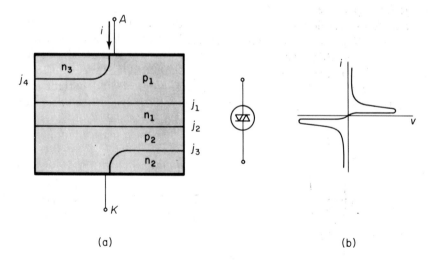

FIGURE 11-8. A bilateral diode switch: (a) schematic of the device configuration and a common circuit symbol; (b) typical *I-V* characteristic.

---

†Circuit configurations for these and other applications of the SCS can be found in the *G.E. Transistor Manual*, General Electric Co., Syracuse, N.Y.

one end (as is j_4) by the metal contact. When j_2 is biased to breakdown, however, a lateral current flows in p_2 biasing the left edge of j_3 into injection, and the device switches. During this operation junction j_4 remains dormant. Because the device is symmetrical, j_4 serves the shorted emitter function when the polarity is reversed (*K* positive with respect to *A*), and j_1 is the junction which is biased to breakdown in initiating the switching operation. If the bilateral diode is constructed properly, the forward and reverse characteristics are symmetrical as shown in Fig. 11-8b.

   Bilateral triode switches (sometimes called *triacs*) can be constructed with SCR characteristics which can be triggered in either the forward- or reverse-bias mode. A good discussion of these devices is presented in the book by Gentry et al. in the reading list for this chapter.

11.2.4 Fabrication and Applications. Many variations of diffusion, alloying, and epitaxial growth are used in the fabrication of p-n-p-n devices. The type of fabrication process depends largely on the power rating and intended use of the device. We can gain some insight into SCR fabrication by considering the example of Fig. 11-9a, which employs a combination

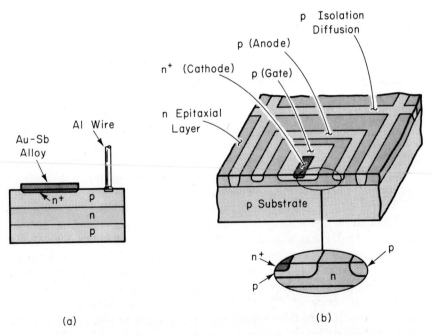

(a)   (b)

FIGURE 11-9. Typical SCR structures: (a) alloy-diffused power SCR; (b) planar diffused SCR (the oxide and metallization layers are omitted for clarity).

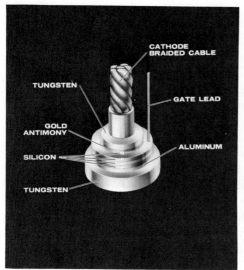
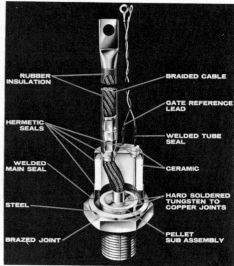

FIGURE **11-10.** A high-current SCR, fabricated by the alloy-diffused method of Fig. 11-9a: (a) mounting of the anode and cathode regions to tungsten disks; (b) cutaway view of the encapsulated device. Illustrations courtesy of General Electric Company, from *GE Silicon Controlled Rectifier Manual*, fourth edition, Syracuse, New York (1967).

of alloying and diffusion. In this example, a p-n-p structure is formed by diffusing p regions into both sides of an n-type Si wafer. The cathode is then formed by an n-type alloyed junction (using an alloy such as Au–Sb), and contact is made to the gate region by alloying an Al wire to the top p region. Contact is made to the anode terminal by alloying the anode p region to a metal substrate with an Al preform (Fig. 11-10a). For high-current devices the anode is attached to a heavy copper stud, and the cathode is contacted by a large cable. In high-current operation, heat is carried away from the junction by the massive metal substrate. In the device shown in Fig. 11-10, tungsten disks are soldered to the anode and cathode regions to take advantage of the similarity of the coefficient of thermal expansion between tungsten and Si. The entire device is hermetically sealed in a housing which provides protection from the atmosphere and from thermal and mechanical shock (Fig. 11-10b). Devices with this type of mounting can be rated at several hundred amperes in the conducting state. Of course, SCR devices intended for small-signal applications can be made in simpler and smaller packages.

Among the many other device configurations used in SCR fabrication, the planar structure of Fig. 11-9b illustrates a small-signal device which can be used in IC's. In this example the p regions can be obtained during the transistor base diffusion and the n-type cathode region can be made during the emitter diffusion of neighboring transistors.

Applications of SCR's and other four-layer devices are quite varied and extend into many fields of electronics, switching, and control. As a simple

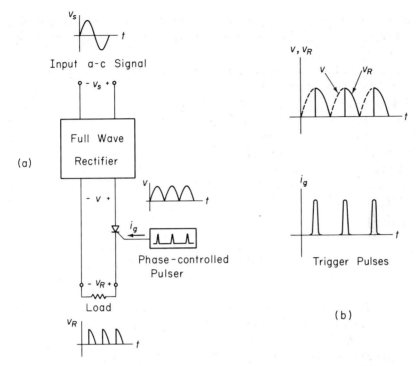

FIGURE 11-11. Example of the use of an SCR to control the power delivered to a load: (a) schematic diagram of the circuit; (b) waveforms of the delivered signal and the phase-variable trigger pulse.

example, let us consider the problem of delivering variable power to a load from a constant line source (Fig. 11-11). The load may be the heater windings of a furnace, a light bulb, or another circuit. The amount of power delivered to the load during each half-cycle depends on the switching of the SCR. If pulses are delivered to the gate near the beginning of each half-cycle, essentially the full power of the input is delivered to the load. On the other hand, if the trigger pulses are delayed, the SCR does not turn on until later in the half-cycle. As a result, the amount of power delivered to the load can be varied from almost full power to no power.

Many examples of SCR and other four-layer device applications can be found in the reading list of this chapter and in the current literature.

READING LIST

Special report: "The Unijunction Transistor," *Electronics*, vol. 38, no. 12, pp. 87–110, 14 June 1965.

L. S. Senhouse, Jr., "An Epitaxial Planar Structure for the Unijunction Transistor," *IEEE Transactions on Electron Devices*, vol. ED-16, pp. 161–165, February 1969.

Special issue on high-power semiconductor devices, *Proceedings of the IEEE*, vol. 55, no. 8, August 1967.

F. E. Gentry, F. W. Gutzwiller, N. Holonyak, Jr., and E. E. von Zastrow, *Semiconductor Controlled Rectifiers: Principles and Application of p-n-p-n Devices*. Englewood Cliffs, N. J.: Prentice-Hall, Inc., 1964.

S. M. Sze, *Physics of Semiconductor Devices*. New York: John Wiley & Sons, Inc., 1969, pp. 310–340.

GE Silicon Controlled Rectifier Manual (4th ed.). Syracuse, N.Y.: General Electric Company, 1967.

PROBLEMS

11.1 (a) Write an expression for the switching voltage V_P for the unijunction transistor of Fig. 11-1 and expressions for the power delivered by E_{bb} before and after switching.

(b) Discuss the effects of junction location, base doping, and sample dimensions on these quantities. Is Si preferred over Ge for a UJT? Why?

11.2 (a) In the p-n-p-n diode (Fig. 11-5a), the junction j_3 is forward biased during the forward-blocking state. Why, then, does the forward bias provided by the gate-to-cathode voltage in Fig. 11-6 cause switching?

(b) Sketch the energy band diagrams for the p-n-p-n diode in equilibrium; in the forward-blocking state; and in the forward-conducting state.

(c) Sketch the excess minority carrier densities in regions n_1 and p_2 when the p-n-p-n diode is in the forward-conducting state.

11.3 Use schematic techniques such as those illustrated in Fig. 8-22 to describe the hole flow and electron flow in a p-n-p-n diode for the forward-blocking state and for the forward-conducting state. Explain the diagrams and be careful to define any new symbols (e.g., those representing EHP generation and recombination).

11.4 Using the coupled transistor model, rewrite Eqs. (11-3) to include avalanche multiplication in j_2, and show that Eq. (11-6) is valid for the p-n-p-n diode.

NEGATIVE CONDUCTANCE
MICROWAVE DEVICES

12

We have discussed a number of devices which are useful in microwave circuits, such as the step-recovery diode, varactor, and tunnel diode. We also studied specially designed high-frequency transistors which can provide amplification and other functions at the lower microwave frequencies. However, transit time and other effects limit the application of transistors beyond the 10^9-Hz range. Therefore, other devices are required to perform electronic functions such as amplification and d-c to microwave power conversion at higher frequencies.

Several important devices for high-frequency applications use the instabilities which occur in semiconductors. One interesting and useful instability is the *acoustoelectric effect*,[†] involving the high-frequency interaction between electrons and the lattice vibrations of certain crystals. There are other types of instabilities, including an important class which involves *negative conductance*. Here we shall concentrate on two of the most commonly used negative conductance devices: *IMPATT* diodes, which depend on a combination of impact ionization and transit time effects, and *Gunn* diodes, which depend on the transfer of electrons from a high-mobility state to a low-mobility state. Each is a two-terminal device which can be operated in a negative conductance mode to provide amplification or oscillation at microwave frequencies in a proper circuit.

We have discussed the use of negative resistance devices in amplification and oscillation in our study of the tunnel diode (Section 6.2.3). In the case of the tunnel diode, the *I–V* characteristic exhibits a clear negative resistance

---

[†]See, for example, R. Bray, "A Perspective on Acoustoelectric Instabilities," *IBM J. Res. Develop.*, vol. 13, pp. 487–493, September 1969. For a general review of high-frequency elastic waves, see R. M. White, "Surface Elastic Waves," *Proc. IEEE*, vol. 58, no. 8, pp. 1238–1276, August 1970.

region. However, the functions of amplification and oscillation can occur when the local a-c current density is properly out of phase with the local a-c electric field within the device. In this chapter we shall investigate how negative conductance arises and how it can be used in microwave applications.

12.1 Impact Avalanche Transit Time (IMPATT) Devices

In this section we describe a type of microwave negative conductance device which operates by a combination of avalanche multiplication and transit time effects. Diodes with simple p-n junction structure, or with variations on that structure, are reverse biased to avalanche breakdown, and an a-c voltage is superimposed on the d-c bias. The carriers generated by the avalanche process are swept through a drift region to the terminals of the device. We shall see that the a-c component of the resulting current can be approximately 180° out of phase with the applied voltage under proper conditions of bias and device configuration, giving rise to negative conductance and oscillation in a resonant circuit. IMPATT devices can convert d-c to microwave a-c signals with high efficiency and are very useful in the generation of microwave power for many applications.

12.1.1 The Read Diode. The original suggestion for a microwave device of the IMPATT type was made by W. T. Read and involved an n^+-p-i-p^+ structure such as that shown in Fig. 12-1. Although IMPATT operation can be obtained in simpler structures, the Read diode is best suited for illustration of the basic principles. The device consists essentially of two regions: (1) the n^+-p region at which avalanche multiplication occurs and (2) the i (essentially intrinsic) region through which generated holes must drift in moving to the p^+ contact. Similar devices can be built in the p^+-n-i-n^+ configuration, in which electrons resulting from avalanche multiplication drift through the i region.

Although detailed calculations of IMPATT operation are complicated and generally require computer solutions, the basic physical mechanism is simple. Essentially, the device operates in a negative conductance mode when the a-c component of current is negative over a portion of the cycle during which the a-c voltage is positive, and vice-versa. The negative conductance occurs because of two processes, causing the current to lag behind the voltage in time: (1) a delay due to the avalanche process and (2) a further

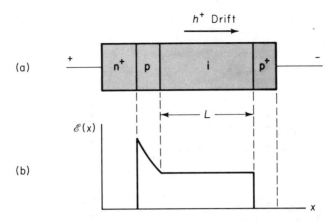

FIGURE 12-1. The Read diode: (a) basic device configuration; (b) electric field distribution in the device under reverse bias.

delay due to the transit time of the carriers across the drift region. If the sum of these delay times is approximately one-half cycle of the operating frequency, negative conductance occurs and the device can be used for oscillation and amplification.

From another point of view, the a-c conductance is negative if the a-c component of carrier flow drifts opposite to the influence of the a-c electric field. For example, with a d-c reverse bias on the device of Fig. 12-1, holes drift from left to right (in the direction of the field) as expected. Now, if we superimpose an a-c voltage such that $\mathscr{E}$ decreases during the negative half-cycle, we would normally expect the drift of holes to decrease also. However, in IMPATT operation the drift of holes through the i region actually increases while the a-c field is decreasing. To see how this happens, let us consider the effects of avalanche and drift for various points in the cycle of applied voltage (Fig. 12-2).

To simplify the discussion, we shall assume that the p region is very narrow and that all the avalanche multiplication takes place in a thin region near the n^+-p junction. We shall approximate the field in the narrow p region by a uniform value. If the d-c bias is such that the critical field for avalanche $\mathscr{E}_a$ is just met in the n^+-p space charge region (Fig. 12-2a), avalanche multiplication begins at $t = 0$. Electrons generated in the avalanche move to the n^+ region, and holes enter the i drift region. We assume the device is mounted in a resonant microwave circuit so that an a-c signal can be maintained at a given frequency. As the applied a-c voltage goes positive, more and more holes are generated in the avalanche region. In fact, the pulse of holes (dotted line) generated by the multiplication process continues to grow as long as the electric field is above $\mathscr{E}_a$ (Fig. 12-2b). It can be shown that the particle

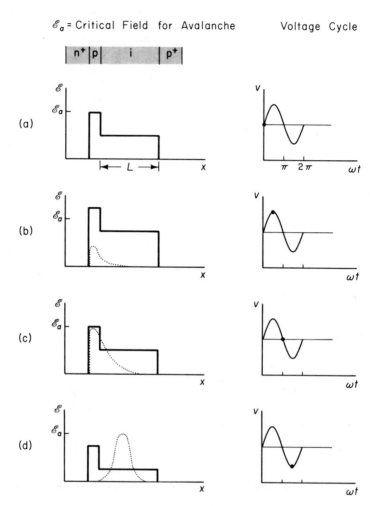

$\mathscr{E}_a$ = Critical Field for Avalanche   Voltage Cycle

FIGURE 12-2. Time dependence of the growth and drift of holes during a cycle of applied voltage for the Read diode: (a) $\omega t = 0$; (b) $\omega t = \pi/2$; (c) $\omega t = \pi$; (d) $\omega t = 3\pi/2$. The hole pulse is sketched as a dotted line on the field diagram.†

current due to avalanche increases exponentially with time while the field is above the critical value. The important result of this growth is that the hole pulse reaches its peak value not at $\pi/2$ when the voltage is maximum, but at π (Fig. 12-2c). Therefore, there is a phase delay of $\pi/2$ inherent in the avalanche process itself. A further delay is provided by the drift region. Once

---

†General features after De Loach, referenced in reading list.

the avalanche multiplication stops ($\omega t > \pi$), the pulse of holes simply drifts toward the p^+ contact (Fig. 12-2d). But during this period the a-c terminal voltage is negative. Therefore, the dynamic conductance is negative, and energy is supplied to the a-c field.

If the length of the drift region is chosen properly, the pulse of holes is collected at the p^+ contact just as the voltage cycle is completed, and the cycle then repeats itself. The pulse will drift through the length L of the i region during the negative half-cycle if we choose the transit time to be one-half the oscillation period

(12-1)
$$\frac{L}{v_d} = \frac{1}{2}\frac{1}{f}, \qquad f = \frac{v_d}{2L}$$

where f is the operating frequency and v_d is the drift velocity for holes.[†] Therefore, for a Read diode the optimum frequency is one-half the inverse transit time of holes across the drift region v_d/L. In choosing an appropriate resonant circuit for this device, the parameter L is critical. For example, taking $v_d = 10^7$ cm/sec for Si, the optimum operating frequency for a device with an i region length of 5 μm is $f = 10^7/2(5 \times 10^{-4}) = 10^{10}$ Hz. Negative resistance is exhibited by an IMPATT diode for frequencies somewhat above and below this optimum frequency for exact 180° phase delay. A careful analysis of the small-signal impedance shows that the minimum frequency for negative conductance varies as the square root of the d-c bias current for frequencies in the neighborhood of that described by Eq. (12-1).

12.1.2 Other IMPATT Structures.

Although the Read diode of Fig. 12-1 displays most directly the operation of IMPATT devices, simpler structures can be used, and in some cases they may be more efficient. Negative conductance can be obtained in simple p-n junctions or in p-i-n devices. In the case of the p-i-n, most of the applied voltage occurs across the i region, which serves as a uniform avalanche region and also as a drift region. Therefore, the two processes of delay due to avalanche and drift, which were separate in the case of the Read diode, are distributed within the i region of the p-i-n. This means that both electrons and holes participate in the avalanche and drift processes.

An important and highly efficient mode of operation is illustrated in Fig. 12-3. If a large current is suddenly applied to the sample, the displacement current $\epsilon \, d\mathscr{E}/dt$ can be so large that the point $\mathscr{E} > \mathscr{E}_a$ propagates through

---

[†]In general, v_d is a function of the local electric field. However, these devices are normally operated with fields in the i region sufficiently large that holes drift at their scattering limited velocity (Fig. 3-25). For this case the drift velocity does not vary appreciably with the a-c variations in the field.

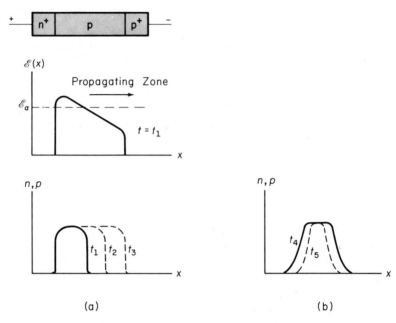

FIGURE 12-3. TRAPATT cycle: (a) electron-hole plasma created by a rapidly propagating field; (b) depletion of the plasma.

the device faster than the maximum carrier velocity. Therefore, an *avalanche zone* moves through the i region, filling it with an electrostatically neutral electron–hole population, or *plasma*. The plasma is created very rapidly, before the carriers have a chance to drift appreciably. The presence of these carriers in large numbers greatly reduces the terminal voltage across the diode. The recovery portion of the cycle resembles that of the step-recovery diode (Section 6.1.2). The electron–hole plasma collapses as holes drift to the right and electrons drift to the left. As the EHP plasma is depleted during the recovery transient, the terminal voltage builds up to a high value and the current goes toward a low value. In a proper resonant circuit, the cyclic buildup and discharge of the plasma gives rise to efficient microwave power generation. Since the electron–hole plasma is initiated during the transit of an avalanche zone, this type of operation is generally called a *TRAPATT* cycle, for *TRA*pped *P*lasma *A*valanche *T*riggered *T*ransit.

IMPATT and TRAPATT diodes are useful for generation of microwave power because of their small size, reliability, and other features typical of solid state devices. Furthermore, these devices can be built in well-established materials such as Si, Ge, or GaAs, using fairly standard fabrication techniques. One important drawback is that the noise inherent in impact ionization processes can interfere with the signal in some frequency ranges.

12.2 The Gunn Effect and Related Devices

Microwave devices which operate by the *transferred electron* mechanism are often called *Gunn diodes* after J. B. Gunn, who first demonstrated one of the forms of oscillation. As we shall see, there are many modes of operation for these devices. In the transferred electron mechanism, the conduction electrons of some semiconductors are shifted from a state of high mobility to a state of low mobility by the influence of a strong electric field. Negative conductance operation can be achieved in a diode† for which this mechanism applies, and the results are varied and useful in microwave circuits.

First, we shall describe the process of electron transfer and the resulting change of mobility. Then we shall consider some of the modes of operation for diodes using this mechanism.

12.2.1 The Transferred Electron Mechanism.
In Section 3.4.3 we discussed the nonlinearity of mobility at high electric fields. In most semiconductors the carriers reach a scattering limited velocity, and the velocity vs. field plot saturates at high fields (Fig. 3-25). In some materials, however, the energy of electrons can be raised by an applied field to the point that they transfer from one region of the conduction band to another, higher-energy region. For some band structures, negative conductivity can result from this electron transfer. To visualize this process, let us recall the discussion of energy bands in Section 3.1.4. The simplified band diagrams we usually draw are good approximations when the conduction electrons exist near the minimum energy of the conduction band. However, in the more complete band diagram, electron energy is plotted vs. the propagation vector **k**, as in Fig. 3-6. It can be shown that the **k** vector is proportional to the electron momentum in the vector direction (Prob. 3.2); therefore, energy bands such as those in Fig. 3-6 are said to be plotted in *momentum space.*

A simplified band diagram for GaAs is shown in Fig. 12-4 for reference; some of the detail has been omitted in this diagram to isolate the essential features of electron transfer between bands. In n-type GaAs the valence band is filled, and the *central valley* (or *minimum*) of the conduction band at $\mathbf{k} = 0$ normally contains the conduction electrons. There is a set of *subsidiary minima* (sometimes called *satellite valleys*) at higher energy,‡ but these

---

†These devices are called diodes, since they are two-terminal devices. No p-n junction is involved, however. Gunn effect and related devices utilize bulk instabilities, which do not require junctions.

‡We have shown only one satellite valley for convenience; there are other equivalent valleys for different directions in **k**-space. The effective mass ratio of 1.2 refers to the combined satellite valleys.

minima are many kT above the central valley and are normally unoccupied. Therefore, the direct band gap and the energy bands centered at $\mathbf{k} = 0$ are often used to describe the conduction processes in GaAs. This was true of our discussion of GaAs lasers in Section 7.4, for example. The presence of the satellite valleys is crucial to the Gunn effect, however. If the material is subjected to an electric field above some critical value (about 3000 V/cm), the electrons in the central valley of Fig. 12-4 gain more energy than the 0.36 eV separating the valleys; therefore, there is considerable scattering of electrons into the higher-energy satellite valley.

Once the electrons have gained enough energy from the field to be transferred into the higher-energy valley, they remain there as long as the field is greater than the critical value. The explanation for this involves the fact that the combined effective density of states for the upper valleys is much greater than for the central valley (by a factor of about 70). Although we shall not prove it here, it seems reasonable that the probability of electron scattering between valleys should depend on the density of states available in each case, and that scattering from a valley with many states into a valley with few states would be unlikely. As a result, once the field increases

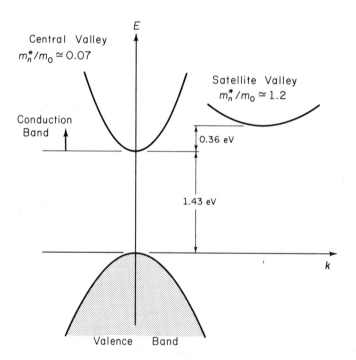

FIGURE 12-4. Simplified band diagram for GaAs, illustrating the lower and upper valleys in the conduction band.

above the critical value, most conduction electrons in GaAs reside in the satellite valleys and exhibit properties typical of that region of the conduction band. In particular, the effective mass for electrons in the higher valley is almost twenty times as great as in the central valley, and the electron mobility is much lower. This is an important result for the negative conductivity mechanism: As the electric field is increased, the electron velocity increases until a critical field is reached; then the electrons *slow down* with further increase in field. The electron transfer process allows electrons to gain energy at the expense of velocity over a range of values of the electric field. Taking current density as $qv_d n$, it is clear that current also drops in this range of increasing field, giving rise to a negative differential conductivity $dJ/d\mathscr{E}$.

A possible dependence of electron velocity vs. electric field for a material capable of electron transfer is shown in Fig. 12-5. For low values of field, the

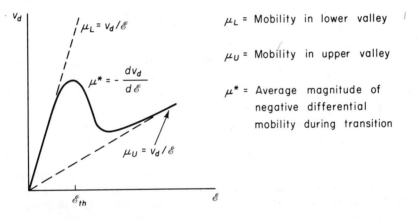

FIGURE 12-5. A possible characteristic of electron drift velocity vs. field for a semiconductor exhibiting the transferred electron mechanism.

electrons reside in the lower valley of the conduction band, and the mobility ($\mu_L = v_d/\mathscr{E}$) is high and constant with field. For high values of field, electrons transfer to the satellite valleys, where their velocity is smaller and their mobility lower. Between these two states is a region of negative slope on the v_d vs. $\mathscr{E}$ plot, indicating a negative differential mobility $dv_d/d\mathscr{E} = -\mu^*$.

The existence of a drop in mobility with increasing electric field and the resultant possibility of negative conductance were predicted by Ridley and Watkins and by Hilsum several years before Gunn demonstrated the effect in GaAs. The mechanism of electron transfer is therefore often called the Ridley–Watkins–Hilsum mechanism. This negative conductivity effect depends only on the bulk properties of the semiconductor and not on junction or surface effects. It is therefore called a *bulk negative differential conductivity* (*BNDC*) effect.

12.2.2 Formation and Drift of Space Charge Domains.

If a sample of GaAs is biased such that the field falls in the negative conductivity region, space charge instabilities result, and the device cannot be maintained in a d-c stable condition. To understand the formation of these instabilities, let us consider first the dissipation of space charge in the usual semiconductor. It can be shown from treatment of the continuity equation that a localized space charge dies out exponentially with time in a homogeneous sample with positive resistance (Prob. 12.2). If the initial space charge is Q_0, the instantaneous charge is

$$(12\text{-}2) \qquad\qquad Q(t) = Q_0 e^{-t/\tau_d}$$

where $\tau_d = \epsilon/\sigma$ is called the *dielectric relaxation time*. Because of this process, random fluctuations in carrier density are quickly neutralized, and space charge neutrality is a good approximation for most semiconductors in the usual range of conductivities. For example, the dielectric relaxation time for a 1.0 Ω-cm Si or GaAs sample is approximately 10^{-12} sec.

Equation (12-2) gives a rather remarkable result for cases in which the conductivity is negative. For these cases τ_d is negative also, and *space charge fluctuations build up* exponentially in time rather than dying out. This means that normal random fluctuations in carrier density can grow into large space charge regions in the sample. Let us see how this occurs in a GaAs sample biased in the negative conductivity regime. The velocity–field diagram for n-type GaAs is illustrated in Fig. 12-6a. If we assume a small shift of electron density in some region of the device, a dipole layer can form as shown in Fig. 12-6b. Under normal conditions this dipole would die out quickly. However, under conditions of negative conductivity the charge within the dipole, and therefore the local electric field, builds up as shown in Fig. 12-6c. Of course, this buildup takes place in a stream of electrons drifting from the cathode to the anode, and the dipole (now called a *domain*) drifts along with the stream as it grows. Eventually the drifting domain will reach the anode, where it gives up its energy as a pulse of current in the external circuit.

During the initial growth of the domain, an increasing fraction of the applied voltage appears across it, at the expense of voltage across the rest of the bar. As a result, it is unlikely that more than one domain will be present in the bar at a time; after the formation of one domain, the electric field in the rest of the bar quickly drops below the threshold value for negative conductivity. If the bias is d-c, the field outside the moving domain will stabilize at a positive conductivity point such as A in Fig. 12-6a, and the field in the domain will stabilize at the high-field value B.

Let us follow the motion of a single domain as illustrated by Fig. 12-6. A small dipole forms from a random noise fluctuation (or at a permanent *nucleation site* such as a crystal defect, a doping inhomogeneity, or the cathode itself), and this dipole grows and drifts down the bar as a domain. During the early stages of domain development, we can assume a uniform electric field in the bar, except just at the small dipole layer. If the field is in the negative mobility region, such as point C of Fig. 12-6a, the slightly higher field within the dipole results in a lower value of electron drift velocity inside the dipole than outside. As a result, electrons on the right (downstream) of the domain drift away, while electrons pile up on the left (upstream) side. This causes the accumulation and depletion layers of the domain to grow,

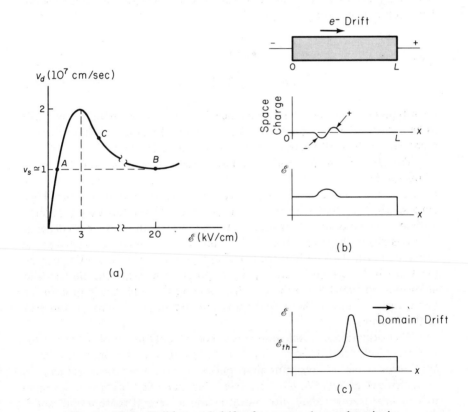

FIGURE 12-6. Buildup and drift of a space charge domain in GaAs: (a) velocity-field characteristic for n-type GaAs; (b) formation of a dipole; (c) growth and drift of a dipole for conditions of negative conductivity.

thereby further increasing the electric field in the domain. This is obviously a runaway process, in which the electric field within the domain grows while that outside the domain decreases. A stable condition is realized when the domain field increases to point B in Fig. 12-6a, and the field outside drops to point A. When this condition is met, the electrons drift at a constant velocity v_s everywhere, and the domain moves down the bar without further growth.

In this discussion we have assumed that the domain has time to grow to its stable condition before it drifts out of the bar. This is not always the case; for example, in a short bar with a low density of electrons, a dipole can drift the length of the bar before it develops into a domain. We can specify limits on the electron density n_0 and sample length L for successful domain formation by requiring the transit time (L/v_s) to be greater than the dielectric relaxation time (absolute value) in the negative mobility region. This requirement gives

(12-3)
$$\frac{L}{v_s} > \frac{\epsilon}{q\mu^* n_0}$$

$$Ln_0 > \frac{\epsilon v_s}{q\mu^*} \simeq 10^{12} \text{ cm}^{-2}$$

for n-type GaAs, where the average negative differential mobility† is taken to be $-100 \text{ cm}^2/\text{V-sec}$. Therefore, for successful domain formation there is a critical product of electron density and sample length.

The type of domain motion we have described here was the first mode of operation observed by Gunn. In the observation of current vs. voltage for a GaAs sample, Gunn found a linear ohmic relation up to a critical bias, beyond which the current came in sharp pulses. The pulses were separated in time by an amount proportional to the sample length. This length dependence was due to the transit time L/v_s required for a domain nucleated at the cathode to drift the length of the bar. Gunn performed an interesting experiment in which he used a tiny capacitive probe to measure the electric field at various positions down the bar. By scanning the field distribution in the bar at various times in the cycle, he was able to plot out the growth and drift of the domains.

The formation of stable domains is not the only mode of operation for transferred electron devices. Nor is it the most desirable mode for most applications, since the resulting short pulses of current are inefficient sources of microwave power. We have discussed this mode first because it is simple to visualize; now we shall turn our attention to several more useful modes of operation.

---

†This is a rather crude approximation, since μ^* is not a constant but varies considerably with field; the negative dielectric relaxation time therefore changes with time as the domain grows.

12.2.3 Modes of Operation in Resonant Circuits. The simple drift of stable domains discussed above is attainable in resistive-loaded circuits with d-c bias, but this mode is not really typical of microwave applications. Negative conductivity devices are usually operated in resonant circuits, such as high-Q resonant microwave cavities. Since the applied voltage varies in time, we can expect the device behavior to depend on both the amplitude and the frequency of the voltage variations.

Although we cannot consider all of the possible modes of operation here, we can discuss several of the most important. Figure 12-7 summarizes several cases for which the bias voltage varies in amplitude and frequency. The plot of electron velocity vs. field for GaAs is repeated in Fig. 12-7a as a reference. When the field rises above the threshold value $\mathscr{E}_{th}$, negative conductance exists until a stable domain is formed with $v_d = v_s$ in both the high- and low-field regions. Domains drift along the sample until they reach the anode or until the low-field value drops below the *sustaining field* $\mathscr{E}_s$ required to maintain v_s (Prob. 12.4). We note that creation of a domain requires a larger field outside the domain ($\mathscr{E}_{th}$) than that needed to maintain it ($\mathscr{E}_s$). As the bias is reduced, the domain width decreases since smaller voltages can be maintained across it. Finally, the domain collapses (is *quenched*) when the field outside drops below $\mathscr{E}_s$, and a new domain cannot nucleate until the field is raised above $\mathscr{E}_{th}$ again. This distinction between the threshold and the sustaining fields is important in several resonant modes.

For a steady bias such that the average field is above the threshold value $\mathscr{E}_{th}$, stable (Gunn) domains form and drift toward the anode (Fig. 12-7b). This is the mode discussed in the previous section; the current waveform is composed of a series of spikes which appear with a frequency equal to the inverse transit time v_s/L. Obviously, this mode is not appropriate for efficient conversion of d-c to microwave power; a resonant circuit mode is more desirable.

An a-c *transit time (Gunn)* mode of operation can be achieved in a resonant circuit if the oscillation period is chosen to be essentially equal to the transit time. Assuming the field never drops below the sustaining field $\mathscr{E}_s$, a domain can form and propagate to the anode during each cycle (Fig. 12-7c). Again, the efficiency is low since the current is collected only when the domain reaches the anode. The best efficiency for this mode is attained when the n_0L product is chosen such that the domain width equals about half the sample length; for this case the current is collected during most of the negative half-cycle of the voltage. In this mode the d-c to microwave conversion efficiency can be about ten per cent under ideal conditions.

Another method for increasing the efficiency is to collect the domain early in the negative half-cycle of the voltage and delay the formation of a new domain. For example, if the transit time is chosen such that the domain is collected while $\mathscr{E} < \mathscr{E}_{th}$ in Fig. 12-7d, a new domain cannot form until

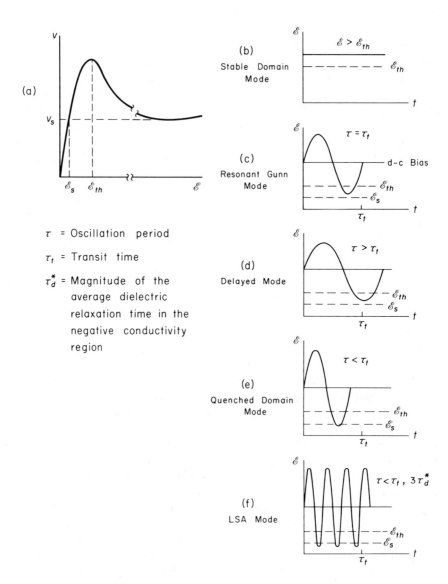

FIGURE 12-7. Modes of operation: (a) velocity-field characteristic for n-type GaAs, with threshold and sustaining fields for domains indicated; (b) d-c biasing for stable domain propagation; (c) bias at the transit time frequency for resonant Gunn mode; (d) delayed or inhibited mode; (e) quenched domain mode; (f) LSA mode.

the field rises above threshold again. During the portion of the negative half-cycle that no domain exists, there is an ohmic component of current higher than that flowing during the domain motion (Prob. 12.5). The efficiency of this *delayed domain mode* or *inhibited mode* is better than that of the simple transit time mode. If the bias drops below threshold early during the negative half-cycle and the domain is collected soon thereafter, the efficiency of the device can theoretically be about twenty per cent. A further advantage of this mode is that the oscillation frequency can be determined within certain limits by the resonant circuit rather than by the transit time.

If the bias field drops below $\mathscr{E}_s$ during the negative half-cycle (Fig. 12-7e), the domain collapses before it reaches the anode. In this *quenched domain mode* the operating frequency must be higher than the transit time frequency so that the domain can be quenched before it is collected. In this mode also the frequency can be designated by the choice of resonant cavity, within the required limits.

One of the most efficient and useful modes of operation does not involve the formation of domains. In the *limited space charge accumulation (LSA) mode*, the frequency is so high that domains have insufficient time to form while the field is above threshold (Fig. 12-7f). As a result, most of the sample is maintained in the negative conductance state during a large fraction of the voltage cycle. We recall this is not possible in the other modes, since most of the sample drops back to the positive conductance state once a domain forms. The LSA mode is actually the simplest mode of operation; since domains do not form, electrons drift from cathode to anode through a sample which displays negative conductance and an essentially uniform field during much of the cycle. In this mode the frequency, determined by the resonant circuit, is much higher than the transit time frequency. This is a distinct advantage in achieving efficient conversion of power in the upper microwave range. The primary requirement for this mode is that the frequency be high enough that domains have insufficient time to form while the signal is above threshold. However, it is also necessary that any accumulation of electrons near the cathode have time to collapse while the signal is below threshold. Therefore, the oscillation period τ should be no more than several times larger than the magnitude of the dielectric relaxation time in the negative conductance regime, τ_d^*, but must be much larger than the positive resistance low-field value τ_d. Although this condition seems restrictive, LSA diodes can be operated over a very wide range of frequencies by proper choice of doping density. The efficiency of the LSA mode can be as high as 20 per cent, and LSA diodes can be operated at very high frequencies.

There are other modes of operation, including an *amplification mode* in which negative conductance is utilized without domain formation. In sub-critically doped samples ($n_0 L \ll 10^{12}$), there are too few carriers for domain formation within the transit time. Therefore, amplification of signals near the transit time frequency can be accomplished.

To summarize the various modes of operation, we can construct an approximate mode diagram (Prob. 12.6) as shown in Fig. 12-8. In this diagram we plot the product of frequency and sample length fL vs. the product of doping and length n_0L for n-type GaAs. This type of figure helps in visualizing the rather complex interrelationships of resonant frequency, dielectric relaxation time, and transit time which give rise to the various modes. The first step in constructing such a diagram is to draw a vertical line at $n_0L = 10^{12}$, where the transit time equals the negative conductivity dielectric relaxation time τ_d^*. Domains are possible to the right of this line but not to the left. This is the first of two lines which separate regions of domain formation from regions with no domains. Next we can identify the fL product corresponding to the sustaining drift velocity ($v_s = 10^7$ cm/sec). Along this horizontal line the frequency equals the inverse transit time ($f = v_s/L$). In the neighborhood of this line we can expect Gunn domains and the amplification mode.

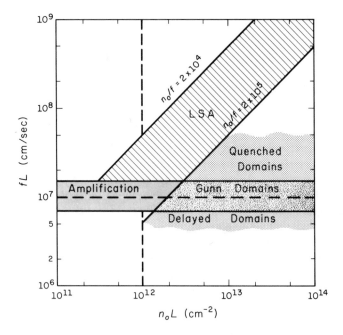

FIGURE 12-8. Mode diagram for n-type GaAs.†

---

†General features after J. A. Copeland, "LSA Oscillator Diode Theory," *J. Appl. Phys.*, vol. 38, pp. 3096–3101, July 1967.

To identify the approximate LSA region, we recall that the oscillation period τ must be much greater that τ_d, but no greater than several τ_d^*. Using $\mu_n = 5000$ and $\mu^* = 100 \text{ cm}^2/\text{V-sec}$, we find† (Prob. 12.6) that $\tau = \tau_d$ when $n_0/f = 1.4 \times 10^3$ and $\tau = \tau_d^*$ when $n_0/f = 7 \times 10^4$. In defining the LSA boundaries, we can choose the low-frequency limit such that the period τ is about $3\tau_d^*$. This gives the boundary $n_0/f = 2 \times 10^5$, indicating the limit for domain formation; this is the second line dividing the domain and no-domain regions. In the upper frequency limit for LSA of Fig. 12-8, we have taken $n_0/f = 2 \times 10^4$ as a somewhat arbitrary value to satisfy the requirement $\tau \gg \tau_d$. These two boundaries are the values quoted by Copeland as satisfying experimental evidence as well as falling within the general range predicted theoretically.

When $\tau \gg \tau_d^*$ (to the right of the $n_0/f = 2 \times 10^5$ line), domains have time to form while the field is above threshold. For frequencies near the inverse transit time we expect Gunn domains propagating to the anode during each cycle; for lower frequencies the domain can be collected during the negative half-cycle while $\mathscr{E} < \mathscr{E}_{th}$ (delayed domain); for higher frequencies domains can be quenched before they reach the anode.

We have omitted quite a few subtleties of mode behavior in Fig. 12-8, and the boundaries are more complicated than this; however, the basic division of the diagram should help to bring together the discussion of this section. It is clear from this diagram that the operation of devices related to the Gunn effect is complex. On the other hand, this complexity gives rise to a rich field of device applications.

12.2.4 Fabrication. Devices utilizing the Gunn effect and its variations can be made in a number of materials which have appropriate band structures. Although GaAs is the most common material, transferred electron effects have been observed in InP, CdTe, ZnSe, GaAsP, and other materials. The band structure of some materials can be altered to exhibit properties appropriate for electron transfer. For example, the energy bands of InAs can be distorted by the application of pressure to the crystal, such that a set of satellite valleys becomes available for electron transfer, although these upper valleys are too far above the lower valley at normal pressures. We have discussed the device behavior in terms of GaAs, since this material can be prepared with good purity and is most widely used in microwave applications.

---

†As in Eq. (12-3), we have approximated the dielectric relaxation times and mobilities as discrete positive and negative resistance values. Actually, the electric field sweeps through the v_d–$\mathscr{E}$ diagram of Fig. 12-7a periodically so that these quantities are complex functions of time. A careful analysis would use a proper average of these quantities over the oscillation period.

Gunn diodes and related devices are simple structures in principle, since they are basically homogeneous samples with ohmic contacts on each end. In practice, however, considerable care must be taken in fabricating and mounting workable devices. In addition to the obvious requirements on doping density, carrier mobility, and sample length implied by Fig. 12-8, there are important problems with contacts, heat sinking, and parasitic reactances of the packaged device.

The samples must have high mobility, few lattice defects, and homogeneous doping in the range giving carrier concentrations $n_0 \simeq 10^{13}$–10^{16} cm^{-3}. Devices can be made from GaAs bulk samples cut from an ingot, but it is more common to use ingot material as a substrate for an epitaxial layer, which serves as the active region of the device. The material properties of vapor-grown GaAs epitaxial layers are generally superior to bulk samples, and the precise control of layer thickness is helpful in these devices, which require exact sample lengths. In a typical configuration, an n-type epitaxial layer about 10 μm thick is grown on an n$^+$ substrate wafer which is perhaps 100 μm thick. The substrate serves as one of the contacts to the active region. A thin n^{++} layer is grown on top of the n region, so that an n$^+$-n-n^{++} sandwiched structure results. External contacts can be made by evaporating a thin layer of In or Sn and plating Au or Ni on each surface, followed by a brief alloying step in a hydrogen atmosphere. The wafer is divided into individual devices by cutting or cleaving (giving a cube structure) or by selective etching (giving a mesa structure). Each device is mounted with the n^{++} side down on a copper stud or other heat sink, so that the active region can dissipate heat to the mount in one direction and to the substrate layer in the other direction. Then the substrate side can be contacted by a wire or pressure contact. In other configurations, planar fabrication techniques can be used to produce lateral devices in an n-type epitaxial layer grown on a high-resistivity substrate.

Removal of heat is a very serious problem in these devices. The power dissipation may be 10^7 W/cm^3 or greater (Prob. 12.7), giving rise to considerable heating of the sample. As the temperature increases, the device characteristics vary because of changes in n_0 and mobility. As a result of such heating effects, these devices seldom reach their theoretical maximum efficiency. Pulsed operation allows better control of heat dissipation than does continuous operation, and efficiencies near the theoretical limits can sometimes be achieved in the pulsed mode. The LSA mode is particularly suitable for microwave power generation because of its relatively high efficiency and high operating frequencies. If the application does not require continuous operation, peak powers of hundreds of watts can be achieved in pulses of microwave oscillation.

Many variations can be found on the basic principles presented here. For example, a bar supporting domains can be shaped so that its cross section varies in a predetermined way. The output waveform for such a device reflects the shape of the sample. This interesting *functional oscillator* effect can be used to produce a wide variety of waveforms. Additional flexibility of application can be found by adding biasing contacts along the length of the sample to further control space charge formation and propagation.

READING LIST

H. KROEMER, "Negative Conductance in Semiconductors," *IEEE Spectrum,* vol. 5, no. 1, pp. 47–56, January 1968.

B. R. PAMPLIN, "Negative Differential Conductivity Effects in Semiconductors," *Contemporary Physics,* vol. 11, no. 1, pp. 1–19, January 1970.

G. GIBBONS and H. T. MINDEN, "Avalanche and Gunn-Effect Microwave Oscillators," *Solid State Technology,* vol. 13, no. 2, pp. 37–48, February 1970.

Special issues of the *IEEE Transactions on Electron Devices:* "Bulk Effect and Transit-Time Devices," vol. ED-13, no. 1, January 1966; this issue contains "flip-page movies" of the dynamics of several Gunn related modes; by the same title, vol. ED-14, no. 9, September 1967; "Wave Interactions in Solids," vol. ED-17, no. 3, March 1970.

Special issue on "Instabilities in Semiconductors," *IBM Journal of Research and Development,* vol. 13, no. 5, September 1969.

B. C. DE LOACH, JR., "Avalanche Transit-Time Microwave Diodes," pp. 464–496, and M. UENOHARA, "Bulk Gallium Arsenide Devices," pp. 497–536 in *Microwave Semiconductor Devices and their Circuit Applications.* ed. H. A. WATSON. New York: McGraw-Hill, Inc., 1969.

I. B. BOTT and W. FAWCETT, "The Gunn Effect in Gallium Arsenide," in *Advances in Microwaves,* vol. 3. ed. L. YOUNG. New York: Academic Press, Inc., 1968, pp. 223–300.

S. M. SZE, *Physics of Semiconductor Devices.* New York: John Wiley & Sons, Inc., 1969, Ch. 5, "Impact-Avalanche Transit-Time Diodes (IMPATT DIODES)," pp. 200–260, and Ch. 14, "Bulk-Effect Devices," pp. 731–787.

PROBLEMS

12.1 (a) Calculate the ratio N_U/N_L of the effective density of states in the upper valleys to the effective density of states in the lower valley of the GaAs conduction band (Fig. 12-4). (Continued)

(Continued 12.1)

(b) Assuming a Boltzmann distribution $n_U/n_L = (N_U/N_L) \exp(-\Delta E/kT)$, calculate the ratio of the density of conduction band electrons in the upper valley to the density in the lower valley in equilibrium at 300°K.

(c) As a rough calculation, assume an electron at the bottom of the central valley has kinetic energy kT. After it is promoted to the satellite valley, what is its approximate equivalent temperature?

12.2 (a) Use Poisson's equation, the continuity equation, and the definition of current density in terms of the gradient of electrostatic potential to relate the time variation of space charge density ρ to the conductivity σ and the dielectric constant ϵ of a material, neglecting recombination.

(b) Assuming a space charge density ρ_0 at $t = 0$, show that $\rho(t)$ decays exponentially with a time constant equal to the dielectric relaxation time τ_d.

(c) Given a sample of thickness L and area A, calculate the inherent RC time constant if the conductivity is σ and the dielectric constant is ϵ.

12.3 Assuming that n_L electrons/cm³ are in the lower (central) valley of the GaAs conduction band at time t and n_U are in the upper valley, show that the criterion for negative differential conductivity ($dJ/d\mathscr{E} < 0$) is

$$\frac{\mathscr{E}(\mu_L - \mu_U)\dfrac{dn_L}{d\mathscr{E}} + \mathscr{E}\left(n_L\dfrac{d\mu_L}{d\mathscr{E}} + n_U\dfrac{d\mu_U}{d\mathscr{E}}\right)}{n_L\mu_L + n_U\mu_U} < -1$$

where μ_L and μ_U are the electron mobilities in the lower and upper valleys, respectively. *Note:* $n_0 = n_U + n_L$. Discuss the conditions for negative differential conductivity, assuming the mobilities are approximately proportional to $\mathscr{E}^{-1}$.

12.4 Explain why space charge in a domain is quenched when the electric field outside the domain drops below the sustaining field $\mathscr{E}_s$ (Fig. 12-7a).

12.5 Explain why the current increases during the period after a domain is collected (Fig. 12-7d) or quenched (Fig. 12-7e) and before a new domain is formed.

12.6 Draw a mode diagram for n-type GaAs (Fig. 12-8); show the calculations and explain each step in forming the diagram.

12.7 We wish to estimate the d-c power dissipated in a GaAs Gunn diode. Assume the diode is 5 μm long and operates in the stable domain mode.

(Continued 12.7)

    (a) What is the minimum electron density n_0? What is the time between current pulses?

    (b) Using data from Fig. 12-6a, calculate the power dissipated in the sample per unit volume when it is biased just below threshold, if n_0 is chosen from the calculation of part (a). In general, does operation at a higher frequency result in greater power dissipation?

APPENDIX I. DEFINITIONS OF COMMONLY USED SYMBOLS†

a	Chapter 1: unit cell dimension (Å); Chapter 8: metallurgical channel half-width for an FET (cm)
a, b, c	basis vectors
A	area (cm²)
A_v	voltage gain
$\mathscr{B}$	magnetic flux density (Wb/cm²)
B	base transport factor for a BJT
B, E, C	base, emitter, collector of a BJT
c	speed of light (cm/sec)
C_j	junction capacitance (F)
C_s	charge storage capacitance (F)
D_n, D_p	diffusion constant for electrons, holes (cm²/sec)
D, G, S	drain, gate, source of an FET
e	Napierian base
e^-	electron
$\mathscr{E}$	electric field strength (V/cm)
E	energy (J, eV)‡; battery voltage (V)
E_a, E_d	acceptor, donor energy level (J, eV)
E_c, E_v	conduction band, valence band edge (J, eV)
E_F	equilibrium Fermi level (J, eV)
E_g	band gap energy (J, eV)
E_i	intrinsic level (J, eV)
E_r, E_t	recombination, trapping energy level (J, eV)
$f(E)$	Fermi–Dirac distribution function
F_n, F_p	quasi-Fermi level for electrons, holes (J, eV)
g, g_{op}	EHP generation rate, optical generation rate (cm⁻³-sec⁻¹)
g_m	mutual transconductance (Ω⁻¹)
G, K, P	grid, cathode, plate of a triode
h	Planck's constant (J-sec, eV-sec); Chapter 8: FET channel half-width (cm)
$h\nu$	photon energy (J, eV)
h, k, l	Miller indices

---

†This list does not include some symbols which are used only in the section where they are defined. Units are given in common semiconductor usage, involving cm where appropriate; it is important to note, however, that calculations should be made in the MKS system in some formulas.

‡In the Boltzmann factor $\exp(-\Delta E/kT)$, ΔE can be expressed in J or eV if k is expressed in J/°K or eV/°K, respectively.

h^+	hole
i, I†	current (A)
I(subscript)	inverted mode of a BJT
i_B, i_C, i_E	base, collector, emitter current in a BJT (A)
I_{CO}, I_{EO}	magnitude of the collector, emitter saturation current with the emitter, collector open (A)
I_{CS}, I_{ES}	magnitude of the collector, emitter saturation current with the emitter, collector shorted (A)
I_{DS}	channel current in an FET, directed from drain to source (A)
i_P	plate current in a triode (A)
j	$\sqrt{-1}$
J	current density (A/cm²)
k	Boltzmann's constant ($J/°K$, $eV/°K$)
k	wave vector (cm⁻¹)
k_d	distribution coefficient
K	$4\pi\epsilon_0$ (F/cm)
l, L	length (cm)
$\bar{l}$	mean free path for carriers in random motion (cm)
m, m^*	mass, effective mass (kg)
m_n^*, m_p^*	effective mass for electrons, holes (kg)
m_0	rest mass of the electron (kg)
M	avalanche multiplication factor
m, n	integers; exponents
n	density of electrons in the conduction band (cm⁻³)
n	n-type semiconductor material
n_i	intrinsic density of electrons (cm⁻³)
n_n, n_p	equilibrium density of electrons in n-type, p-type material (cm⁻³)
n_0	equilibrium density of electrons (cm⁻³)
N(subscript)	normal mode of a BJT
N_a, N_d	density of acceptors, donors (cm⁻³)
N_a^-, N_d^+	density of ionized acceptors, donors (cm⁻³)
N_c, N_v	effective density of states at the edge of the conduction band, valence band (cm⁻³)
p	density of holes in the valence band (cm⁻³)
p	p-type semiconductor material

---

†See note at the end of this list.

p	momentum (kg-m/sec)
p_i	intrinsic hole density (cm^{-3}) $= n_i$
p_n, p_p	equilibrium density of holes in n-type, p-type material (cm^{-3})
p_0	equilibrium hole density (cm^{-3})
q	magnitude of the electronic charge (C)
Q	operating point; quality factor of a resonant circuit
Q_+, Q_-	total positive, negative charge (C)
Q_n, Q_p	charge stored in an electron, hole distribution (C)
r, R	resistance (Ω)
R_H	Hall coefficient (cm^3/C)
t	time (sec)
$\dagger$	sample thickness (cm)
$\bar{t}$	mean free time between scattering collisions (sec)
t_{sd}	storage delay time (sec)
T	temperature (°K)
$v, V\dagger$	voltage (V)
V	potential energy (J)
$\mathcal{V}$	electrostatic potential (V)
V_{CB}, V_{EB}	voltage from collector to base, emitter to base in a BJT (V)
V_{DS}, V_{GS}	voltage from drain to source, gate to source in an FET (V)
V_{GK}, V_{PK}	voltage from grid to cathode, plate to cathode in a triode (V)
$\mathcal{V}_n, \mathcal{V}_p$	electrostatic potential in the neutral n, p material (V)
V_0	contact potential (V)
V_P	Chapter 8: pinch-off voltage for an FET; Chapter 11: forward breakover voltage for a UJT or SCR (V)
$\mathbf{v}, \mathbf{v}_d$	velocity, drift velocity (cm/sec)
w	sample width (cm)
W	depletion region width (cm)
W_b	base width in a BJT, measured between the edges of the emitter and collector junction depletion regions (cm)
x	distance (cm)
x_n, x_p	distance in the neutral n region, p region of a junction, measured from the edge of the transition règion (cm)
x_{n0}, x_{p0}	penetration of the transition region into the n region, p region, measured from the metallurgical junction (cm)
Z	atomic number; dimension in z-direction (cm)

$\dagger$See note at the end of this list.

α	emitter-to-collector current amplification factor in a BJT
α	optical absorption constant (cm^{-1})
α_r	recombination coefficient (cm^3/sec)
β	base-to-collector current amplification factor in a BJT
γ	emitter injection efficiency; in a p-n-p, the fraction of i_E due to the hole current i_{Ep}
δ, Δ	incremental change
$\delta n, \delta p$	excess electron, hole density (cm^{-3})
$\Delta n_p, \Delta p_n$	excess electron, hole density at the edge of the transition region on the p side, n side (cm^{-3})
$\Delta p_C, \Delta p_E$	excess hole density in the base of a BJT, evaluated at the edge of the transition region of the collector, emitter junction (cm^{-3})
$\epsilon, \epsilon_r, \epsilon_0$	permittivity, relative dielectric constant, permittivity of free space (F/cm); $\epsilon = \epsilon_r \epsilon_0$
λ	wavelength of light (μm, Å)
μ	mobility ($cm^2/V\text{-}sec$)
ν	frequency of light (sec^{-1})
ρ	resistivity ($\Omega\text{-}cm$)
σ	conductivity ($\Omega\text{-}cm$)$^{-1}$
τ_d	dielectric relaxation time (sec)
τ_n, τ_p	recombination lifetime for electrons, holes (sec)
τ_t	transit time (sec)
ϕ	flux density ($cm^2\text{-}sec$)$^{-1}$
ψ, Ψ	time-independent, time-dependent wave function
ω	angular frequency (sec^{-1})
$\langle \ \rangle$	average of the enclosed quantity

---

†For d-c voltage and current, capital symbols with capital subscripts are used; lower-case symbols with lower-case subscripts represent a-c quantities; lower-case symbols with capital subscripts represent total (a-c + d-c) quantities. See, for example, Fig. 8-3. For voltage symbols with double subscripts, V is positive when the potential at the point referred to by the first subscript is higher than that of the second point. For example, V_{GS} is the potential difference $V_G - V_S$.

APPENDIX II. PHYSICAL CONSTANTS AND CONVERSION FACTORS†

Avogadro's number	$N_A = 6.02 \times 10^{23}$ molecules/mole
Boltzmann's constant	$k = 1.38 \times 10^{-23}$ J/°K
	$= 8.62 \times 10^{-5}$ eV/°K
Electronic charge (magnitude)	$q = 1.60 \times 10^{-19}$ C
Electronic rest mass	$m_0 = 9.11 \times 10^{-31}$ kg
Permittivity of free space	$\epsilon_0 = 8.85 \times 10^{-14}$ F/cm
	$= 8.85 \times 10^{-12}$ F/m
Planck's constant	$h = 6.63 \times 10^{-34}$ J-sec
	$= 4.14 \times 10^{-15}$ eV-sec
Room temperature value of kT	$kT = 0.0259$ eV
Speed of light	$c = 2.998 \times 10^{10}$ cm/sec

Prefixes:

1 Å (angstrom) $= 10^{-8}$ cm	milli-,	m- $= 10^{-3}$
1 μm (micron) $= 10^{-4}$ cm	micro-,	μ- $= 10^{-6}$
1 mil $= 10^{-3}$ in.	nano-,	n- $= 10^{-9}$
2.54 cm $= 1$ in.	pico-,	p- $= 10^{-12}$
1 eV $= 1.6 \times 10^{-19}$ J	kilo-,	k- $= 10^{3}$
	mega-,	M- $= 10^{6}$
	giga-,	G- $= 10^{9}$

A wavelength λ of 1 μm corresponds to a photon energy of 1.24 eV.

---

†Since cm is used as the unit of length for many semiconductor quantities, caution must be exercised to avoid unit errors in calculations. When using quantities involving length in formulas which contain quantities measured in MKS units, it is usually best to use all MKS quantities. Conversion to standard semiconductor usage involving cm can be accomplished as a last step. Similar caution is recommended in using J and eV as energy units.

APPENDIX III. PROPERTIES OF SEMICONDUCTOR MATERIALS

	E_g (eV)	μ_n (cm²/V-sec)	μ_p (cm²/V-sec)	ρ (Ω-cm)	Transition	Doping	Lattice	a (Å)	ϵ_r	Density (g/cm³)	Melting point (°C)
Si	1.11	1350	480	2.5×10^5**	i	n,p	D	5.43	11.8	2.33	1415
Ge	0.67	3900	1900	43	i	n,p	D	5.66	16	5.32	936
SiC(α)	2.86	500		10^{10}	i	n,p	W	3.08	10.2	3.21	2830
AlP	2.45	80		10^{-5}	i	n,p	Z	5.46		2.40	2000
AlAs	2.16	180		0.1	i	n,p	Z	5.66	10.9	3.60	1740
AlSb	1.6	200	300	5	i	n,p	Z	6.14	11	4.26	1080
GaP	2.26	300	150	1	i	n,p	Z	5.45	11.1	4.13	1467
GaAs	1.43	8500	400	4×10^8**	d	n,p	Z	5.65	13.2	5.31	1238
GaSb	0.7	5000	1000	0.04	d	n,p	Z	6.09	15.7	5.61	712
InP	1.28	4000	100	8×10^{-3}	d	n,p	Z	5.87	12.4	4.79	1070
InAs	0.36	22600	200	0.03	d	n,p	Z	6.06	14.6	5.67	943
InSb	0.18	10^5	1700	0.06	d	n,p	Z	6.48	17.7	5.78	525
ZnS	3.6	110		10^{10}	d	n	Z,W	5.409	8.9	4.09	1650†
ZnSe	2.7	600		10^9	d	n	Z	5.671	9.2	5.65	1100†
ZnTe	2.25		100	100	d	p	Z	6.101	10.4	5.51	1238†
CdS	2.42	250		10^3	d	n	W,Z	4.137	8.9	4.82	1475
CdSe	1.73	650	15		d	n	W	4.30	10.2	5.81	1258
CdTe	1.58	1050	100	10^{10}	d	n,p	Z	6.482	10.2	6.20	1098
PbS	0.37	575	200	5×10^{-3}	i	n,p	H	5.936	161	7.6	1119
PbSe	0.27	1000	1000	10^{-3}	i	n,p	H	6.147	280	8.73	1081
PbTe	0.29	1600	700	10^{-2}	i	n,p	H	6.452	360	8.16	925

All values at 300°K.   *intrinsic resistivity.   †vaporizes.

Definitions of symbols: ρ is resistivity of high-purity material; i is indirect; d is direct; D is diamond; Z is zinc blende; W is wurtzite; H is halite (NaCl). Values of mobility and resistivity are for material of available purity; these values are considered approximate (exception: Si and GaAs resistivities are extrapolated to intrinsic material). Most of the values in this table were taken from publications of the Electronic Properties Information Center (EPIC), Hughes Aircraft Co., Culver City, California; also, M. Neuberger, "III–V Semiconducting Compounds–Data Tables," published with permission from Plenum Publishing Corporation, copyright 1970.

Crystals in the wurtzite structure are not described completely by the single lattice constant given here, since the unit cell is not cubic. Several II–VI compounds can be grown in either the zinc blende or wurtzite structures.

Many values quoted here are approximate or uncertain, particularly for the II–VI and IV–VI compounds.

443

APPENDIX IV. DERIVATION OF THE DENSITY OF STATES IN THE CONDUCTION BAND

In this derivation we shall consider the conduction band electrons to be essentially free. Constraints of the particular lattice can be included in the effective mass of the electron at the end of the derivation. For a free electron, the three-dimensional Schrödinger wave equation becomes

$$(\text{IV-1}) \qquad -\frac{\hbar^2}{2m}\nabla^2\psi = E\psi$$

where ψ is the wave function of the electron and E is its energy. The form of the solution to Eq. (IV-1) is

$$(\text{IV-2}) \qquad \psi = (\text{const.})\, e^{j\mathbf{k}\cdot\mathbf{r}}$$

We must describe the electron in terms of a set of boundary conditions within the lattice. A common approach is to use periodic boundary conditions, in which we quantize the electron energies in a cube of material of side L. This can be accomplished by requiring that

$$(\text{IV-3}) \qquad \psi(x + L, y, z) = \psi(x, y, z)$$

and similarly for the y- and z-directions. Thus our wave function can be written as

$$(\text{IV-4}) \qquad \psi_n = A \exp\left[j\frac{2\pi}{L}(\mathbf{n}_x x + \mathbf{n}_y y + \mathbf{n}_z z) \right]$$

where the $2\pi\mathbf{n}/L$ factor in each direction guarantees the condition described by Eq. (IV-3), and A is a normalizing factor. Substituting ψ_n into the Schrödinger equation (IV-1), we obtain

$$(\text{IV-5}) \qquad -\frac{\hbar^2}{2m} A\nabla^2 \exp\left[j\frac{2\pi}{L}(\mathbf{n}_x x + \mathbf{n}_y y + \mathbf{n}_z z) \right]$$

$$= EA \exp\left[j\frac{2\pi}{L}(\mathbf{n}_x x + \mathbf{n}_y y + \mathbf{n}_z z) \right]$$

$$(\text{IV-6}) \qquad E_n = \frac{\hbar^2}{2m}\left(\frac{2\pi}{L}\right)^2 (\mathbf{n}_x{}^2 + \mathbf{n}_y{}^2 + \mathbf{n}_z{}^2) = \frac{\hbar^2 \mathbf{n}^2}{2mV^{2/3}}$$

where V is the volume of the cube and $\mathbf{n}^2 = \mathbf{n}_x{}^2 + \mathbf{n}_y{}^2 + \mathbf{n}_z{}^2$ denotes the energy level.

Now let us find the number of allowed energy states per unit volume N as a function of energy. The number of electron states below the energy level $\mathbf{n}$ can be determined as shown in Fig. IV-1 by calculating the volume occupied in $\mathbf{n}$-space. Here we assume that the electron energy levels are filled in a spherical volume about the origin and the number of states within the

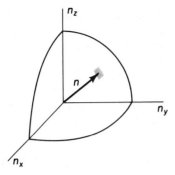

FIGURE IV-1. Portion of a sphere in **n**-space.

volume is large. This allows us to consider an essentially continuous distribution of states in **n**-space.

If we count the electron states per energy level in accordance with the Pauli principle, the number of states with energy level less than **n** is twice the volume in **n**-space, $2(4\pi/3)\mathbf{n}^3$. Thus the total number of states in the volume V can be expressed as

$$(\text{IV-7}) \qquad NV = \frac{8\pi}{3}\mathbf{n}^3$$

From Eq. (IV-6), the energy of the **n**'th energy level is

$$(\text{IV-8}) \qquad E_n = \frac{h^2\mathbf{n}^2}{2mV^{2/3}} = \frac{h^2}{2mV^{2/3}}\left(\frac{NV}{8\pi/3}\right)^{2/3} = \frac{h^2}{2m}(3\pi^2 N)^{2/3}$$

If **n** is some maximum value $\mathbf{n}_F$, such that all states below this level are filled and all states above it are empty, then the energy E_F as obtained from Eq. (IV-8) is the Fermi energy at 0°K.

Denoting the density of allowed states by $N(E)$, we can use Eq. (IV-8) to write the total number of states per unit volume as

$$(\text{IV-9}) \qquad N = \int N(E)\,dE = \frac{1}{3\pi^2}\left(\frac{2mE}{h^2}\right)^{3/2}$$

Differentiating Eq. (IV-9) we obtain

$$(\text{IV-10}) \qquad N(E) = \frac{1}{3\pi^2}\left(\frac{3}{2}\right)\left(\frac{2m}{h^2}\right)^{3/2}E^{1/2} = \frac{1}{2\pi^2}\left(\frac{2m}{h^2}\right)^{3/2}E^{1/2}$$

for the density of states as a function of energy.

To include the probability of occupation of any energy level E, we use the Fermi–Dirac distribution function

$$(\text{IV-11}) \qquad f(E) = \frac{1}{e^{(E-E_F)/kT} + 1}$$

The density of electrons in the range dE is given by the product of the density of allowed states in that range and the probability of occupation. Thus the density of occupied electron states N_e in dE is

$$(\text{IV-12}) \qquad\qquad N_e \, dE = N(E) f(E) \, dE$$

We may calculate the concentration of electrons in the conduction band at a given temperature by integrating Eq. (IV-12) across the band

$$(\text{IV-13}) \qquad n = \int_0^\infty N(E) f(E) \, dE = \frac{1}{2\pi^2} \left(\frac{2m}{\hbar^2}\right)^{3/2} e^{E_F/kT} \int_0^\infty E^{1/2} e^{-E/kT} \, dE$$

In this integration we have referred the energies in the conduction band to the band edge (E_c taken as $E = 0$). Furthermore, we have taken the function $f(E)$ to be

$$(\text{IV-14}) \qquad\qquad f(E) = e^{(E_F - E)/kT}$$

for energies such that $(E - E_F) \gg kT$.
The integral in Eq. (IV-13) is of the standard form

$$(\text{IV-15}) \qquad\qquad \int_0^\infty x^{1/2} e^{-ax} \, dx = \frac{\sqrt{\pi}}{2a\sqrt{a}}$$

Thus Eq. (IV-13) gives

$$(\text{IV-16}) \qquad\qquad n = 2\left(\frac{2\pi mkT}{h^2}\right)^{3/2} e^{E_F/kT}$$

If we refer to the bottom of the conduction band as E_c instead of $E = 0$, the expression for the electron density is

$$(\text{IV-17}) \qquad\qquad n = 2\left(\frac{2\pi m_n^* kT}{h^2}\right)^{3/2} e^{(E_F - E_c)/kT}$$

which corresponds to Eq. (3-15). We have included constraints of the lattice through the effective mass of the electron in the crystal, m_n^*.

APPENDIX V. SOLID SOLUBILITIES OF IMPURITIES IN Si AND Ge†

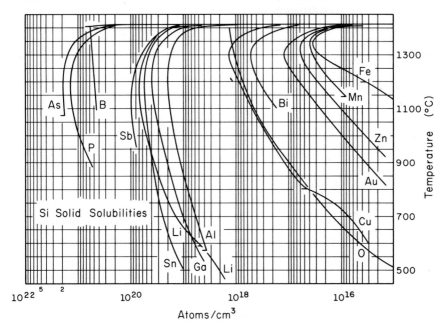

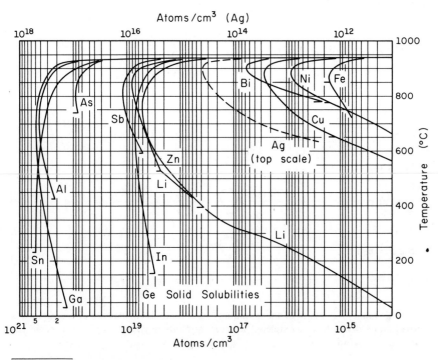

†From F. A. Trumbore, "Solid Solubilities of Impurity Elements in Si and Ge," *Bell System Technical Journal*, vol. 39, no. 1, pp. 205–233, January 1960, copyright 1960, The American Telephone and Telegraph Co., reprinted by permission. Alterations have been made to include later data.

APPENDIX VI. COMMON CIRCUIT SYMBOLS FOR SOLID STATE DEVICES†

Diodes

p-n diode, p-i-n, IMPATT

Breakdown ("Zener") diode

Bidirectional Zener diode

Step-recovery (snap) diode

Tunnel diode

Photodiode, solar cell

Light-emitting diode (LED)

Transistors

Bipolar p-n-p

Bipolar n-p-n

---

†This table gives many of the commonly used symbols for devices discussed in this book. Standardization is incomplete, however, and other symbols are often used in the electronics literature.

Transistors (Continued)

Unijunction (n-base)

Unijunction (p-base)

JFET (n-channel)

JFET (p-channel)

Depletion   Enhancement

IGFET (n-channel)

IGFET (p-channel)

Four-Layer and Related Devices

p-n-p-n (Shockley) diode

SCR

Diac   Triac

Bilateral switch

SCS

INDEX

Abrupt junction, *see* Step junction
Absorption, optical, 95–99, 253
  constant (α), 98
  free carrier, 96
a-c equivalent circuit:
  BJT, 351
  JFET, 292
  triode, 283
  tunnel diode, 225
Acceptor:
  action in semiconductors, 68–69
  atoms, 68, 111, 266
  doping profile, 143, 200, 205–206, 316, 341
  ionization energy in Si, 111
  level, 68
Accumulation layer:
  in a BNDC device, 427
  in an FET, 293
Acoustoelectric effect, 417
Activator, 101–102
Admittance, a-c, 191
Air isolation (IC), 388–389
AlGaAs, 267–268
Alkali halides, 53
Alloy:
  die bond, 381
  preform, 140, 142, 309, 311
  process, 140–141, 229–230
  station, 141–142
Alloyed junction:
  BJT, 308–311
  diode, 140–142
  SCR, 413–414
  tunnel diode, 229–230
  UJT, 401
Alpha (α) of a BJT:
  calculation, 324

Alpha (α) of a BJT (*Contd.*):
  definition, 306
  dependence on current, 345–346, 405–406
  dependence on voltage, 342–343
Aluminum (Al):
  acceptor in Si, 68, 111
  alloy in Si, 11, 401, 414
  contact pads, 377
  metallization, 146–147, 376–377, 391–393
  wire bonding, 382–383
Ambipolar drift and diffusion, 134
Amorphous solids, 3–4
Amplification factor, 283, 306
Amplification mode (BNDC devices), 431–432
Amplifier:
  a-c equivalent circuit, 282–284
  graphical analysis, 281–282, 307–308
  transistor, 307–308
  triode, 281–284
  tunnel diode, 224–225
Angular momentum, 32–33
Annealing, 149
Anode:
  gate in an SCS, 411
  of a p-n-p-n diode, 403
  of a thermionic diode, 278–279
  of a triode, 279–280
Antimony (Sb), 18, 67, 111, 229, 372, 396, 401, 447
Arrays:
  diode, 244–246
  IC, 368–369
  solar cell, 236
Aspect ratio, 375
Astable circuit, 223
Asymmetrically doped junction, 169–171

Cadmium selenide (CdSe):
platelet laser, 272
properties, 443
thin-film transistor, 300
Cadmium sulfide (CdS):
photodetector, 117–118
properties, 443
thin-film transistor, 300
Capacitance:
abrupt junction, 187–191
arbitrary junction, 216–217, 248
charge storage, 191
depletion layer, 187
diffusion, 187
in a BJT, 351–353
junction, 187–189
linearly graded junction, 205
MOS, 376
parasitic, 283, 357, 364
Capacitor, integrated, 366, 376
see also Varactor
Capture of electrons and holes in
recombination, 101, 109, 196–197
Carrier concentration, 70–81
constant product, 76, 115
equilibrium, 74–77
excess, *see* Excess carriers
gradient, 119
intrinsic, 77–78
relation to doping, 79
temperature dependence, 77–80
Carrier injection, *see* Injection, carrier
Carrier recombination, *see* Recombination,
carrier
Carriers:
majority and minority, 70
see also Electrons, Holes, Majority
carrier, Minority carrier
Cathode:
p-n-p-n diode, 403
thermionic diode, 278–279
triode, 279–280
gate (SCS), 411
Cathode-ray tube (CRT), 103–104
Cathodoluminescence, 103–104
Centers, recombination and trapping,
100–101, 108–114, 186–197, 231–232,
242
Channel (FET):
conductance, 291, 297
diffused, 293–294
induced, 293–299
metallurgical, 286, 289–290
pinch-off, 288, 290, 298–299
width, 287, 290, 297
Channeling, 149
Charge, surface (MOS), 295–296, 298–299
Charge carriers, 61–70

Charge control model:
p-n junction, 167–169, 181–186, 189–191
transistor, 330–331, 335–339, 352–353
Charge-coupled device (CCD), 300–301
Charge storage:
capacitance, 191, 352
in a BJT, 330–331, 335–339, 352–353
in a p-n junction, 167–169, 181–186,
189–191
Circuit models:
BJT, 328–329, 333, 351
diode, 208
JFET, 292–293
triode, 282–283
tunnel diode, 225
UJT, 399
Circuit symbols, 448–449
Cleavage:
along crystal planes, 10–12
in injection laser fabrication, 267
Closed-tube diffusion, 144–145
Coactivator, 101–102
Coherent light, 252
Collector (BJT):
breakdown, 344–345
characteristics, 303, 307, 343
contact, 309–311, 357, 371–373
injection, *see* Inverted mode
region, 303, 357
resistance, 351, 372
Color:
of light emission, 241–242
of materials, 96–97
Common-base configuration (BJT), 303,
343–344
Common-emitter configuration (BJT), 307,
343–344
Compensation, 80–81, 139–140, *see also*
Self-compensation
Complementary error function (erfc), 143,
186, 205–206
Complex conjugate, 37
Composite transistor, 372–373
Compound semiconductors, 2, 443
Computer:
devices, 300, 390–394
use in IC design, 379–380, 394
Conduction band, 56–61
effective density of states, 75, 444–446
electron density, 74–81
Conduction processes, *see* Drift, Diffusion
Conductivity, 81–86
intrinsic, 94
temperature dependence, 87
type, 67
Conductivity modulation, 199, 211, 398
Conductors, 366, 376–377
Confinement in laser junctions, 267–268